61

Springer Series in Chemical Physics

Edited by J. P. Toennies

Springer
Berlin
Heidelberg
New York
Barcelona
Budapest
Hong Kong
London
Milan
Paris
Santa Clara
Singapore
Tokyo

Springer Series in Chemical Physics

Editors: Vitalii I. Goldanskii Fritz P. Schäfer J. Peter Toennies

Managing Editor: H. K.V. Lotsch

40 **High-Resolution Spectroscopy of Transient Molecules**
By E. Hirota

41 **High Resolution Spectral Atlas of Nitrogen Dioxide 559–597 nm**
By K. Uehara and H. Sasada

42 **Antennas and Reaction Centers of Photosynthetic Bacteria**
Structure, Interactions, and Dynamics
Editor: M. E. Michel-Beyerle

43 **The Atom-Atom Potential Method**
Applications to Organic Molecular Solids
By A. J. Pertsin and A. I. Kitaigorodsky

44 **Secondary Ion Mass Spectrometry SIMS V**
Editors: A. Benninghoven, R. J. Colton, D. S. Simons, and H. W. Werner

45 **Thermotropic Liquid Crystals, Fundamentals**
By G. Vertogen and W. H. de Jeu

46 **Ultrafast Phenomena V**
Editors: G. R. Fleming and A. E. Siegman

47 **Complex Chemical Reaction Systems**
Mathematical Modelling and Simulation
Editors: J. Warnatz and W. Jäger

48 **Ultrafast Phenomena VI**
Editors: T. Yajima, K. Yoshihara, C. B. Harris, and S. Shionoya

49 **Vibronic Interactions in Molecules and Crystals**
By I. B. Bersuker and V. Z. Polinger

50 **Molecular and Laser Spectroscopy**
By Zu-Geng Wang and Hui-Rong Xia

51 **Space-Time Organization in Macromolecular Fluids**
Editors: F. Tanaka, M. Doi, and T. Ohta

52 **Multiple-Photon Laser Chemistry**
By. R.V. Ambartzumian, C. D. Cantrell, and A. Puretzky

53 **Ultrafast Phenomena VII**
Editors: C. B. Harris, E. P. Ippen, G. A. Mourou, and A. H. Zewail

54 **Physics of Ion Impact Phenomena**
Editor: D. Mathur

55 **Ultrafast Phenomena VIII**
Editors: J.-L. Martin, A. Migus, G. A. Mourou, and A. H. Zewail

56 **Clusters of Atoms and Molecules**
Salvation and Chemistry of Free Clusters, and Embedded, Supported and Compressed Clusters
Editor: H. Haberland

57 **Radiationless Transitions in Polyatomic Molecules**
By E. S. Medvedev and V. I. Osherov

58 **Positron Annihilation in Chemistry**
By O. E. Mogensen

59 **Soot Formation in Combustion**
Mechanisms and Models
Editor: H. Bockhorn

60 **Ultrafast Phenomena IX**
Editors: P. F. Barbara, W. H. Knox, G. A. Mourou, and A. H. Zewail

61 **Gas Phase Chemical Reaction Systems**
Experiments and Models 100 Years After Max Bodenstein
Editors: J. Wolfrum, H.-R. Volpp, R. Rannacher, and J. Warnatz

62 **Ultrafast Phenomena X**
Editors: P. F. Barbara, J. G. Fujimoto, W. H. Knox, and W. Zinth

Volumes 1–39 are listed at the end of the book

J. Wolfrum · H.-R. Volpp
R. Rannacher · J. Warnatz (Eds.)

Gas Phase Chemical Reaction Systems

Experiments and Models 100 Years After Max Bodenstein

Proceedings of an International Symposion, held at the "Internationales Wissenschaftsforum Heidelberg", Heidelberg, Germany, July 25–28, 1995

With 148 Figures

Springer

Professor Dr. J. Wolfrum
Dr. H.-R. Volpp
Physikalisch-Chemisches Institut
Universität Heidelberg
Im Neuenheimer Feld 253
D-69120 Heidelberg, Germany

Professor Dr. J. Warnatz
Interdisziplinäres Zentrum
für Wissenschaftliches Rechnen
Universität Heidelberg
Im Neuenheimer Feld 368
D-69120 Heidelberg, Germany

Professor Dr. R. Rannacher
Institut für Angewandte Mathematik
Universität Heidelberg
Im Neuenheimer Feld 293
D-69120 Heidelberg, Germany

Cataloging-in-Publication Data applied for

Die Deutsche Bibliothek - CIP-Einheitsaufnahme

Gas phase chemical raction systems : experiments and models 100 years after Max Bodenstein ; proceedings of an international symposion, held at the Internationales Wissenschaftsforum Heidelberg, Heidelberg, Germany, July 25 - 28, 1995 / J. Wolfrum ... (ed.). - Berlin ; Heidelberg ; New York ; Barcelona ; Budapest ; Hong Kong ; London ; Milan ; Paris ; Santa Clara ; Singapore ; Tokyo : Springer, 1996
(Springer series in chemical physics ; Vol. 61)
ISBN 3-540-61662-4
NE: Wolfrum, Jürgen [Hrsg.]; Internationales Wissenschaftsforum <Heidelberg>; GT

ISSN 0172-6218

ISBN-13: 978-3-642-80301-7 e-ISBN-13: 978-3-642-80299-7
DOI: 10.1007/978-3-642-80299-7

Softcover reprint of the hardcover 1st edition 1996

SPIN: 10545060 54/3144 3 2 1 0 - Printed on acid-free paper

Preface

This volume consists of edited papers presented at the International Symposion *Gas Phase Chemical Reaction Systems: Experiments and Models 100 Years After Max Bodenstein*, held at the Internationales Wissenschaftsforum Heidelberg (IWH) in Heidelberg during July 25–28, 1995.

The intention of this symposion was to bring together leading researchers from the fields of reaction dynamics, kinetics, catalysis and reactive flow modelling to discuss and review the advances in the understanding of chemical kinetics about 100 years after *Max Bodenstein's* pioneering work on the "hydrogen iodine reaction", which he carried out at the Chemistry Institute of the University of Heidelberg. The idea to focus in his doctoral thesis [1] on this reaction was brought up by his supervisor *Victor Meyer* (successor of *Robert Bunsen* at the Chemistry Institute of the University of Heidelberg) and originated from the non-reproducible behaviour found by *Bunsen* and *Roscoe* in their early photochemical investigations of the H_2/Cl_2 system [2] and by *van't Hoff* [3], and *V. Meyer* and co-workers [4] in their experiments on the slow combustion of H_2/O_2 mixtures.

Whereas earlier work by *Hautefeuille* [5] and *Lemoine* [6] on the "behaviour of hydrogen iodine gas in the heat" and the "photosensitivity of hydrogen iodine" was more of a qualitative character, it was *Bodenstein's* systematic studies on the thermal [7] and light-induced [8] decomposition of hydrogen iodine, where he combined experimental reaction kinetics studies with mathematical analysis and modelling, that paved the way for modern quantitative treatment of chemical systems of practical interest. Chemical processes occurring in industrial applications, such as in engine combustion or the chemical vapour deposition of diamond, usually consist of a large number of elementary chemical reactions coupled with transport and diffusion phenomena and very often involve heterogeneous reaction steps. Therefore, the development of reliable computational methods for numerical modelling and simulation of such processes necessitates a close and well-coordinated interaction across the disciplinary boundaries of mathematics, physics, chemistry, and engineering.

In the past, two workshops held here in Heidelberg [9,10] paid tribute to this multidisciplinary challenge, which was also recognized by the foundation of the *Interdisziplinäres Zentrum für Wissenschaftliches Rechnen (IWR)* in 1987 and the subsequent establishment of the *Sonderforschungbereich (SFB) 359: Reaktive Strömungen, Diffusion und Transport* at the University of Heidelberg in 1993. It was the encouraging experience and enthusiasm within the research work of the SFB 359 that led to the idea of this symposion, which we hope has achieved its aims, both to remember the historical impact *Max Bodenstein* had on the development of chemical kinetics and to allow for intense discussions between scientists who actively work on different aspects of the quantitative investigation and modelling of chemical reaction systems.

Overall about 100 scientists from 10 countries (England, France, Germany, Hungary, India, Israel, Italy, Sweden, Taiwan, and the U.S.A.) participated, presenting and discussing a total of 23 papers and 21 posters and making this symposion a very stimulating one. We hope that this book will transfer a flavour of this stimulating atmosphere to a wider audience.

The "*Bodenstein family tree*" depicted in Fig. I was created – with the enthusiastic help of the participants – during the symposion in order to illustrate the scientific connections (rather than strict teacher-student relationships) originating from *Max Bodenstein's* pioneering work. We by no means intend this tree to be an exhaustive representation of the kinetics community.

In addition, for those readers who are interested in more historical details of *Max Bodenstein's* life we would like to recommend the beautiful article by *Erika Cremer* [11], in which she describes his scientific life which led him from Heidelberg (where he received his doctoral degree in 1893), *via* Berlin, Göttingen, back to Heidelberg (where he published his Habilitation in 1899 [12]), and *via* Leipzig, Hannover finally again to Berlin, where he became the successor of *Nernst* in 1923.

This book is divided into six parts (Parts I–VI), reflecting the diversity of the topics discussed during the symposion.

Part I is devoted to experimental studies aiming at the elucidation of the microscopic dynamics of elementary reactions by employing laser and molecular beam techniques. The paper that opens this volume was presented by *A.H. Zewail* and describes the fascinating possibilities femtosecond lasers have offered in chemistry – actually leading to a new branch known as "laser femtochemistry", in which reactive events can nowadays be investigated in real-time (fs) and with atomic (Å) resolution. The contributions by *H.-R. Volpp* and *J. Wolfrum*, as well as that by *F.F. Crim et al.* describe experiments in which the influence of selective reagent excitation on the reactivity and product-channels of gas phase elementary reactions is studied by using the laser "pump-and-probe" technique. *R.N. Zare, D.W. Chandler* and co-workers report on bimolecular reaction product imaging studies and give a comprehensive description of the reconstruction technique necessary to derive quantum state-resolved three-dimensional scattering information from two-dimensional ion-images of the reaction products. *C.B. Moore et al.* present results from unimolecular reaction dynamics studies near the dissociation threshold, which confirm the fundamental hypothesis of statistical transition state theory. Finally, the contributions by *Y.T. Lee et al.*, and *G.G. Volpi, P. Casavecchia et al.* present results from reactive scattering studies of bimolecular three- and four-atom reactions using the cross-molecular-beam method.

Part II is the theoretical counterpart of Part I. The paper by *D.G. Truhlar et al.* gives an overview of theoretical achievements made in the dynamical treatment of the Cl + H_2 reaction and presents recent results from quantum scattering studies using a new *ab initio* potential energy surface, which allows for a direct comparison with experimental results from *G.G. Volpi, P. Casavecchia et al.*. *M. Baer* and co-workers present a review on their recent quantum mechanical calculations for triatomic and tetraatomic systems, which for the H + H_2O reaction can be directly compared with the experimental results obtained by *H.-R. Volpp* and *J. Wolfrum.* Using the quasiclassical trajectory method, *G.C. Schatz et al.* studied mode specificity for two different four-atom gas phase reactions which are of importance in combustion chemistry. Employing the same technique *T. Raz* and *R.D. Levine* simulated an "exotic" regime of reaction dynamics called "cluster impact-induced chemistry", in which high-barrier processes like the four-center $H_2 + I_2 \rightarrow 2$ HI collision mechanism – as suggested by *Max Bodenstein* for the gas phase reaction – can take place, but in this case inside an impact-heated Xe cluster.

Part III starts with a summary of the current status of the Bodenstein ($H_2 + I_2$) "text book reaction" given by *J.B. Anderson*, who presented experimental and theoretical results which suggest that at low temperatures the reaction proceeds by a direct bimolecular reaction involving vibrationally excited I_2 molecules and a termolecular reaction mechanism, $H_2 + I + I \rightarrow HI + HI$. Both are among the possible reactions already suggested by *Max Bodenstein* about hundred years ago. In the following paper *J.V. Michael* reports on recent advances made in the direct measurement of high-temperature bimolecular rate constants, while the paper presented by *I.W.M. Smith* deals with the other extreme: reaction kinetics investigations at ultra-low temperatures.

Part IV again is a theoretical one, in which different approaches for the calculation of state-specific and thermal rate data are described. The article by *A.F. Wagner* presents a new approach to describe the influence of hindered rotations on recombination/dissociation kinetics in the framework of transition state theory. In the papers by *D.C. Clary* and *G. Nyman* an approximate quantum mechanical method is described and used to calculate thermal rate coefficients for gas phase reactions of interest in atmospheric chemistry which involve polyatomic molecules. Finally, different approaches to describe vibrational relaxation of diatoms in thermal collisions are discussed by *E.E. Nikitin.*

Part V includes papers presented in the "heterogeneous reactions"-session, which was opened by *G. Ertl*, who started by highlighting *Max Bodenstein's* impact on the field of heterogeneous chemical catalysis and then continued by presenting some fascinating examples from a rich variety of phenomena, like oscillatory and chaotic kinetics, which were observed in the catalytic oxidation of carbon monoxide on Pt(110) single crystal surfaces. The paper by *D.M. Golden et al.* describes experimental studies on the chemical interaction between gaseous species with both atmospherically relevant liquid surfaces and soot particles, and emphasizes the importance of heterogeneous processes in the chemical balance of the stratosphere. The following contribution by *J. Warnatz, F. Behrendt* and co-workers reviews the current status achieved in the numerical simulation of heterogeneous reaction systems. Hydrocarbon ignition and its chemical kinetic modelling is discussed with reference to a wide range of experimental and practical configurations (e.g., internal combustion engines) in the paper presented by *C.K. Westbrook.*

Part VI is devoted to the modelling of turbulence, reactive flows and complex chemical reaction systems. The paper presented by *M. Baum* describes the possibilities of direct numerical simulation (DNS) methods with detailed chemical reaction kinetics for the modelling of turbulent combustion. The following paper by *Vit.A. Volpert et al.* reviews recent experimental and theoretical results on the stability of reaction fronts. In the article by *J.A. Miller* and *P. Glarborg* an improved chemical kinetic model for the selective non-catalytic reduction in the "Thermal De-NO_x" process is presented. The final contribution of the symposion was given by *U. Maas*, in which he describes a new approach for the systematic and mathematically correct simplification of chemical kinetics starting from a detailed reaction mechanism. This work points to a promising way of including even complex chemical kinetics into reactive flow simulation codes for practical applications.

Although it is almost impossible to cover all aspects of this rapidly developing research field in one book, we feel that the articles in this volume together with the list of references given, reflect the diversity and importance, as well as the exciting challenge the investigation of gas phase reaction systems still offers.

To thank those who contributed their efforts and energy to make the current symposion a success is a special pleasure for us. Among the first to be mentioned is *Mrs. Sylvia Boganski*; without her work behind the scenes the organization of the symposion would not have been possible. Our thanks are due to our students *Ralph Tadday*, *Lüko Willms* and in particular to *Thomas Laurent* and to *Dr.*

Rajesh K. Vatsa (on sabbatical leave from BARC Bombay), who were a great help during the whole conference. The very pleasant atmosphere created by *Mrs. Dr. Th. Reiter* and the staff of the IWH as well as the support from *Peter Hochstein, Werner Weis* and the technical staff of the Physikalisch-Chemisches Institut (PCI) in organizing the poster session also deserves special thanks.

We are very grateful to the "Deutsche Forschungsgemeinschaft" and the SFB 359 at the University of Heidelberg, whose financial support made this symposion possible. The partial financial support received from the Alexander von Humboldt-Stiftung *via* the donation of a 1993 Max-Planck research award is also gratefully acknowledged. Finally, we want to thank *Dr. R.A. Brownsword* (EC HCM fellow at the PCI) for his help in the technical editing and *Dres. H.K.V. Lotsch* and *W. Skolaut* at Springer-Verlag as well as *Mrs. Ch. Pendl* for their constant advice during the preparation of this volume.

Heidelberg,
June 1996

J. Wolfrum
H.-R. Volpp
R. Rannacher
J. Warnatz

References

[1] M. Bodenstein und V. Meyer: Ber. dtsch. chem. Ges. **26**, 1146 (1893); M. Bodenstein: *ibid.*, **26**, 2603 (1893), II Mitteil.

[2] Bunsen, Roscoe: Pogg. Ann. **117**, 536 (1862).

[3] Van't Hoff: Études de dynamique chimique, p. 50 ff. (Amsterdam 1884).

[4] V. Meyer, Krause, Askenasy: Ann. d. Chem. **264**, 85 (1891); *ibid.* **269**, 85 (1892).

[5] Hautefeuille: Compt. rend. **64**, 608 (1867).

[6] Lemoine: Ann. chim. phys. (5) **12**, 145 (1877).

[7] M. Bodenstein: Z. phys. Chem. **13**, 56 (1894).

[8] M. Bodenstein: Z. phys. Chem. **22**, 23 (1897).

[9] K.H. Ebert, P. Deuflhard, and W. Jäger (Eds.), ***Modelling of Chemical Reaction Systems,*** Springer Series in Chem. Phys. Vol. **18** (Berlin, Heidelberg: Springer-Verlag, 1980).

[10] J. Warnatz and W. Jäger (Eds.), ***Complex Chemical Reaction Systems: Mathematical Modelling and Simulation***, Springer Series in Chem. Phys. Vol. **47** (Berlin, Heidelberg: Springer-Verlag, 1987).

[11] E. Cremer: *Max Bodenstein in memoriam*, in: Ber. dtsch. chem. Ges. **100**, XCV–CXXVI (1967) (incl. list of Max Bodenstein's publications).

[12] M. Bodenstein: ***Gasreaktionen in der chemischen Kinetik,*** Habilitationsschrift, Universität Heidelberg (Leipzig, Engelmann, 1899).

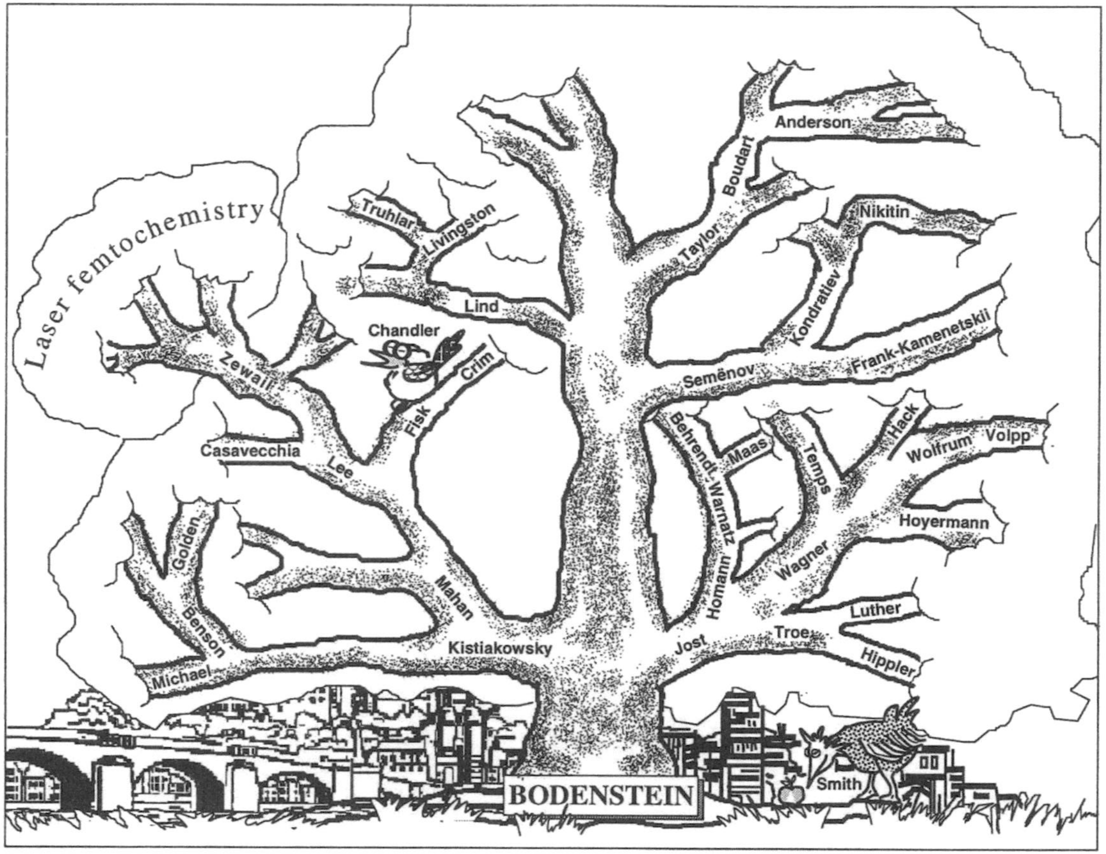

Figure I: "Bodenstein family tree" – Prepared during the conference.

Contents

Part VI Modelling of Flow, Turbulence and Complex Chemical Reactions

Part I

Microscopic Dynamics of Elementary Reactions: Experiment

Femtochemistry and Max Bodenstein's Impact

A.H. Zewail
Arthur Amos Noyes Laboratory of Chemical Physics,
California Institute of Technology, MailCode 127-72,
Pasadena, California 91125, U.S.A.

Abstract

This article gives a summary of the presentation made in tribute to Max Bodenstein at the conference entitled "Gas Phase Chemical Reaction Systems: Experiments and Models 100 Years after Max Bodenstein", in Heidelberg, during the period July 25 to 28, 1995.

1 Introduction

In 1894 Max Bodenstein (1871–1942), at the University of Heidelberg, published a landmark paper [1] which has played an important role in the development of gas-phase chemical kinetics. In these investigations [1,2], he reported the rate measurements for the "text-book" example of the hydrogen-iodine reaction

$$H_2 + I_2 \rightarrow HI + HI \quad (1)$$

and the reverse hydrogen-iodine decomposition reaction,

$$2\,HI \rightarrow H_2 + I_2, \quad (2)$$

Bodenstein's work triggered significant other developments in the understanding of elementary reaction mechanisms. It raised fundamental questions pertinent to Arrhenius' concept of the "activated state", Hinshelwood's and Kistiakowsky's ideas of the nature of activation by collisions, and the application by Eyring of absolute rate theory (just developed) to Bodenstein's reactions. Besides his scientific impact, he also spawned a new generation of "quantitative kinetists", among them George Kistiakowsky who did his Ph.D. work in Bodenstein's group (in Berlin) and continued this tradition in the United States at Harvard University.
Bodenstein's era enjoyed numerous successes and fundamental contributions both by him and by other giants of the time. Among them, see Fig. 1, are J. H. van't Hoff, S. Arrhenius, W. Ostwald, W. Nernst, C.N. Hinshelwood, N.N. Semënov, M. Polanyi, H. Eyring, R. Norrish,.... Our students will recognize these names in their textbooks on kinetics and dynamics, and they should enjoy learning the rich historical evolution and dynamics that led to such discoveries and achievements. Today's celebration of Max Bodenstein's contributions is a celebration of an era which has impacted many fields.

The field of femtochemistry has its roots in these contributions. The theoretical insights of van't Hoff and Arrhenius before the turn of the century, of Polanyi and Eyring in the thirties, and the experimental approach to quantitative kinetics by Bodenstein and others, promoted the desire for elemental description of the dynamics. At the heart of this problem are three essentials: the theoretical foundation describing the time scale for bond breaking and bond making, the experimen-

Springer Series in Chemical Physics, Volume 61
Gas Phase Chemical Reaction Systems
Eds.: J. Wolfrum, H.-R. Volpp, R. Rannacher, and J. Warnatz

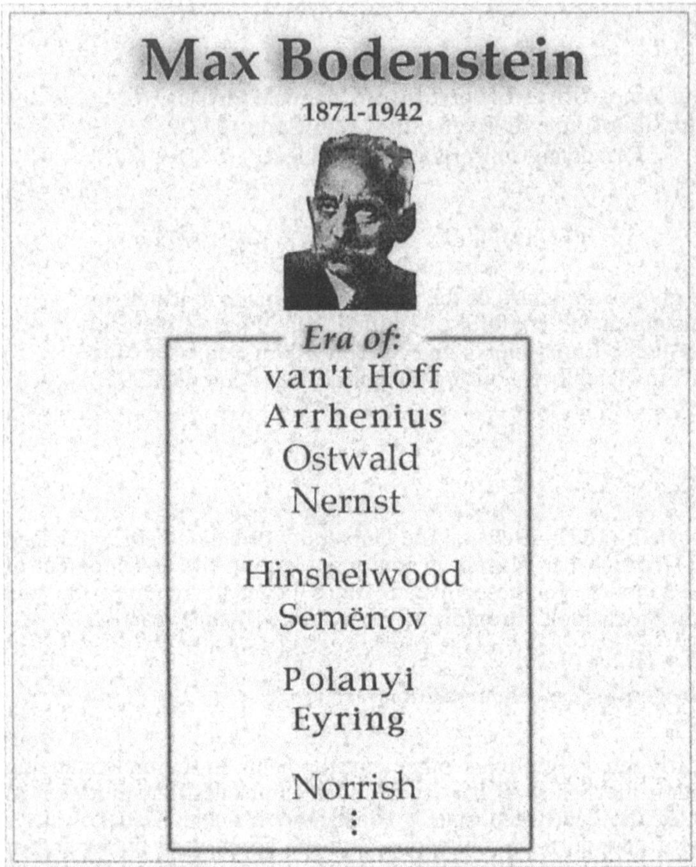

Figure 1: Max Bodenstein and the era of chemical kinetics.

tal advancement of new methods for tracking the elementary steps, and the development of concepts for observing and describing the atomic motions during an elementary process.

In 1931, four years after the breakthrough of the Heitler-London quantum treatment of the H + H problem, Eyring and Polanyi provided the first potential energy surface describing the motion of three atoms (Fig. 2):

$$H_\alpha + H_\beta - H_\gamma \rightarrow H_\alpha - H_\beta + H_\gamma$$

With this type of potential, J. O. Hirschfelder, Eyring and B. Topley (1936) provided the first trajectory calculations for the same reaction. Although in those days the time steps were often expressed in atomic units (!), the steps of the trajectory were in femtoseconds (Fig. 2). The work led to a clearer picture of the nature of Arrhenius' activated state, what is now termed the *transition state*.

Since this early theoretical work in reaction dynamics, the "arrow of time" (Fig. 3) in kinetics has continued to search for better resolution, reaching the femtosecond (fs) resolution almost half a century later. Experimentally, an important stride was made in this search for improved time resolution in kinetic measurements around the middle of the century. The contributions by Norrish and G.

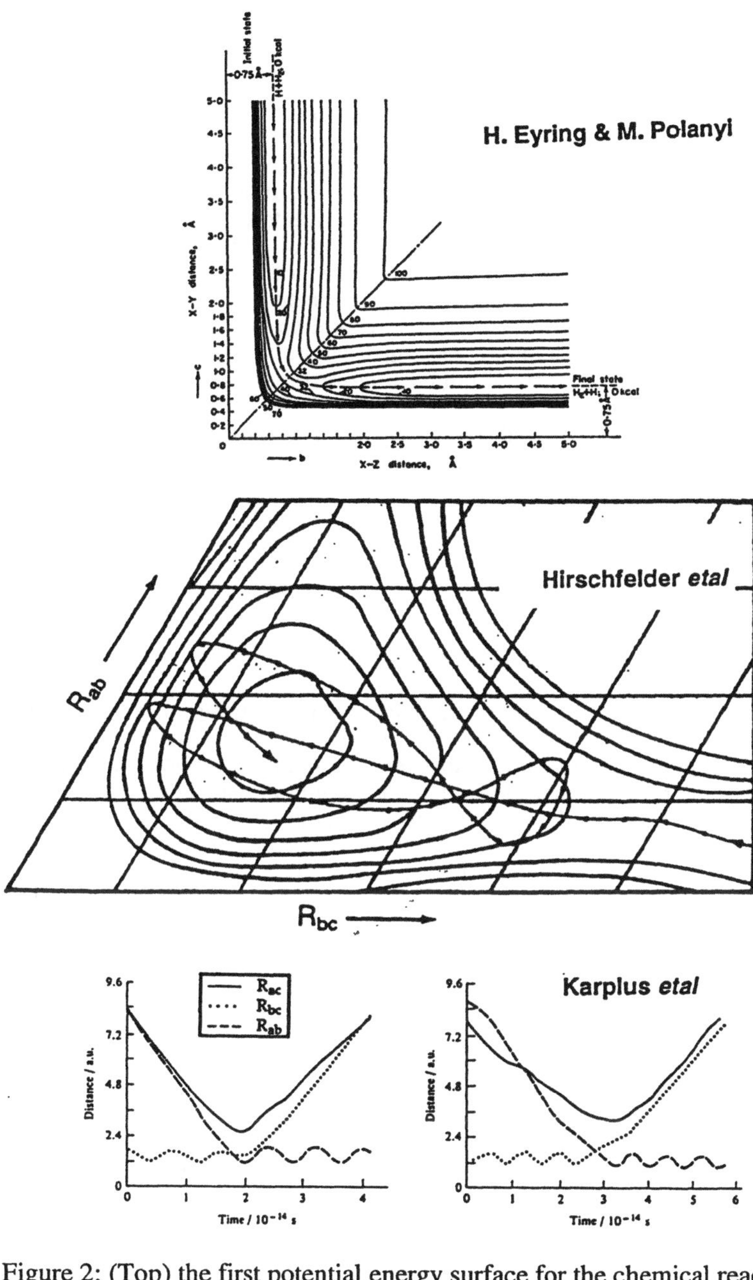

Figure 2: (Top) the first potential energy surface for the chemical reaction $H + H_2$; (middle) the first trajectory calculations showing the fs steps for the same reaction; and (bottom) the first trajectory calculation on realistic potentials of $H + H_2$ emphasizing the time scale of the transition state (~ 10 fs) - See text and ref. [3] for more details.

Porter (flash photolysis) and by M. Eigen (relaxation methods) made it possible to reach the millisecond and then the microsecond time resolution. In the sixties, the development of molecular beams and chemiluminescence method by D.R. Herschbach, Y.T. Lee and J.C. Polanyi, introduced a time clock, from studies of angular and product-state distributions, in the studies of elementary reactions.

With the advent of femtosecond lasers, it became possible to observe in real time the actual motion of nuclei and to study the elementary mechanisms pictured by Bodenstein in his description of gas-phase reactions. In all branches of femtochemistry, this study of elementarity is basic and is due to the inherent resolution achieved in femtochemical studies. Since the velocity of atoms in reactions is ~1 km/sec, with 10 fs resolution the distance scale reached is ~ 0,1 Å, the atomic scale of motion. As discussed below, this ability to create such localized, coherent wave packets with the atomic scale of distance resolution was part of the development of quantum mechanics as a theoretical construct, but was not an experimental reality until the development of the required time resolution of motion in atoms, molecules, and reactions.

The talk, which is summarized here, focuses on examples of femtochemical studies from the myriad of applications studies in different phases and around the world. Of particular interest are reactions involving multiple-center transition states, and, of historical interest, those related to Bodenstein and his fellow giants. The experimental approach in femtochemistry is detailed elsewhere [3] and will not be discussed here.

2 Concepts in Femtochemistry

In 1926, Erwin Schrödinger introduced the idea of a wave group in order to make a natural connection between quantum and classical descriptions [4]. One year later, P. Ehrenfest published his famous theorem [5], the correspondence principle, outlining the regime for the transition from a quantum to classical description – the quantum expectation values behave classically in the classical limit.

The use of wave groups or wave packets in physics, and certainly in chemistry, was limited to a few theoretical examples in the applications of quantum mechanics. The solution of the time-dependent Schrödinger equation for a particle in a box, or for a harmonic oscillator, and the elucidation of the uncertainty principle by superposition of waves are two of these examples. However, essentially all theoretical problems are presented as solutions in the time-independent frame picture. In part, this practice is due to the desire to start from a quantum-state description. But, more importantly, it was due to the lack of *experimental* ability to synthesize wave packets.

Even on the picosecond (1×10^{-12} second) time scale, the molecular systems are in eigenstates, and there is only one evolution, the change of *population* with time from that state. Thus, with this time resolution, which opened up numerous applications in chemistry and biology, one is mainly concerned with kinetics, not dynamics, but now on the picosecond time scale.

On the femtosecond time scale, an entirely new domain emerges. First, a wave packet can be prepared, as the temporal resolution is sufficiently short to "freeze" the nuclei at a given internuclear separation. Put in another way, the time resolution is much shorter than the vibrational (and rotational) motions such that the wave packet is prepared, highly localized with a de Broglie wave length of ~ 0.1 Å, with the structure frozen. Second, this synthesis is not in violation of the

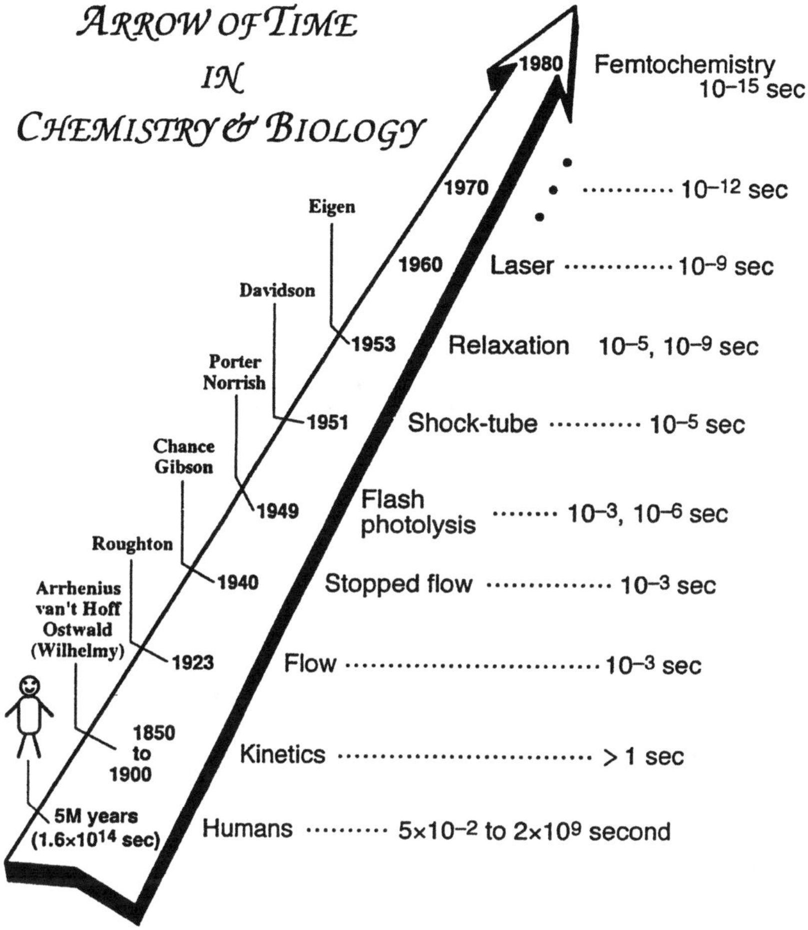

Figure 3: The "arrow of time" in chemistry and biology drawn to describe some strides in real-time studies [3]. For other developments, such as molecular beam and chemiluminescence studies in the kinetics of elementary reactions, see text.

uncertainty principle, as the key here is the *coherent* preparation of the system. It has been shown in, e.g., a 2-atom system (iodine) that because its width is < 0.2 Å, the wave packet oscillates spatially and executes distance changes between 2 to 5 Å depending on the energy. The preparation and probing is done coherently and only as such can one see the bond stretch and compress, and the molecule rotate. For non-reactive and reactive systems, the same picture applies. Third, because of this coherent synthesis, the transition from *kinetics* to *dynamics* is made, as one is able to monitor the evolution at the atomic resolution of motion with all nuclei "glued" together!

There is a fourth important point: Ehrenfest's classical limit is actually reached on this time scale for molecular systems. The spreading of the wave packet turned out to be not a problem, contrary to anticipation, and we now know why [3]. This can easily be seen by considering the motion of a Gaussian packet in

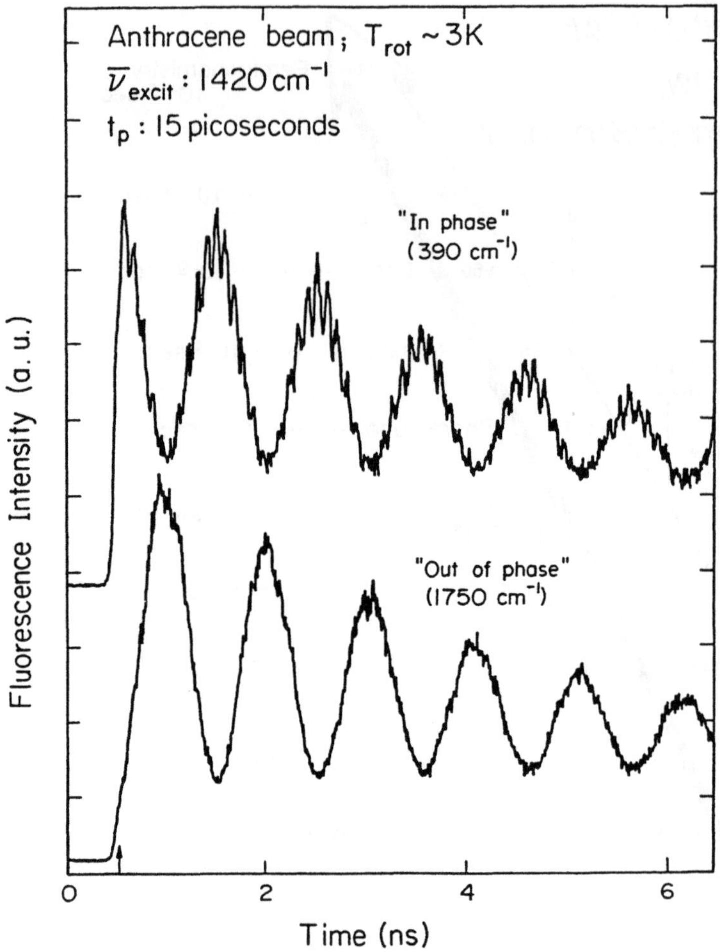

Figure 4: The first coherent, vibrational wave packet observed in a large molecule (anthracene) with 66 vibrational degrees of freedom in a molecular beam [see publications in ref. 3]. Observations in all classes of reactions followed since this 1980 work, but with femtosecond instead of the picosecond resolution [see ref. 3 and Fig. 5].

free space. The wave packet disperses, but not significantly on the fs time scale. The dispersion time is given by

$$\tau_d = 2m\{\Delta R^2(t = 0)\}\hbar^{-1}$$

and is in the ps time domain for most molecular systems [3]. Note that $\Delta R(t = 0)$ relates to the momentum ΔP by the uncertainty relationship. Thus, the fs resolution provides the required ΔP (ΔE), which in turn gives the sub-Ångström ΔR resolution in complete accord with the uncertainty principle. As mentioned before, the key is the *coherent* nature of the experiment at preparation and probing and the time resolution, which is *much shorter* than the vibrational and rotational times of the motion.

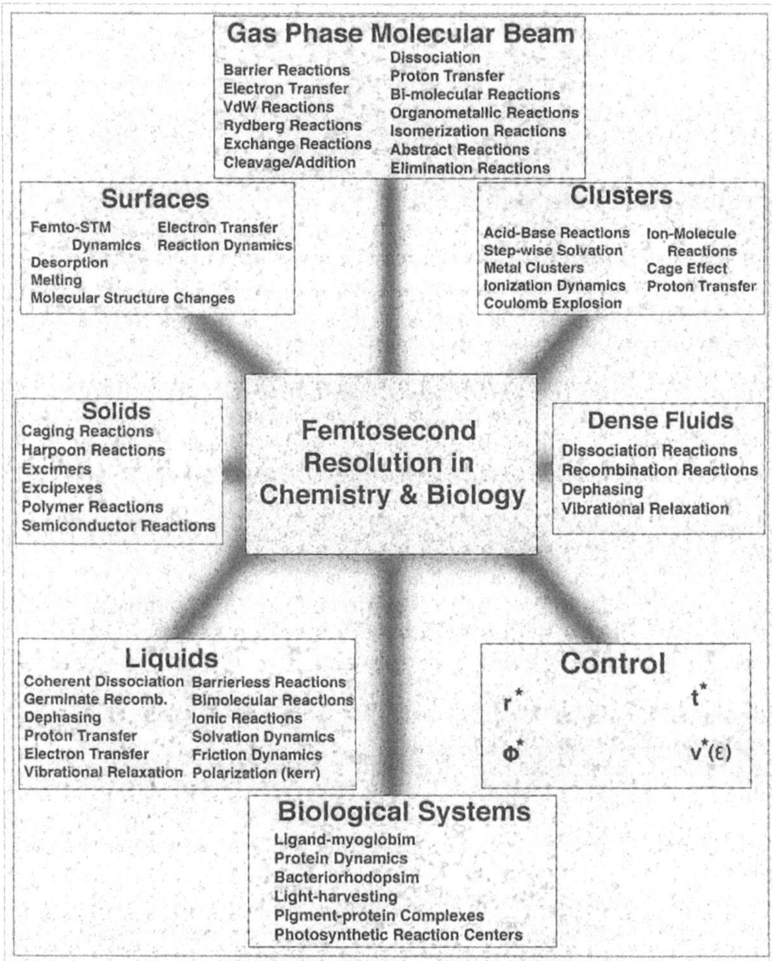

Figure 5: Applications of the femtosecond resolution in chemistry and biology, covering the different phases of matter and the new direction for the control of chemical dynamics [See, e.g., ref. 8, 9].

These concepts provide a temporal image of the dynamics. We can speak of the motions in time in the same way we conceive of them. In one sense, we now can speak of classical reactions, molecules and atoms [6]. Even in complex molecules these wave packets can be prepared and studies. In fact, the first experimental trigger of these ideas was made (1980) on an *isolated large* molecule with 66 vibrational degrees of freedom (anthracene) in a molecular beam (Fig. 4).

On the theoretical side, especially for molecular systems, a major step forward was made when E. Heller reformulated the time dependent picture for applications in spectroscopy and J. Kinsey and D. Imre described their novel dynamical Raman experiments in terms of the wave packet theory. The progress was greatly helped by advances made in the theoretical execution and speed of computation made by R. Kosloff and subsequently by many others. There is a parallelism between the experimental diversity of applications in different areas (Fig. 5) and the impressive theoretical applications to many experiments and systems. J. Manz, who has played a significant role in this field, has given an overview to the recent progress made in the two volumes he edited with L. Wöste [7].

3 Examples from Femtochemistry

Here, we will not give a review of the phases and systems studies so far (Fig. 5). Instead, we will provide references to the examples discussed in the lecture for our own work on the different reactions spanning elementary and complex systems. Details of the findings can be found in the original references given.

3.1 Benchmark Systems for Elementary, Unimolecular, and Bimolecular Reactions:

a) Real-Time Femtosecond Probing of "Transition States" in Chemical Reactions.
M. Dantus, M.J. Rosker, and A.H. Zewail
J. Chem. Phys. **87**, 2395 (1987)

Femtosecond Real-Time Probing of Reactions. I. The Technique.
M.J. Rosker, M. Dantus, and A.H. Zewail
J. Chem. Phys. **89**, 6113 (1988)

b) Femtosecond Real-Time Observation of Wave Packet Oscillations (Resonance) in Dissociation Reactions.
T. S. Rose, M. J. Rosker, and A. H. Zewail
J. Chem. Phys. **88**, 6672 (1988)

Femtosecond Real-Time Probing of Reactions. IV. The Reactions of Alkali Halides.
T. S. Rose, M. J. Rosker, and A. H. Zewail
J. Chem. Phys. **91**, 7415 (1989)

c) Femtochemistry of the Reaction: $IHgI^* \rightarrow [IHg\cdots I]^{\ddagger *} \rightarrow HgI+I$.
R. M. Bowman, M. Dantus, and A. H. Zewail
Chem. Phys. Lett. **156**, 131 (1989)

Femtosecond Real-Time Probing of Reactions. V. The Reaction of IHgI.
M. Dantus, R. M. Bowman, M. Gruebele, and A. H. Zewail
J. Chem. Phys. **91**, 7437 (1989)

Femtochemistry of the Reaction of IHgI: Theory Versus Experiment.
M. Gruebele, G. Roberts, and A. H. Zewail
Philos. Trans. Roy. Soc. London A **332**, 223 (1990)

d) Real-Time Picosecond Clocking of the Collision Complex in a Bimolecular Reaction: The Birth of OH from $H + CO_2$.
N. F. Scherer, L. R. Khundkar, R. B. Bernstein, and A. H. Zewail
J. Chem. Phys. **87**, 1451 (1987)

Real-Time Clocking of Bimolecular Reactions: Applications to H + CO_2.
N. F. Scherer, C. Sipes, R. B. Bernstein, and A. H. Zewail
J. Chem. Phys. **92**, 5239 (1990)

Femtosecond Real-Time Probing of Reactions. VIII. The Bimolecular Reaction of Br + I_2.
I. R. Sims, M. Gruebele, E. D. Potter, and A. H. Zewail
J. Chem. Phys. **97**, 4127 (1992)

Now we consider more complex systems, with examples from organic, inorganic, and physical chemistry.

3.2 Pericyclic Reactions: Woodward-Hoffmann's Classic

The Validity of The "Diradical" Hypothesis: Direct Femtosecond Studies of the Transition-State Structures.
S. Pedersen, J. L. Herek, and A. H. Zewail
Science **266**, 1359 (1994)

3.3 Norrish's Reactions

Direct Femtosecond Observation of the Transient Intermediate in the α-Cleavage Reaction of $(CH_3)_2CO$ to $2CH_3$ + CO: Resolving the Issue of Concertedness.
S. K. Kim, S. Pedersen, and A. H. Zewail
J. Chem. Phys. **103,** 477 (1995)

3.4 Elimination Reactions

a) Picosecond Monitoring of a Chemical Reaction in Molecular Beams: Photofragmentation of R-I $\rightarrow R^{\ddagger}$ + I.
J. L. Knee, L. R. Khundkar, and A. H. Zewail
J. Chem. Phys. **83**, 1996 (1985)

b) Picosecond Photofragment Spectroscopy. IV. Dynamics of Consecutive Bond Breakage in the Reaction $C_2F_4I_2 \rightarrow C_2F_4 + 2I$.
L. R. Khundkar and A. H. Zewail
J. Chem. Phys. **92**, 231 (1990)

3.5 Bodenstein-type Reactions

a) Kinetic-Energy, Femtosecond Resolved Reaction Dynamics: Modes of Dissociation (in Idobenzene) from Time-Velocity Correlations.
P. Y. Cheng, D. Zhong, and A. H. Zewail
Chem. Phys. Lett. **237**, 399 (1995)

b) Femtosecond, Velocity-Gating of Complex Structures in Solvent Cages.
P. Y. Cheng, D. Zhong, and A. H. Zewail
J. Phys. Chem., submitted for publication

3.6 Organometallic Reactions

Femtochemistry of Organometallics: Dynamics of Metal–Metal and Metal–Ligand Bond Cleavage in $M_2(CO)_{10}$.
S. K. Kim, S. Pedersen, and A. H. Zewail
Chem. Phys. Lett. **233**, 500 (1995)

3.7 Charge-Transfer Reactions

a) Transition States of Charge-Transfer Reactions: Femtosecond Dynamics and the Concept of Harpooning in the Bimolecular Reaction of Benzene with Iodine.
P. Y. Cheng, D. Zhong, and A. H. Zewail
J. Chem. Phys. **103**, 5153 (1995)

b) Microscopic Solvation and Femtochemistry of Charge-Transfer Reactions: The Problem of Benzene(s)-Iodine Binary Complexes and Their Solvent Structures.
P. Y. Cheng, D. Zhong, and A. H. Zewail
Chem. Phys. Lett. **242**, 368 (1995)

3.8 Acid-Base Reactions

a) Real-Time Probing of Reactions in Clusters.
J. J. Breen, L. W. Peng, D. M. Willberg, A. Heikal, P. Cong, and A. H. Zewail
J. Chem. Phys. **92**, 805 (1990)

b) Solvation Ultrafast Dynamics of Reactions: VIII. Acid-Base Reactions in Finite-sized Clusters of Naphthol in Ammonia, Water and Piperidine.
S. K. Kim, J. J. Breen, D. M. Willberg, L. W. Peng, A. Heikal, J. A. Syage, and A. H. Zewail
J. Phys. Chem. **99**, 7421 (1995)

3.9 Barrier-Crossing Reactions

a) Picosecond Dynamics and Photoisomerization of Stilbene in Supersonic Beams. II. Reaction Rates and Potential Energy Surface.
J. A. Syage, P. M. Felker, and A. H. Zewail
J. Chem. Phys. **81**, 4706 (1984)

b) Rates of Photoisomerization of trans-Stilbene in Isolated and Solvated Molecules: Experiments on the Deuterium Isotope Effect and RRKM Behavior.
P. M. Felker and A. H. Zewail
J. Chem. Phys. **89**, 5402 (1985)

c) Microscopic Friction and Solvation in Barrier Crossing: Isomerization of Stilbene in Size-Selected Hexane Clusters.
A. A. Heikal, S. H. Chong, J. S. Baskin, and A. H. Zewail
Chem. Phys. Lett. **242**, 380 (1995)

4 Concluding Remarks

I hope that this talk has succeeded in describing some of the continued and exciting research in molecular reaction dynamics over the 100 years since Max Bodenstein's publication. I am personally delighted to take part in this historic conference; I now can relax after giving this opening lecture and enjoy the contributions on the program which I look forward to listening to. I wish to give special thanks to Professor Dr. Jürgen Wolfrum and Dr. Hans-Robert Volpp for their generous care and excellent organization. Finally, without the contributions of members of my research group, mentioned during the lecture and in the above references, I would not be able to present to you this story on femtochemistry.

5 Acknowledgments

This work was supported by the Air Force Office of Scientific Research and by the National Science Foundation.

6 References

[1] M. Bodenstein, Z. phys. Chem. **13**, 56 (1894).
[2] M. Bodenstein, Z. phys. Chem. **22**, 1 (1897); *ibid.*,**29**, 295 (1899).
[3] A. H. Zewail, ***Femtochemistry: Ultrafast Dynamics of the Chemical Bond, Vol. I & II***; and articles therein.
[4] E. Schrödinger, Ann. Phys. **79,** 489 (1926).
[5] P. Ehrenfest, Z. Phys. **45**, 455 (1927).
[6] B. Garraway, K.-A. Suominen, Reports on Progress in Physics **58**, 365 (1995).
[7] J. Manz, L. Wöste (Eds.), ***Femtosecond Chemistry, Vol. 1,2***, VCH, Weinheim (1995).
[8] B. Kohler, J. L. Krause, F. Raksi, K. R. Wilson, V. V. Yakovlev, R. M. Whitnell, Y. J. Yan, Acc. Chem. Res. **28**, 133 (1995).
[9] J. C. Polanyi, A. H. Zewail, Acc. Chem. Res. **28**, 119 (1995).

Laserspectroscopic Studies of Bimolecular Elementary Reaction Dynamics in the Gas Phase

H.-R. Volpp and J. Wolfrum
Physikalisch-Chemisches Institut, Universität Heidelberg
Im Neuenheimer Feld 253, 69120 Heidelberg, Germany

Abstract

In the present article we give an overview of recent work carried out in our laboratory in order to study microscopic details of bimolecular gas phase reactions at the molecular level using the laser photolysis / laser-induced fluorescence (LP/LIF) "pump-and-probe" technique. In particular, we will focus on the following three- and four-atom reactions: $H + O_2 \rightarrow O + OH$, $H + CO_2 \rightleftharpoons CO + OH$ and $H + H_2O \rightleftharpoons H_2 + OH$, each of them playing an important role in atmospheric and combustion chemistry. In recent years, these reactions have become prototype systems in the development of full-dimensional quantum mechanical reactive scattering methods and the computation of the necessary accurate *ab initio* potential energy surfaces. We shall present absolute reactive cross sections and nascent OH product vibrational and rotational fine-structure state distributions, measured over a wide range of collision energies to investigate in detail the influence of reagent translational excitation on reactivity and reaction dynamics. The experimental results allow comparison with quasiclassical and recent quantum mechanical scattering calculations on *ab initio* potential energy surfaces.

1 Introduction

Since the days of summer 1894, when Max Bodenstein carried out in Heidelberg his pioneering studies on "the decomposition of the hydrogen iodine gas in the (sun) light", which finally led him to the conclusion that the decomposition of the hydrogen iodine is caused by the light in such a way that "every light wave of suitable wavelength" decomposes one HI molecule it hits *via* a simple $HI \rightarrow H + I$ unimolecular mechanism [1], the development of flash photolysis [2], chemiluminescence [3], molecular beam techniques [4], and in particular the development of lasers with their high temporal, spectral and spatial resolution dramatically expanded the experimental possibilities for the investigation of thermal and state-to-state reaction kinetics [5] and actually paved the way for detailed studies of the microscopic dynamics of uni- and bimolecular elementary reactions [6–10].

Nowadays, femtosecond laser "pump-and-probe" techniques – as pioneered by Ahmed Zewail and his group at the California Institute of Technology [11] – can provide us with fascinating series of snap shots of how reactive systems behave while passing over the transition state [12] and nanosecond laser "pump-and-probe" techniques allow to measure asymptotic scalar and vectorial quantities of the reactive collision; e.g. nascent quantum state distributions of reaction products [13], absolute reaction cross sections [14], state-specific rates [15], and stereodynamical correlations [16], respectively.

Springer Series in Chemical Physics, Volume 61
Gas Phase Chemical Reaction Systems
Eds.: J. Wolfrum, H.-R. Volpp, R. Rannacher, and J. Warnatz

In the present article we will concentrate on a set of gas phase elementary reactions (1–3), which play a central role in combustion [17] and atmospheric chemistry [18], and which have in recent years become important benchmark systems for comparison between the results of experimental reaction dynamics studies and quasiclassical, approximate and – most recently – exact quantum mechanical reactive scattering calculations.

For the following reactions,

$$H + O_2 \rightarrow O + OH \qquad \Delta H_0 = 69.5 \text{ kJ/mol}, \qquad (1)$$

$$H + H_2O \rightarrow H_2 + OH \qquad \Delta H_0 = 62 \text{ kJ/mol}. \qquad (2)$$

$$H + CO_2 \rightarrow CO + OH \qquad \Delta H_0 = 102 \text{ kJ/mol}, \qquad (3)$$

global ground state potential energy surfaces, based on *ab initio* calculations, are available [19,20,21]. Because all of the reactions (1-3) have considerable reaction barriers, it is necessary to generate highly translationally excited H atoms in order to make studies on the dynamics of reactive collisions possible. In the studies which we will describe in the following, H atoms with well-defined translational energies were generated by UV laser photolysis of HX-type precursor molecules. OH radicals produced in the reactions (1–3) were detected by means of laser-induced fluorescence (LIF) under single-collision conditions with quantum state resolution, allowing the determination of the nascent vibrational and rotational fine-structure distribution of the OH products, as well as the measurement of the absolute reaction cross section *via* a calibration method. We shall also briefly discuss results from experiments with velocity-aligned H atoms in which vector correlation were determined.

In addition, the dynamics of the reverse reactions (-2) and (-3) were investigated using a similar method. In this case translationally excited OH radicals were generated by laser photolysis of H_2O_2 and the H atoms produced in the reactions were detected by means of VUV-LIF at the Lyman-α-transition. We will compare the experimental results with results from dynamical simulations e.g. quasiclassical trajectory (QCT) calculations and – if possible – with the results obtained by applying quantum scattering (QMS) methods.

2 Experimental Method: The Laser Photolysis / Laser-Induced Fluorescence (LP/LIF) "pump-and-probe" Technique

Following early experiments in which highly translationally excited reactants for chemical studies were generated by nuclear reactions [22] or flash-lamp photolysis of appropriate precursor compounds [23], it was Quick and Tiee [24] who first used the LP/LIF method for a study of the reactions (1,3). In these experiments, translationally "hot" H atoms were generated by laser irradiation of static HBr/O_2 and HBr/CO_2 mixtures and the time history of OH production was monitored. Shortly after this, for the same reactions, nascent OH rotational state distributions measured under single collision conditions in a flow system were reported by Kleinermanns and Wolfrum [25], who also introduced a method to measure absolute reaction cross sections for the gas phase reactions (1–3). Further LP/LIF–studies were performed in a molecular beam containing $CO_2\cdots HX$ van der Vaals complexes to investigate the influence of reactant orientation (as defined by the different $CO_2\cdots HX$ geometries for X = HS, Cl, Br, I) on reactivity and the OH product state distributions of reaction (3) [26]. Zewail and co-workers carried out the first real-time dynamics measurements of a bimolecular reactive collision [27]. In these studies, reaction (3) was photo-initiated within a $CO_2\cdots HI$ complex (containing the two potential reagents H and CO_2 in close proximity) using a picosecond UV "pump"-pulse (to photodissociate selectively the HI) and the formation of OH was detected by a delayed picosecond "probe"-pulse *via* LIF. By stepwise changing the delay between the"pump" and "probe"-pulse it was possible to measure the time evolution of the OH appearance, which is a direct measure of the lifetime of the $HOCO^{\dagger}$ reaction intermediate. More experimental details about this real-time method can be found in Ref. 11. We will focus in the following on the nanosecond laser "pump-and-probe" technique, which we used to investigate the gas phase reaction dynamics of bimolecular reactive collisions.

2.1 Reaction Dynamics Studies Employing Translationally Excited H Atom Reagents and OH Product Detection

The experimental setup used to study the reactions H + $O_2/H_2O/CO_2$ is outlined in detail elsewhere [28], and general construction principles of flow systems to be used in combination with laser detection techniques can be found e.g. in [29], so only a brief description will be given in the following.

The experiments were carried out in a Teflon coated quartz cell equipped with long sidearms in which special baffle systems were included, to keep scattered light (produced mainly by the photolysis laser pulse in the flow cell windows) out of the fluorescence collection optics. Translationally excited H atoms were generated by pulsed UV laser photolysis of the HX-type precursor (X = HS, Cl, Br, I) continuously flowing through the reactor together with the stable reagent (O_2, CO_2, and H_2O, respectively) in order to avoid accumulation of reaction products. All experiments were carried out at room temperature, at a total pressure of typically 40-100 mTorr with an HX:reagent ratio between 1:7 and 1:10. The photolysis wavelengths used were: 193 nm for HCl, HBr and H_2S, 248 nm for HI and H_2S, and 266 nm for HI. With the above precursor-wavelength combinations, H atoms with well defined center-of-mass (c.m.) frame collision energies ranging from 1 eV up to 2.6 eV could be obtained. The photolysis laser beam was provided by an excimer laser operating with ArF (193 nm) or KrF mixture (248 nm). The fourth harmonic of a Nd:YAG laser was used to photodissociate HI at 266 nm.

Typically 60-150 ns after the photodissociation "pump" pulse, nascent OH product radicals were probed by a second copropagating UV "probe" laser beam by LIF in the $A^2\Sigma^+ - X^2\Pi$ system as schematically depicted in Fig. 1. The probe beam with a bandwidth of 0.2 cm^{-1} was provided by a frequency-doubled dye laser, pumped by a XeCl excimer laser and was tunable in the wavelength region 305–320 nm in order to scan OH LIF excitation spectra. Line positions for the different OH fine-structure transitions were taken from Ref. 30. Both lasers were operated at 20 Hz and each data point in a LIF excitation spectrum was obtained by averaging the collected fluorescence signal over 40–80 laser shots. The measured LIF spectra were stored on a computer for further data processing. OH vibrational and rotational fine-structure state distributions were determined by numerical integration of

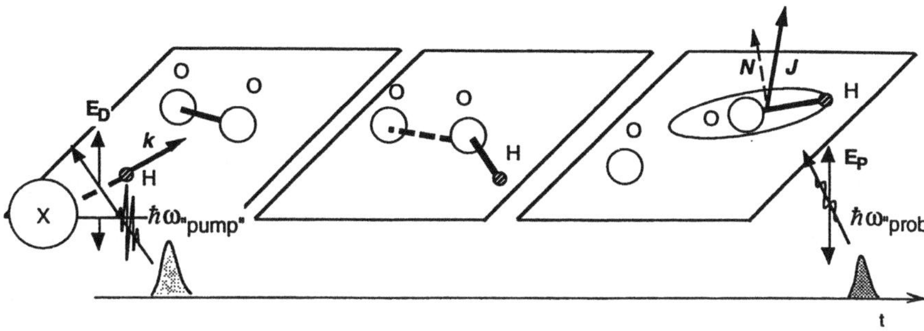

Figure 1: Schematic describtion of the LP/LIF "pump-and-probe" method.

the spectral lines and normalization with respect to the corresponding Einstein coefficients of absorption [31].

For the absolute reaction cross section measurements a calibration method was used, in which the photodissociation of H_2O_2 was utilized as a well characterized OH radical source in order to calibrate the unknown OH number densities produced in the reactions. The entire experimental procedure is described in detail in Ref. 28. For vector correlation studies using velocity-aligned H atoms the photolysis laser beam could be linearly polarized by means of a 10-plate Brewster-angle stack-polarizer. In these experiments the OH probe laser beam was also completely linearly polarized using a Glan-Taylor polarizer. A $\lambda/2$ plate inserted into the laser beam path allowed to adjust the probe laser polarization vector (denoted as $\mathbf{E_P}$ in Fig.1) to any desired angle with respect to the direction of the photodissociation laser (denoted as $\mathbf{E_D}$ in Fig.1) [32]. The way e.g. the OH total angular momentum center-of-mass frame alignment parameter $A_0^{(2)}$(c.m.) = 2 $\langle P_2(\boldsymbol{J}\cdot\boldsymbol{k})\rangle$ – the brackets imply an expectation value averaged over all angles – can be derived from OH LIF intensity measurements for different pump/probe geometries and polarization alignments is described in Ref. 33. $\langle P_2(\boldsymbol{J}\cdot\boldsymbol{k})\rangle$ can range from –0.5 to +1 and is a measure of the alignment of the OH total angular momentum $\boldsymbol{J}$ with respect to the initial H atom fragment recoil velocity vector $\boldsymbol{k}$. In the limiting case $\langle P_2(\boldsymbol{J}\cdot\boldsymbol{k})\rangle = -0.5$, $\boldsymbol{J}$ is perpendicular to $\boldsymbol{k}$ while in the other limit, $\langle P_2(\boldsymbol{J}\cdot\boldsymbol{k})\rangle = +1$, $\boldsymbol{J}$ is aligned parallel to $\boldsymbol{k}$. A experimental method based on the detailed analysis of OH Doppler line shapes was used by Hall and co-workers in their H + O_2 studies [34]. Basic principles of state-resolved stereodynamics experiments and the analytical procedure used in their analysis and interpretation are described in detail elsewhere [16].

2.2 Reaction Dynamics Studies Employing Translationally Excited OH Reagents and H Atom Product Detection

The experimental setup used to study the reactions OH + CO/H_2 is depicted schematically in Fig. 2. Due to the need to directly connect the VUV light generation cell to the reactor, a crossed laser beam arrangement had to be used. The experiments were carried out at room temperature in mixtures of H_2O_2 and CO and H_2, respectively, flowing through a stainless steel flow cell pumped by a rotary pump with a flow rate high enough to prevent accumulation of reaction products. Typically the H_2O_2:CO, H_2O_2:H_2 ratio was between 1:8 and 1:10 with the total cell-pressure being 50-120 mTorr.

Translationally excited OH radicals were generated by laser photolysis of

H_2O_2 at two different wavelengths (248 and 193 nm) allowing dynamics investigations at two markedly different c.m. collision energies. The UV photodissociation dynamics of H_2O_2 has been characterized in great detail in previous studies where the energy partitioning into the OH vibrational, rotational and relative translational degree of freedom was determined [35]. Absorption cross sections and OH quantum yields were also measured [36]. In Ref. 37 it has been found that at a photodissociation wavelength of 193 nm almost all OH radicals are produced in the vibrational ground state with a Gaussian-like rotational state distribution centred at $K \approx 12$ with a FWHM of $\Delta K \approx 7$. The quantum number $K = N + \Lambda$ (where Λ denotes the projection of the electronic orbital momentum on the OH internuclear axis), as defined in Hund's case b [38], is used to denote the $OH(^2\Pi)$ rotational states. In order to determine the energy partitioning we measured the OH vibrational and rotational state distribution at a wavelength of 248 nm and found that here also OH is exclusively produced in the vibrational ground state with a Gaussian OH(v = 0) rotational state distribution centred at $K \approx 7$ with a FWHM of $\Delta K \approx 8$. In accordance with the results from the analysis of measured OH Doppler profiles the fraction of the available energy which appears as relative OH translational energy in the laboratory frame in the H_2O_2 photolysis was determined to be $f_{trans} = 0.82$ at 248 nm and $f_{trans} = 0.83$ at 193 nm [37]. From these values the collision energy distributions in the OH–reagent c.m. system can be calculated [38,39].

The VUV probe laser beam, used to detect the produced H atoms by means of one-photon LIF at the Lyman–α transition ($\lambda_{VUV} = 121.567$ nm), was generated by a resonant sum-difference frequency mixing scheme, $\omega_{VUV} = 2\omega_R - \omega_T$, in a krypton/argon gas mixture. The output of the dye laser A (Fig. 2), operated at a fixed wavelength of 425.1 nm was, after being frequency doubled in a BBO II crystal, in resonance with the two-photon transition 4p–5p(1/2,0) in krypton ($\lambda_R = 212.55$ nm). Dye laser B was tuned in the wavelength range $\lambda_T = 844$–846 nm in order to measure H atom Doppler profiles. As depicted in Fig. 2, both laser beams (λ_R, λ_T) were focused in a cell containing the krypton/argon mixture at a typical total pressure of 200 mbar, in order to generate the VUV radiation. Both dye lasers were simultaneously pumped by a XeCl excimer laser.

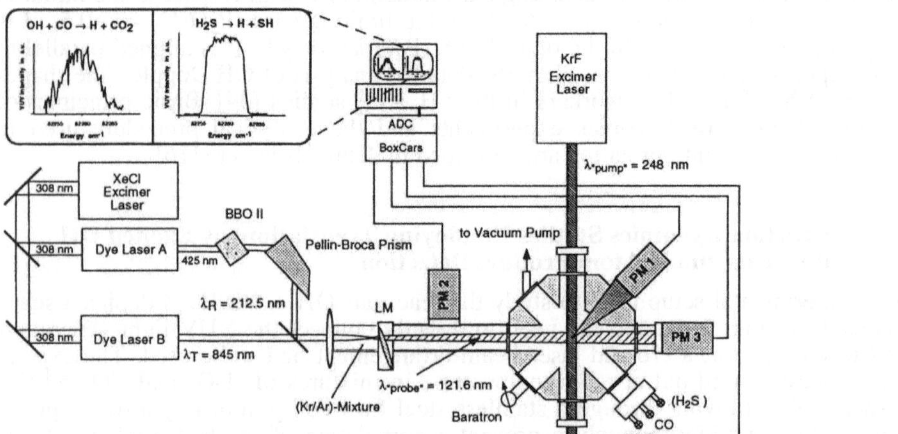

Figure 2: Schematic diagram of the experimental "pump-and-probe" setup as used in the studies of the $OH + CO \rightarrow H + CO_2$ reaction in which H atoms are detected by VUV-LIF.

The generated VUV probe light was separated from the fundamental laser beams by a lens monochromator (LM in Fig. 2) to avoid photochemical perturbations due to the fundamental UV laser radiation. The photolysis laser beam, with an intensity of typically 2–4 mJ/cm^2, was aligned perpendicularly with respect to the VUV probe laser beam and adjusted in order to provide an optimal overlap in the viewing region of the "solar blind" LIF photomultiplier (denoted as PM 1 in Fig. 2), which was equipped with a band pass filter. Another photomultiplier of the same type (PM 2) was used to monitor the VUV probe laser intensity. The photomultiplier signals as well as the signal from the photodiode, which was used to monitor the photolysis laser intensity behind the reaction cell, were fed into a 4-channel boxcar system and transferred, via an analogue to digital converter, to a microcomputer where the H atom LIF signal was normalized to both the VUV probe and the photolysis laser intensity. To improve the S/N ratio, each point of the recorded H spectral lines from reaction OH + CO and from the H_2S photolysis has been obtained by averaging over 30 laser shots. The temporal synchronization of the experiment was performed by a pulse generator. Delay times between the photolysis and the VUV probe laser pulse were 60–120 ns, measured using a fast oscilloscope.

3 The H + $O_2 \rightarrow$ O + OH Reaction

The reaction

$$H(^2S) + O_2(^3\Sigma_g^-) \rightarrow O(^3P) + OH(^2\Pi) \quad (1)$$

is of fundamental importance in combustion chemistry as a chain branching step in the ignition of H_2/O_2 mixtures and in the oxidation of hydrocarbons [40]. Already in 1928 – shortly after Semenov [41] developed the general theory of chain reactions [42] – Haber and Bonhoeffer [43] suggested in the course of their systematic spectroscopic studies of flame processes the first chain reaction mechanism for the oxidation of hydrogen, in which the reaction H + O_2 was considered as a main chain branching step. A number of subsequent experiments carried out at Haber's institute in Berlin aimed at proving this hypothesis [44]. Since then, many experimental studies on the thermal kinetics [45] and the microscopic dynamics of reaction (1) were carried out [25a,46]. In addition, a number of theoretical investigations were performed using statistical [17a], quasiclassical [19a,46b,47] and recently quantum mechanical methods [48].

In the reaction dynamics experiments [46] it was observed that with increasing collision energy the nascent OH(v = 0) product rotational state distributions continuously change from a statistical one at $E_{c.m.}$ = 1.0 eV to a non-statistical one at $E_{c.m.}$ = 2.5 eV. This behaviour could be well reproduced by QCT calculations and was attributed to a change of the underlying reaction mechanism from a HO_2^* complex forming mechanism at lower energies to a direct one at intermediate and higher collision energies [46b,47b]. A analysis of measured vibrational and OH (v = 1) rotational distributions further suggests that with increasing collision energy the formation of highly rotationally excited OH in the vibrational ground state is favoured and that the amount of energy available for relative translation of the recoiling OH and O products increases with collision energy [46g]. At all collision energies investigated, a preference in the population of the symmetric $\Pi(A')$ state of the OH products was observed [46a-d]. This preference for the $\Pi(A')$ population – as depicted in

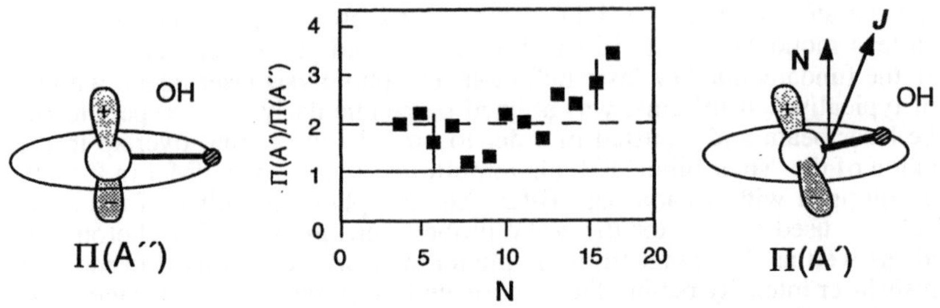

Figure 3: Population ratio of the Λ-components (averaged over the spin-orbit components) of OH radicals produced in the reaction $H + O_2$ at $E_{c.m.} = 1.9$ eV.

Fig. 3 – was attributed to a planar reaction mechanism. Here the reaction occurs preferentially for collisions where the H atom velocity vector lies in the plane defined by the three atoms (see Fig. 1), leaving the lobes of the singly occupied OH π-orbital – which point along the direction of the broken bond – in the plane of the three atoms. A QCT study in which the degree of planarity of reactive collisions was evaluated further supports this interpretation [46b]. In Fig. 3. the orientation and symmetry of the OH π-orbital (as defined in the Hund's case b limit) with respect to the plane of rotation is depicted schematically for the two different Λ-components.

In addition, experiments with velocity aligned H atoms (generated by linear polarized photodissociation of H_2S at 193 nm) carried out by Hall and co-workers [34] showed a strong perpendicular alignment of the total OH angular momentum $\boldsymbol{J} = \boldsymbol{K} \pm \boldsymbol{S}$ (depending on the spin-orbit state) with respect to the final OH velocity vector for high-J Π(A′) states. For the similar high-J of the Π(A′′) component a considerably lower $\boldsymbol{J}$ polarization was observed. Taking into consideration the strongly forward-peaked differential cross section, this is consistent with the value of $\langle P_2(\boldsymbol{J}\cdot\boldsymbol{k})\rangle = -(0.11 \pm 0.04)$ we determined for the alignment of $\boldsymbol{J}$ with respect to the initial H atom fragment recoil velocity vector $\boldsymbol{k}$ at high-J Π(A′′) states at $E_{c.m.} = 2.2$ eV [49]. However, this value is considerably lower than the value determined in earlier measurements at $E_{c.m.} = 2.6$ eV, in which OH fluorescence intensity ratios were observed, which suggests a completely perpendicular angular momentum polarization ($\boldsymbol{J} \perp \boldsymbol{k}$: $\langle P_2(\boldsymbol{J}\cdot\boldsymbol{k})\rangle \approx -0.5$) for the Π(A′′) component [50].

In Ref. 51, the relative population of the $O(^3P_{j=2,1,0})$ product fine-structure states were determined at different collision energies. The energy dependence in the change of the population of the $O(^3P_{j=2,1,0})$ fine-structure states from a diabatic one at $E_{c.m.} = 1.6$ eV to an adiabatic one at $E_{c.m.} = 2.5$ eV was attributed to a reduction of nonadiabatic coupling in the exit channel due to the increase of translational energy released to the O + OH products with increasing collision energy [46g].

Relative and absolute reaction cross section measurements for reaction (1) were carried out in the energy range 1.0 eV < $E_{c.m.}$ < 2.6 eV. The experimental results as depicted in Fig. 4 show a pronounced maximum at $E_{c.m.} \approx 1.7$ eV, a feature which could not be reproduced by earlier QCT calculations on different *ab initio* PESs [47a-e]. In quantum scattering calculations, where reaction probabilities for total angular momentum equals zero (J = 0) were calculated, it was observed that reactivity is not enhanced by either initial O_2 vibrational or rotational excitation [48a]. On the other hand, it was found that reactivity markedly increases with increasing collision energy, in particular at the opening of the OH(v = 1) product channel at $E_{c.m.} \approx 1.25$ eV [48a]. Recently a new extrapolation method was devel-

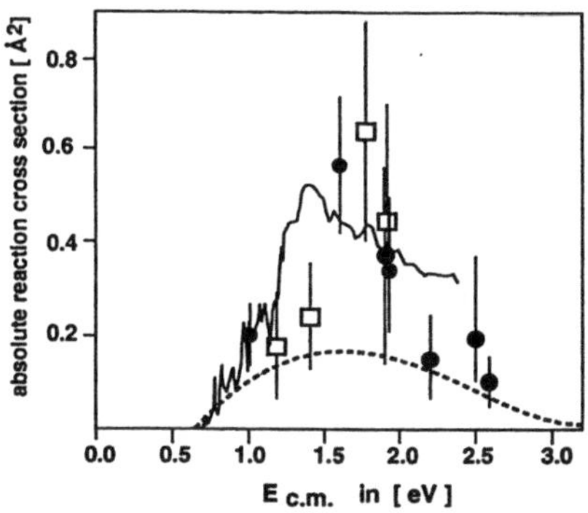

Figure 4: Absolute reaction cross section for $H + O_2 \rightarrow OH + O$ as a function of collision energy. Filled circles [46g] and open squares [46f] are experimental results. Dashed line represent QCT [47a], solid line QCT/QM-QJ [52] results.

oped by Varandas [52], which allows the calculation of the absolute reaction cross section for reaction (1) from the quantum mechanical (J = 0) reaction probability. In these QCT/(quantum mechanical-quadratic J) QM-QJ calculations a relatively sharp maximum of the excitation function, as depicted in Fig. 4, could be observed for the first time. Although the maximum of the theoretical curve is slightly shifted to lower energies compared to the experimental values, the theoretical results are very encouraging. In the QCT/QM-QJ calculations total angular momentum values up to J = 57 had to be included. This, as a consequence, makes rigorous quantum mechanical calculations a formidable task.

Up to now all the $H(^2S) + O_2(^3\Sigma_g^-)$ reactive scattering calculations were carried out on a single-valued $^2A''$ ground state surface which in C_s geometry correlates directly *via* the ground state of the HO_2 radical to the products $O(^3P)$ and $OH(^2\Pi)$ [53a]. However, there are several energetically low-lying electronically excited states which correlate with the same products and could therefore be involved in the real reaction process [53b]. The true ground state surface should have two conical intersections, one for C_{2v} and one for $C_{\infty v}$ geometry. Recently, using a DIM (diatomics in molecules) model, a global PES was derived which includes both of these conical intersections [54]. Most recently, accurate 3D quantum scattering calculations were carried out for thermal collision energies in which geometric phase effects due to the C_{2v} conical intersection were included [55]. In these calculations, significant changes due to the geometric phase effect were observed in the state-to-state transition probabilities suggesting that also at higher collision energies the "Berry"-phase could influence the dynamics of reaction (1).

4 The $H + H_2O \rightleftharpoons H_2 + OH$ System

Among four-atom reactions this system has become a prototype test case for the comparison between quantum state resolved dynamics experiments and bimolecular collision theory. In 1973 the first semi-empirical LEPS and BEBO PESs were constructed by Zellner and Smith [56]. The subsequent development of a global H-HOH(A´) ground state PES in 1980 [20b], based on *ab initio* calculations [20a], was favoured by the fact that three of the four atoms involved are H atoms. The

Walch-Dunning-Schatz-Elgertsma (WDSE) PES was since then used in a number of dynamical calculations, and in the following we will compare our experimental results with results obtained in some of the most recent quantum mechanical ones.

4.1 The $H + H_2O \rightarrow H_2 + OH$ Reaction

For this reaction the WDSE PES has a barrier of 0.923 eV and an overall endoergicity of 0.659 eV [20b]. Furthermore, it was found that the reaction proceeds via a bent and planar H–HOH transition state, with the HOH angle being close to that of the free water molecule [57]. Schatz *et al.* performed QCT studies in which the influence of reagent translation and vibrational excitation on reactivity and the product energy partioning was investigated [58], and obtained good agreement with experimental studies by the groups of Crim [59a,b], Zare [59c,d], and Wolfrum [28,60], where the influence of vibrational and translational excitation on the reaction cross section was investigated. These calculations showed that vibrational excitation is much more efficent in promoting reactivity than translational excitation. This behaviour is consistent with the general principle introduced by J.C. Polyani for the effects of translational *versus* vibrational energy on the reactive cross section of endothermic bimolecular reactions with a "late" barrier [61].

Experiments carried out to investigate the influence of translational excitation on the internal state distributions of the OH products [28] revealed further interesting details. In these studies, which covered the collision energy range 1.0 eV < $E_{c.m.}$ < 2.5 eV, it was observed that OH is almost exclusively produced in the vibrational ground state with only very low rotational excitation; typically only a few percent of the available energy were found in the OH rotational degree of freedom. The observed absence of vibrational excitation of the OH product was attributed to a stripping mechanism in which the reaction proceeds adiabatically with respect to the nonreacting bond. This kind of spectator behaviour of the OH bond can be explained by the fact that the OH bond length is almost unchanged in going from the H_2O reagent through the H–HOH transition state to the OH product [57].

Different models have been developed to explain the small rotational excitation of the OH product. An impulsive model was used in Ref. 62, in which the recoil of the H_2 product from the O atom *via* a bent transition-state configuration produces the torque leading to OH product rotational excitation. Within this model, the degree of OH rotational excitation is actually limited by the small mass of the departing H_2 fragment and the small "lever arm" given by the distance between the center of mass of the OH molecule and the O atom. In Ref. 58a, a stripping model was described, where OH rotational excitation is produced by simply mapping H_2O reagent zero–point bend excitation onto the OH product rotation. Following this idea, a more sophisticated four-atom adiabatic-bend Franck-Condon (ABFC) model was developed to study the influence of H-HOH transition-state bending excitation on the final OH and H_2 rotational state distributions [63a]. In this calculation, it was found that H-HOH bending excitation of up to five quanta leads to generally cold OH rotational state distributions, as observed in the experiments for collision energies up to 2.5 eV [28], suggesting that only a small number of low-lying H-HOH bending states are excited during the reaction. In Fig. 5 a) the experimental OH rotational state distribution (averaged over the four fine-structure components) for $E_{c.m.}$ = 1.0 eV is depicted and compared to theoretical results. ABFC results [63a] where obtained by projecting the wave function of the H-HOH bend ground state onto the free rotor base of the OH molecule. The observed reasonable agreement between experiment and the ABFC calculations suggest that at energies close to the reaction threshold H rotational excitation is dominated at least to a large extent simply by the ground bending state of the H-HOH transition state.

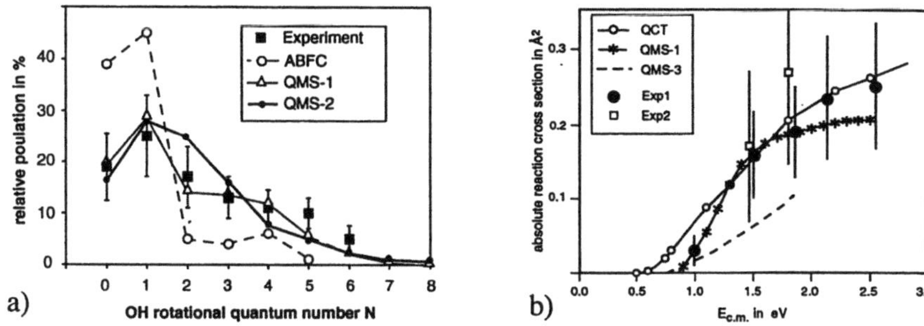

Figure 5: Comparison between experiment and theory for the H + H_2O reaction: a) OH(v = 0) product rotational state distribution at $E_{c.m.}$ = 1.0 eV, b) absolute reactive cross sections. Appreviations are explained in the text.

The QMS-1 ($E_{c.m.}$ = 1.0 eV) and QMS-2 rotational distributions (obtained at a slightly lower collision energy of 0.9 eV) in Fig. 5a represent results from 3D quantum scattering calculations obtained by Nyman and Clary [64], and Szichman and Baer [65], respectively. In addition, the OH(v = 0) rotational state distributions (see [28]) and the absolute reactive cross sections of Nyman and Clary (QMS-1 values in Fig. 5b), and of Schatz and co-workers [58] (QCT values in Fig. 5b) were found to be in reasonable agreement with the measured ones also at the higher energies studied in the experiments [28]. The QMS-3 cross section values in Fig. 5b are recent results from full-dimensional (6D) quantum mechanical (J = 0) calculations carried out by Zhang and Light [66]. Quantum mechanical reactive cross sections obtained by Baer and co-workers for $E_{c.m.} \leq 1.5$ eV are presented in the article by Baer *et. al.* in this volume. More details regarding the theoretical methodology can be found in the articles by Nyman, Clary, and Baer *et al.* in this volume. The fact that the QCT cross sections are slightly larger than the experimental and the QMS cross sections at low collision energies is due probably to the deficiency of classical mechanics in conserving the zero point energy of the reactants.

As in case of reaction (1), the analysis of the measured OH fine-structure state distributions showed a pronounced preference for populating symmetric ("in-plane") Π(A´) Λ-states of the OH products, reflecting the higher cross section for OH(A´) over OH(A´´) formation. This can be explained using a simple classical model based on the analysis of the normal mode eigenvector of the imaginary frequency of the H–HOH transition state, which showed that this mode is mainly an OH stretch [63b]. In this picture, the preference in the population observed for the OH(A´) states simply reflects the fact that a perpendicular ("out-off-plane") collision of H with H_2O (which would lead to the formation of OH in the A´´ state) has less chance to couple with the O–H stretching vibration, which can move the system along the reaction coordinate. A planar approach, on the other hand, has a higher chance to couple with the O–H stretching, resulting in a higher cross section for OH(A´) formation with the unpaired π-orbital lobes in the plane of the reagent approach. A more detailed discussion of fine-structure effects in reactive and nonreactive bimolecular collisions can be found in the papers of Macdonald and Liu [67a].

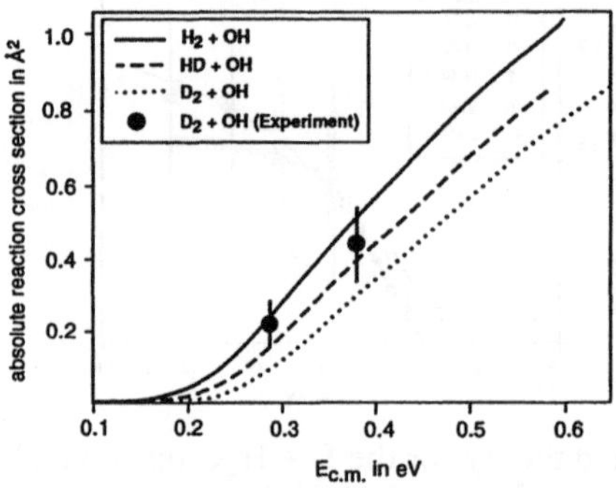

Figure 6.: Absolute reaction cross sections for $D_2 + OH \rightarrow D + HOD$ compared to results from 5D quantum scattering calculations.

4.2 The $H_2 + OH \rightarrow H + H_2O$ Reaction

This diatom-diatom reaction, as one of the most "simple" four-atom reactions, has become a test system towards the development of rigorous quantum reactive scattering methods [68]. The reaction is characterized by a direct mechanism with no potential well being present. This, together with the "light" masses involved, makes this reaction favourable for the application of exact quantum methods [69]. On the experimental side, thermal rate coefficients [70], absolute [60] and differential cross sections [71] have been measured.

In the following we will compare (see Fig. 6) absolute reactive cross sections measured for the isotopically substituted reaction $D_2 + OH \rightarrow D + HOD$ with recent rigorous 5D time-dependent quantum wave-packet calculations [74]. The theoretical values shown in Fig. 6 were obtained for OH and D_2 in the rotational and vibrational ground state, while in the experiment OH is highly rotational excited (see section 2.2) and the D_2 rotational states are populated according to a room temperature ($T_{rot} \approx 300$ K) Boltzmann distribution (in the latter, the most probable rotational state of D_2 is $j = 2$). In Ref. 72 a slightly higher (about a factor of 1.5) total reaction probability was observed for $D_2(j = 2)$ compared to that of $D_2(j = 0)$. The quantum scattering calculations, however, have not yet been extended to such high OH rotational states as are populated in the experiments, which would definitely be necessary for a more detailed comparison between theory and experiment. A comparison between results from approximate quantum scattering calculations [73] and full-dimensional QCT studies and our experiments can be found in Ref. 74.

5 The $H + CO_2 \rightleftharpoons CO + OH$ System

This reaction system is of basic importance for oxidation processes occurring in flames, because in hydrocarbon combustion CO is oxidized to CO_2 almost exclusively by the reaction with OH radicals, and the reaction of hydrogen atoms with CO_2 is an important step in order to establish the water-gas equilibrium. As a result, both reactions have a strong influence e.g. on flame propagation [17a].

5.1 The $H + CO_2 \rightarrow CO + OH$ Reaction

Direct high temperature measurements of the thermal rate coefficient were carried out recently using the shock tube technique [75]. Earlier experimental data were reviewed in Ref. 17a. A number of gas-phase reaction dynamics studies where carried out employing the hot H atom technique. Detailed investigations of inelastic collisions between H atoms and CO_2 were performed by Flynn and Weston and are reviewed in Ref. 76. CO internal state distributions were reported in Ref. 77, and relative reaction cross sections and OH vibrational and rotational state distributions were measured for a variety of collision energies by Wittig and co-workers [78] and analysed by Levine and co-workers [79a] using an information-theoretical approach [79b]. Absolute reaction cross section measurements were also carried out [14a,25b]. Recent measurements of absolute reaction cross sections at $E_{c.m.} = 1.9$, 2.3, 2.6 eV showed an increase of the cross section in the range $E_{c.m.} = 1.9–2.3$ eV (similar to the one observed in Ref. 78), followed by an decrease at higher energies [80]. The new results suggest that the H + CO_2 excitation function exhibits a pronounced maximum located somewhere between 1.9 and 2.6 eV. Such a maximum could so far not be reproduced in QCT calculations [58b] carried out on the Schatz-Fitzcharles-Harding (SFH) PES [21]. In these calculations, strong evidence was found at higher energies for a second reaction pathway besides the HOCO one, which proceeds *via* a transient HCO_2 formation. However, the HCO_2 region of the global SFH-PES is know to be represented in the global fit only with lower accuracy, particularly at high energies [81]. More *ab initio* values for the HCO_2 part of the PES and their inclusion into the global PES would be definitely needed in order to assess more accurately a possible contribution of this reaction pathway to the overall H + CO_2 excitation function at high collision energies.

In recent measurements using velocity aligned H atoms with $E_{c.m.} = 2.3$ eV, a almost isotropic OH total angular momentum distribution, $\langle P_2(\boldsymbol{J}\cdot\boldsymbol{k})\rangle \approx 0$, and a almost completely statistical Λ-doublet state distribution, $\Pi(A')/\Pi(A'') \approx 1$, was observed for the OH products [32]. This, together with the measured OH(v = 1) / OH(v = 0) ratio, which was found to be close to the statistical "prior" value [79b], would be consistent with the interpretation that the reaction proceeds *via* a complex forming mechanism with a lifetime of the order of a rotational period of the collision complex. In contrast to this, in similar experiments for the H + H_2O reaction carried out a comparable collision energy, a strong OH angular momentum alignment, $\boldsymbol{J}\perp\boldsymbol{k}$: $\langle P_2(\boldsymbol{J}\cdot\boldsymbol{k})\rangle = -0.5$, and a highly polarized Λ-doublet state distribution $\Pi(A')/\Pi(A'') \approx 4.5$ for OH(v = 0, K = 9) was observed, which clearly shows that this reaction proceeds *via* a direct mechanism with the H–HOH lifetime being considerably smaller than a rotational period.

5.2 The CO + OH → H + CO_2 Reaction

There have been numerous measurements of the thermal rate coefficient over a wide range of temperatures [82]. The unusual temperature dependence of the thermal rate coefficient was attributed to a complex mechanism in which the reaction proceeds *via* the formation of a $HOCO^\dagger$ reaction intermediate which can decompose at comparable rates either back to OH + CO or to H + CO_2 [56,83]. The overall rate of reaction was found to be pressure dependent [84], and the complete reaction mechanism has been suggested to be:

$$\begin{array}{ccccc} OH + CO & \rightleftharpoons & HOCO^\dagger & \longrightarrow & H + CO_2 \\ & & \downarrow +M & & \\ & & HOCO & & \end{array}$$

The existence of the HOCO radical has been confirmed both in the gas-phase [85a] and in matrices [85b]. State-to-state integral cross sections for OH rotational excita-

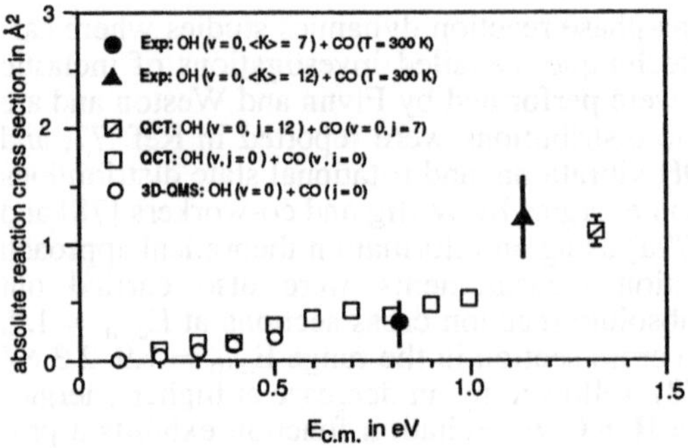

Figure 7.: Absolute reaction cross sections for CO + OH ⟶ H + CO_2 compared to results from QCT and quantum scattering calculations.

tion in collision with CO were measured by Liu and co-workers [67b], and in crossed molecular beam experiments a strong peaking of the reactive differential cross section both in forward and backward directions was found, suggesting the formation of a $HOCO^\dagger$ reaction complex [71]. The vibrational state distribution of CO_2 products were measured using infrared emission [86a] and the diode laser absorption technique [86b].

In state selected experiments, the influence of selective OH [87a] and CO [87b,c] vibrational excitation on the rate of reaction was investigated. Our recent experimental results, as depicted in Fig. 7, represent an extension of these studies towards the investigation of the influence of reagent translational excitation on the absolute reactive cross section [88]. The QCT [89] and approximate quantum scattering results (3D-QMS) [90] depicted in Fig. 7 were obtained using the SFH-PES [21]. In the 3D-QMS calculations the relative motion of the two molecules was restricted to one plane, the CO bond length was held fixed at its equilibrium value and internal rotation of the OH molecule was not allowed. In the QCT study, although all degrees of freedom are treated explicitly, the calculations were restricted to non-rotating OH and CO for $E_{c.m.} \leq 1$ eV. Further QCT calculations with rotationally excited reagents in the vibrational ground state showed that the variation of the cross section following OH and CO rotational excitation is only modest [89], the depicted theoretical values can therefore serve as a reasonable basis for comparison with our experimental values. At the collision energy $E_{c.m.}$ = 1.3 eV the OH and CO internal state distribution in the QCT simulations were similar to the experimental conditions (for CO at $T \approx 300$ K, $j = 7$ is the most probable CO rotational state). In agreement with the QCT data our results show that translational energy actually stimulates reactivity for CO + OH. In the QCT simulation this was found to be due to an increase of "direct reactive processes towards H + CO_2 product formation" in the decay of the $HOCO^\dagger$ intermediate formed in the collision. Here "direct processes" are characterized by a $HOCO^\dagger$ lifetime markedly shorter than several vibrational periods. However, for a more detailed comparison with experiment additional QCT studies which account in more detail for the experimental conditions as well as accurate quantum scattering studies of the OH + CO reaction [91] would be definitely desirable. In addition, state specific calculations using e.g. the statistical adiabatic channel model (SACM) [92], using saddle point parameters – carefully calibrated against the available thermal and state–specific rate data – could lead to new insights, in particular into the accuracy of the so far available PES data.

6 Acknowledgements

We gratefully acknowledge financial support of the Sonderforschungsbereich 359 ("Reaktive Strömungen, Diffusion und Transport"), the Alexander von Humboldt Stiftung, the Max-Planck-Gesellschaft, the European Community, and the Deutsche Forschungsgemeinschaft. Thanks are due to A.J.C. Varandas, D.H. Zhang, J.C. Light, and R.N. Zare for sending pre-prints of their publications and M. Baer, G.C. Schatz and G. Nyman for helpful communications.

7 References

[1] M. Bodenstein, Z. phys. Chem. **22**, 23 (1897).
[2] F.J. Lipscomb, R.G.W. Norrish, and B.A. Thrush, Proc. R. Soc. London Ser. A, **233**, 455 (1956).
[3] J.K. Cashion and J.C. Polanyi, J. Chem. Phys. **30**, 317 (1959).
[4] D.R. Herschbach, Adv. Chem. Phys. **10**, 319 (1966); Y.T. Lee and Y.R. Shen, Phys. Today **33**, 52 (1980); Y.T. Lee, Science **236**, 793 (1987); J. Durup, Laser Chem. **7**, 239 (1987) *"Reviewing major trends in the work of the 1986 Nobel prize winners in chemistry: D.R. Herschbach, J.C. Polanyi and Y.T. Lee"*.
[5] I.W.M. Smith in part III of this volume; M.J. Pilling, I.W.M. Smith (Eds.), ***Modern Gas Kinetics: Theory, Experiment and Application*** (Blackwell Scientific Publications, Oxford, 1987).
[6] A.H. Zewail (Ed.), ***Advances in Laser Chemistry***, Springer Series in Chem. Phys. Vol. **3** (Springer-Verlag, 1978); A. Ben-Shaul, Y. Haas, K.L. Kompa and R.D. Levine, ***Lasers and Chemical Change***, Springer Series in Chem. Phys. Vol. **10** (Springer-Verlag, 1981).
[7] R.D. Levine and R.B. Bernstein, ***Molecular Reaction Dynamics and Chemical Reactivity*** (Oxford University Press, 1987).
[8] M.N.R. Ashfold and J.E. Baggott (Eds.), ***Advances in Gas-Phase Photochemistry and Kinetics: Molecular Photodissociation Dynamics*** (The Royal Chemical Society, 1987).
[9] J. Jortner, R.D. Levine, B. Pulman (Eds.), ***Mode Selective Chemistry*** Vol. **24** (Kluwer Academic Publishers, 1991).
[10] S. Rosenwaks, M. Shapiro (Eds.), ***Lasers in Chemistry***, Special Issue of the Israel J. Chem. Vol. **34** *"in honour of 1993 Wolf Prize Recipient A.H. Zewail"* (Laser Pages Publishing, Jerusalem, 1994).
[11] A.H. Zewail, ***Femtochemistry: Ultrafast Dynamics of the Chemical Bond, I & II***, Vol. **3** (World Scientific Series in 20th Century Chemistry, 1994).
[12] H. Eyring, J. Chem. Phys. **3**, 107 (1935); M.G. Evans and M. Polanyi, Trans. Faraday Soc. **31**, 875 (1935).
[13] K.-H. Gerike, F.J. Comes, and R.D. Levine, J. Chem. Phys. **74**, 6106 (1981); E.E. Marinero, C.T. Rettner, and R.N. Zare, *ibid.* **80**, 4142 (1984); D.A.V. Kliner, D.E. Adelman, and R.N. Zare, *ibid.* **95**, 1648 (1991); K. Honda, M. Takayanagi, T. Nishiya, H. Ohoyama, and I. Hanazaki, Chem. Phys. Lett. **180**, 321 (1991); D.V. Lanzisera, and J.J. Valentini, J. Chem. Phys. **103**, 607 (1995).
[14] a) A. Jacobs, M. Wahl, R. Weller, J. Wolfrum, Chem. Phys. Lett. **158**, 161 (1989); b) A. Jacobs, H.-R.Volpp, and J. Wolfrum, *ibid.* **196**, 249 (1992); c) H.M. Lambert, T. Carrington, S.V. Filseth, and C.M. Sadowski, J. Phys.

Chem. **97**, 128 (1993); c) T. Laurent, P.D. Naik, H.-R.Volpp, J. Wolfrum, T. Arusi-Parpar, I. Bar, and S. Rosenwaks, Chem. Phys. Lett. **236**, 343 (1995); d) T. Laurent, H. Lillich, H.-R.Volpp, J. Wolfrum, A. Melchior, I. Bar, and S. Rosenwaks, Chem. Phys. Lett. **247**, 321 (1995).

[15] I.W.M. Smith in ***Advances in Gas-Phase Photochemistry and Kinetics: Bimolecular Collisions*** (M.N.R. Ashfold, J.E. Baggott Eds., The Royal Chemical Society, 1989); see also the article by I.W.M. Smith in this volume.

[16] a) J.P. Simons, J. Phys. Chem. **91**, 5378 (1987); b) F. Green, G. Hancock, A.J. Orr-Ewing, Faraday Discussion Chem. Soc. **91**, 79 (1991); c) M. Brouard, S.P. Duxon, P.A. Enriquez, and J.P. Simons, J. Chem. Phys. **97**, 7414 (1992); d) M. Brouard, S.P. Duxon, and J.P. Simons in Ref. 10 and references therein; e) W.R. Simpson, T.P. Rakitzis, S.A. Kandel, A.J. Orr-Ewing and R.N. Zare, J. Chem. Phys. **103**, 7313 (1995).

[17] a) J. Warnatz, in ***Combustion Chemistry*** (W.C. Gardiner Ed.; Springer-Verlag, 1984); b) J. Troe, J. Phys. Chem. **90**, 3485 (1986); c) J.A. Miller, R.J. Kee and C.K. Westbrook, Ann. Rev. Phys. Chem. **41**, 345 (1990).

[18] R.P. Wayne, ***Chemistry of Atmospheres*** (Clarendon, Oxford, 1985); J.I. Steinfeld, J.S. Fransisco and W.L. Hase, ***Chemical Kinetics and Dynamics*** (Prentice Hall, Englewood Cliffs, 1989).

[19] a) R.J. Blint, and C.F. Melius, Chem. Phys. Lett. **64**, 183 (1979); b) M.R. Pastrana, L.A.M. Quintales, J. Brañdao, and A.J.C. Varandas; c) V.J. Barclay, C.E. Dateo, and I.P. Hamilton, J. Chem. Phys. **101**, 6766 (1994).

[20] a) S.P. Walch, T.H. Dunning, J. Chem. Phys. **72**, 1303 (1980); b) G.C. Schatz, H. Elgersma, Chem. Phys. Lett. **73**, 21 (1980); c) A.D. Isaacson, J. Phys. Chem. **77**, 3516 (1992); d) T.H. Dunning, L.B. Harding, and E. Kraka in ***Supercomputer Algorithms for Reactivity, Dynamics, and Kinetics of Small Molecules*** (A. Lagàna Ed.; Kluver, Dordrecht, The Netherlands 1989); A new *ab initio* PES has been calculated by H.-J. Werner and co-workers recently (see the article by P. Casavecchia *et al.* in this volume).

[21] G.C. Schatz, M.S. Fitzcharles, L.B. Harding, Faraday Discuss. Chem. Soc. **84**, 359 (1987).

[22] L. Szilard, T.A. Chalmers, Nature **134**, 462 (1934).

[23] R.A. Ogg, R.R. Williams, J. Chem. Phys. **13**, 586 (1945); G.A. Oldershaw, D.A. Porter, Nature **223**, 490 (1969).

[24] C.R. Quick, J.J. Tiee, Chem. Phys. Lett. **100**, 223 (1983).

[25] a) K. Kleinermanns, and J. Wolfrum, Chem. Phys. Lett. **104**, 157 (1984); b) K. Kleinermanns, and J. Wolfrum, J. Chem. Phys. **80**, 1446 (1984).

[26] S.K. Shin, C. Wittig, and W.A. Goddard III, J. Phys. Chem. **95**, 8048 (1991).

[27] N.F. Scherer, L.R. Khundkar, R.B. Bernstein, and A.H. Zewail, J. Chem. Phys. **82**, 1451 (1987).

[28] A. Jacobs, H.-R. Volpp, and J. Wolfrum, J. Chem. Phys. **100**, 1936 (1994).

[29] J.G. Pruett, and R.N. Zare, J. Chem. Phys. **64**, 1774 (1976).

[30] G.H. Dieke, H.M. Crosswhite, J. Quant. Spectr. Rad. Transf. **2**, 97 (1962).

[31] I.L. Chidsey, D.R. Crosley, J. Quant. Spectr. Rad. Transf. **23**, 187 (1980).

[32] A. Jacobs, H.-R. Volpp, and J. Wolfrum, Chem. Phys. Lett. **218**, 51 (1994).

[33] C.H. Greene, and R.N. Zare, J. Chem. Phys. **78**, 6741 (1983).

[34] H.L. Kim, M.A. Wickramaaratchi, X. Zheng, and G.E. Hall, J. Chem. Phys. **101**, 2033 (1994).

[35] M.P. Docker, A. Hodgson, and J.P. Simons, in Ref. 8, pp. 115–137.

[36] G.L. Vaghjiani, A.A. Turnipseed, R.F. Warren, and A.R. Ravishankara, J. Chem. Phys. **96**, 5878 (1992); b) J.W. Schiffman, D.D. Nelson Jr., and D.J. Nesbitt, *ibid.* **98**, 6935 (1993).

[37] A. Jacobs, M. Wahl, R. Weller, and J. Wolfrum, Appl. Phys. B **42** 173 (1987).
[38] G. Herzberg, ***Molecular Spectra and Molecular Structure. Vol. 1. Spectra of Diatomic Molecules*** (Van Nostrand, Princeton, 1945).
[39] W. J. van der Zande, R. Zhang, R.N. Zare, K.G. McKendrick, and J.J. Valentini, J. Phys. Chem. **95**, 8205 (1991); S. Koppe, T. Laurent, P.D. Naik, H.-R. Volpp, J. Wolfrum, T. Arusi-Parpar, I. Bar, and S. Rosenwaks, Chem. Phys. Lett. **214**, 546 (1993).
[40] J. Warnatz, in ***Modelling of Chemical Reaction Systems,*** Springer Series in Chem. Phys. Vol. **18** (K.H. Ebert, P. Deuflhard, W. Jäger Eds., Berlin, Heidelberg, Springer-Verlag, 1980).
[41] N.N. Semenov, Nobel Lecture: *»Some Problems to Chain Reactions and to the Theory of Combustion«* in ***Nobel Lectures Chemistry, 1942–1962*** (Amsterdam, London, New York, 1964).
[42] The expression "chain reaction" was actually coined by Max Bodenstein see E. Cremer, *»Max Bodenstein in memoriam«*, in: Ber. dtsch. chem. Ges. **100**, XCV–CXXVI (1967).
[43] F. Haber, K.F. Bonhoeffer, *»Bandenspektroskopie und Flammenvorgänge«* Sitzungsber. d. Preuss. Akad. d. Wiss., Berlin, 1. März 1928; K.F. Bonhoeffer, F. Haber, Z. phys. Chem. **137**, 263 (1928).
[44] F. Haber, H.D. von Schweinitz, *»Über die Zündung des Knallgases durch Wasserstoffatome«* Sitzungsber. d. Preuss. Akad. d. Wiss., Physik. Math. Kl. 1928, **XXX**, 8; F. Haber, Z. Angew. Chem. **42**, 570 (1929); L. Farkas, P. Goldfinger, und F. Haber, Naturwissenschaften **18**, 266 (1930).
[45] see e.g. the article by J.V. Michael in this volume.
[46] a) K. Kleinermanns, E. Linnebach, J. Wolfrum, J. Chem. Phys. **89**, 2525 (1985); b) K. Kleinermanns, E. Linnebach, M. Pohl, *ibid.* **91**, 2181 (1989); c) M. Bronikowski, R. Zhang, D. J. Rakestraw, R. N. Zare, Chem. Phys. Lett. **156**, 7 (1989); d) A. Jacobs, F.M. Schuler, H.-R. Volpp, M. Wahl, J. Wolfrum, Ber. Bunsenges. Phys. Chem. **94**, 1390 (1990); e) A. Jacobs, H.-R. Volpp, J. Wolfrum, Chem. Phys. Lett. **177**, 200 (1991); f) K. Keßler, K. Kleinermanns, J. Chem. Phys. **97**, 374 (1992); g) S. Seeger, V. Sick, H.-R. Volpp, and J. Wolfrum, in Ref. 10.
[47] a) J.A. Miller, J. Chem. Phys. 74, 5120 (1981); b) K. Kleinermanns, R. Schinke, *ibid.* **80**, 1440 (1984); c) A.J.C. Varandas, *ibid.* **99**, 1076 (1993); d) A.J.C. Varandas, Chem. Phys. Lett. **225**, 18 (1994); e) A.J.C. Varandas, *ibid.* **235**, 11 (1995).
[48] a) R.T. Pack, E.A. Butcher, and G.A. Parker, J. Chem. Phys. **99**, 9310 (1993); b) C. Leforestier, and W.H. Miller, *ibid.* **100**, 733 (1994); c) D.H. Zhang, and J.Z.H. Zhang, *ibid.* **101**, 3671 (1994) d) R.T. Pack, E.A. Butcher, and G.A. Parker, *ibid.* **102**, 5998 (1995).
[49] H.-R. Volpp, and J. Wolfrum, Workshop on Imaging Methods in Molecular Structure and Dynamics, The Weizmann Institute of Science, Israel (1994).
[50] K. Kleinermanns, and E. Linnebach, J. Chem. Phys. **82**, 5012 (1985).
[51] Y. Matsumi, N. Shafer, K. Tonokura, M. Kawasaki, J. Chem. Phys. **95**, 4972 (1991); H.-G. Rubahn, W.J. van der Zande, R. Zhang, M.J. Bronikowski, and R.N. Zare, Chem. Phys. Lett. **186**, 157 (1991).
[52] A.J.C. Varandas, Mol. Phys. **85**(6), 1159 (1995).
[53] a) S.R. Langhoff, and R.L. Jaffe, J. Chem. Phys. **71**, 1824 (1979); b) J. Troe, 22nd Symposium (Int.) on Combustion, The Combustion Institute, Pittsburgh, 1988, pp. 843.
[54] B. Kendrick, and R.T. Pack, J. Chem. Phys. **102**, 1994 (1995); A.J.C. Varandas, and A.I. Voronin (to be published).

[55] B. Kendrick, and R.T. Pack, *Geometric Phase Effects in H + O_2 Scattering I+II*, J. Chem. Phys. in press (1996).
[56] R. Zellner, and I.W.M. Smith, J. Chem. Soc. Faraday Trans. 2, **69**, 1617 (1973).
[57] G.C. Schatz, and H. Elgertsma in ***Potential Energy Surfaces and Dynamics Calculations*** (D.G. Truhlar Ed.; Plenum, New York 1981).
[58] a) G.C. Schatz, J.L. Colton, M.C. Grant, J. Phys. Chem. **88**, 2971 (1984); b) K. Kudla, G.C. Schatz, *ibid.* **95**, 8267 (1991); c) Chem. Phys. Lett. **193**, 507 (1992); d) J. Chem. Phys. **98**, 4644 (1993).
[59] a) F.F. Crim, M.C. Hsiao, J.L. Scott, A. Sinha, R.L. van der Wal, Philos. Trans. R. Soc. London Ser. A: **332**, 259 (1990); b) F.F. Crim, A. Sinha, M.C. Hsiao, J. D. Thoemke in Ref. 9.; c) M.J. Bronikowski, W.R. Simpson, B. Girad, R.N. Zare, J. Chem. Phys. **95**, 8647 (1991); d) M.J. Bronikowski, W.R. Simpson, R.N. Zare, *ibid.* **97**, 2194, 2204 (1993).
[60] S. Koppe, T. Laurent, P.D. Naik, H.-R. Volpp, and J. Wolfrum, Can. J. Chem. **72**, 614 (1994), *"paper dedicated to Professor J.C. Polanyi on the occasion of his 65th birthday"*.
[61] M.H. Mok, and J.C. Polanyi, J. Chem. Phys. **51**, 1451 (1969); J.C. Polanyi, and J.L. Schreiber in ***Physical Chemistry – An Advanced Treatise*** (H. Eyring, D. Henderson, W. Jost Eds.; Academic Press, New York 1974).
[62] K. Kleinermanns, and J. Wolfrum, Appl. Phys. B **34**, 5 (1984).
[63] a) D. Wang, and J.M. Bowman, Chem. Phys. Lett. **207**, 227 (1993); J. Chem. Phys. **98**, 6235 (1993).
[64] G. Nyman, and D.C. Clary, J. Chem. Phys. **100**, 3556 (1994).
[65] H. Szichman, M. Baer (private communication); M. Baer *et al.* this volume.
[66] D.H. Zhang, and J.C. Light, J. Chem. Phys. **104**, 4544 (1996).
[67] a) R.G. Macdonald, K. Liu, J. Chem. Phys. **93**, 2431 (1990); *ibid.* **93**, 2443 (1990); b) D.M. Sonnenfroh, R.G. Macdonald, K. Liu, *ibid.* **93**, 6508 (1991).
[68] see e.g. the article by M. Baer *et al.* in this volume and references therein.
[69] D.H. Zhang and J.Z.H. Zhang, J. Chem. Phys. **99**, 5615 (1993), *ibid.* **100**, 2679 (1994); *ibid.* **101**, 5615 (1994); U. Manthe, T. Seideman, and W.H. Miller, *ibid.* **99**, 10078 (1993); *ibid.* **101**, 4759 (1994); D. Neuhauser, *ibid.* **100**, 9272 (1994).
[70] F.P. Tully, A.R. Ravishankara, J. Phys. Chem. **84**, 3126 (1980); A.R. Ravishankara, J.M. Nicovich, R.L. Thompson, F.P. Tully, *ibid.* **85**, 2498 (1981).
[71] P. Casavecchia *et al.* this volume.
[72] Y. Zhang, D. Zhang, Q. Zhang, D. Wang, D.H. Zhang, and J.H. Zhang, J. Phys. Chem. **99**, 16824 (1995).
[73] D.C. Clary, J. Chem. Phys. **95**, 7298 (1991).
[74] K.S. Bradley, and G.C. Schatz, J. Phys. Chem. **89**, 3788 (1994).
[75] K. Wintergerst, P. Frank in ***Verbrennung und Feuerungen-15. Deutscher Flammentag***, VDI Berichte 922 (VDI Verlag, 1991); see also J.V. Michael, K.P. Lim, Annu. Rev. Phys. Chem. **44**, 429 (1993).
[76] G.W. Flynn, and R.E. Weston, J. Phys. Chem. **98**, 8116 (1993).
[77] J.K. Rice, and A.P. Baronavski, J. Chem. Phys. **94**, 1006 (1990); S.L. Nickolaisen, H.E. Cartland, and C. Wittig, J. Chem. Phys. **96**, 4378 (1992).
[78] G. Rhadhakrishnan, S. Buelow, and C. Wittig, J. Chem. Phys. **84**, 727 (1986); G. Hoffmann, D. Oh, Y. Chen, Y.M. Engel, and C. Wittig, Israel J. Chem. **30**, 115 (1990) *"Special Issue in honour of the 1988 Wolf Prize Recipients R.D. Levine and J. Jortner"*.
[79] a) C. Wittig, Y.M. Engel, and R.D. Levine, Chem. Phys. Lett. **153**, 411 (1988); b) R.D. Levine, J.L. Kinsey in ***Atom-Molecule Collision Theory - A***

Guide for the Experimentalist (R.B. Bernstein Ed., Plenum Press, New York 1979).
[80] S. Koppe, T. Laurent, P.D. Naik, H.-R Volpp, and J. Wolfrum (manuscript in preparation).
[81] G.C. Schatz (private communication).
[82] e.g. a) M. Mozurkewich, S.W. Benson, J. Phys. Chem. **88**, 6429 (1984); b) J. Troe, Ber. Bunsenges. Phys. Chem. **98**, 1399 (1994) and references therein.
[83] J. Troe, J. Chem. Soc. Faraday Trans. **90**, 2303 (1994).
[84] a) G. Paraskevopoulos, and R.S. Irwin, J. Chem. Phys. **80**, 259 (1984); b) A.J. Hynes, P.H. Wine, A.R. Ravishankara, J. Geophys. Res. **91**, 11815 (1986); c) R.A. Stachnik, M.J. Molina, results quoted in Ref. 70; d) D. Fulle, H.F. Harmann, H. Hippler, and J. Troe, J. Chem. Phys. **105**, 983 (1996).
[85] a) H.E. Radford, W. Wei, and T.J. Sears, J. Chem. Phys. **97**, 3989 (1992); b) D.E. Milligan, and M.E. Jacox, *ibid.* **54**, 927 (1971); M.E. Jacox, *ibid.* **88**, 4598 (1988).
[86] a) D.W. Trainor, and C.W. Rosenberg, Jr., Chem. Phys. Lett. **29**, 35 (1974); b) M.J. Frost, J.S. Salh, and I.W.M. Smith, J. Chem. Soc. Faraday Trans. **87**, 1037 (1991).
[87] a) J. Brunning, D.W. Derbyshire, I.W.M. Smith, and M.D. Williams, J. Chem. Soc. **84**, 105 (1988); b) M. Wooldridge, R.K. Hanson, and C.T. Bowman, 25th Symposium (Int.) on Combustion, The Combustion Institute, Pittsburgh, 1994, pp. 741–748; c) T. Dreier, and J. Wolfrum, 18th Symposium (Int.) on Combustion, The Combustion Institute, Pittsburgh, 1981, pp. 801–809.
[88] S. Koppe, T. Laurent, P.D. Naik, H.-R Volpp, and J. Wolfrum, accepted for publication in the proceedings of the 26th Symposium (Int.) on Combustion, 1996.
[89] K. Kudla, G.C. Schatz, and A.F. Wagner, J. Chem. Phys. **95**, 1635 (1991).
[90] D.C. Clary, and G.C. Schatz, J. Chem. Phys. **99**, 4578 (1994).
[91] D.H. Zhang, and J. Z.H. Zhang, J. Chem. Phys. **103**, 6512 (1995).
[92] M. Quack, and J. Troe, Ber. Bunsenges. Phys. Chem. **78**, 240 (1974); J. Troe, Z. Phys. Chem. (Neue Folge) **161**, 209 (1989) *"paper dedicated to Professor H. Gg. Wagner on the occasion of his 60th birthday".*

Spectators and Participants in Vibrational State Controlled Bimolecular Reactions

J.M. Pfeiffer, J.D. Thoemke, R.B. Metz, A. Sinha, M.C. Hsiao, E. Woods, and F.F. Crim
Department of Chemistry, University of Wisconsin-Madison
Madison, Wisconsin 53706, U.S.A.

Abstract
Vibrational excitation provides a means for identifying the degrees of freedom in a molecule that are spectators or participants in a bimolecular reaction. In our study of the reactions of vibrationally excited H_2O with H and Cl atoms and vibrationally excited HCN with H, O, and Cl atoms, we find that in direct reactions the nonreacting bond of a molecule is a spectator to the reaction. When the nonreacting bond participates in the reaction, the reaction mechanism is not direct but proceeds through an intermediate complex.

1 Introduction

The energy requirement for a reaction is a central aspect of chemical kinetics recognized as early as the days of Max Bodenstein. Vibrational excitation plays a crucial role in the course of a chemical reaction, in many cases providing the energy to overcome the barrier to reaction. Detailed studies of the role of vibrations in chemical reactions have progressed from determining the consequences of vibrational excitation in reactions of diatomic molecules with atoms [1] to observing the influence that highly specific excitation has on reactions of triatomic molecules [2-13]. The addition of a single atom to the target molecule creates rich possibilities. Because triatomic molecules have several vibrational modes to excite, their study allows exploration of the role that different vibrations in the same molecule play in a chemical reaction. The simplest view of the means by which vibrations can control a reaction is that excitation of a vibration which becomes the reaction coordinate in a bimolecular reaction should efficiently promote the reaction while other, less coupled vibrations should not.

Several experiments on selected vibrational states of water suggest the utility of this simple view [2-7,10-12]. Exciting O-H stretching vibration states of water promotes its H-atom abstraction reaction ($X + H_2O \rightarrow HX + OH$) with X = H, O, and Cl, consistent with the vibration of the O-H bond becoming the reaction coordinate for breaking that bond and making a new H-X bond. Two experiments suggest that the surviving bond is a spectator in the reaction: the surviving bond retains its initial excitation, preferentially producing OH(v = 1) when it is initially excited and OH(v = 0) when it is not, and reaction of HOD preferentially breaks the O-H bond when it is vibrationally excited and the O-D bond when that is the initially excited bond. Thus, a simple picture in which the non-reacting bond does not participate in the reaction is consistent with many experimental results. Our goal

Springer Series in Chemical Physics, Volume 61
Gas Phase Chemical Reaction Systems
Eds.: J. Wolfrum, H.-R. Volpp, R. Rannacher, and J. Warnatz

here is to explore the limits of this spectator model in reactions of water as well as those of another prototypical triatomic molecule, hydrogen cyanide (HCN).

2 Experimental Approach

The measurements of the effect of vibrations require a means of vibrationally exciting the molecules, a source of reactive atoms, and a detection scheme that probes individual quantum states of the products [2-9]. We use light from a pulsed laser to deposit energy in H_2O or HCN stretching vibrations in the region of two to four quanta of O-H or C-H stretch, respectively. Excitation in the region of two or three quanta of O-H or C-H stretching uses light from a dye laser, either directly or

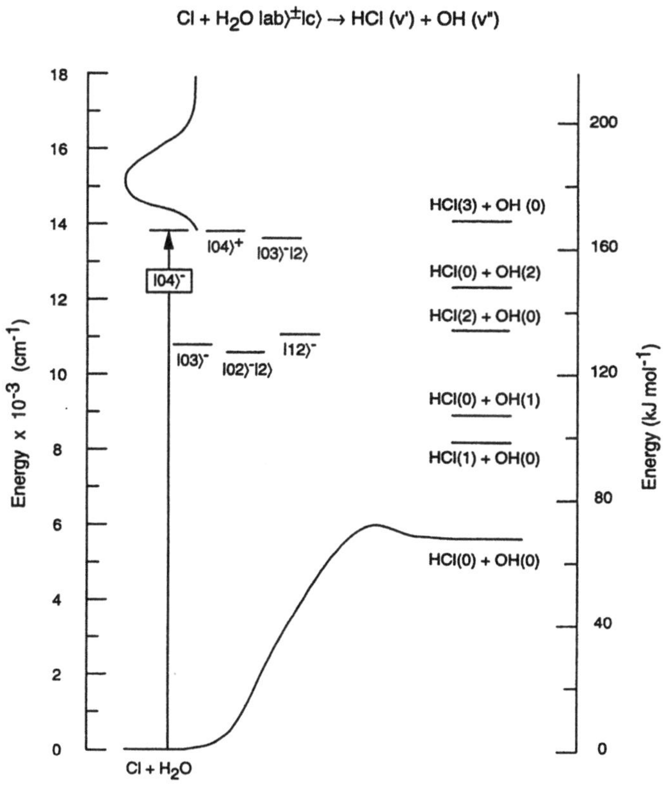

Figure 1: Schematic energy level diagram for the Cl + H_2O reaction. The energy levels on the left indicate the vibrational states in H_2O, and the curve above the $|04>^-$ level reflects the distribution of Cl + H_2O collision energies in our experiment. The energy levels on the right indicate the different vibrational levels of the HCl and OH products.

after stimulated Raman scattering in hydrogen. Preparation of molecules with 4 quanta of light-atom stretching vibration uses a Ti:sapphire laser, which operates in the region of 700-900 nm. Both lasers produce pulses of less than 6 ns with energies ranging from a few mJ to 60 mJ, depending on wavelength, and with bandwidths between 0.02 cm^{-1} and 0.20 cm^{-1}.

The reactive atoms in our experiments come from dissociation in a microwave discharge (H, O) or photolysis (O, Cl). The photolysis experiments use 355 nm light to dissociate either Cl_2 or NO_2 to form Cl or O atoms, respectively. Photolysis of a precursor produces atoms with excess translational energy that potentially alters the reaction rate. To test for artefacts of the translational energy, we create thermal O and Cl atoms and react them with vibrationally excited molecules. We use a microwave discharge of O_2/Ar to make thermal O atoms, and we make thermal Cl atoms by collisionally relaxing photolytic Cl atoms in 1100 mTorr of Cl_2 prior to reaction. In all cases, we see no significant difference between reactions with thermal atoms and those with up to 1670 cm^{-1} of translational energy in the center of mass reference frame. Except in the reaction with thermal Cl atoms, we use low pressures in the gas cell to ensure that few molecules have a collision prior to reaction. At the time delays between excitation and product detection (τ) and the pressures (P) we use ($P\tau \leq 25$ mTorr·μs), fewer than 25% of the molecules have any collision prior to product detection and fewer than 5% have a collision with a potentially reactive atom.

Laser induced fluorescence (LIF) detection with frequency doubled light from a dye laser probes the OH product from the reactions of vibrationally excited water (X + H_2O(|ab>$^{\pm}$|c>) → HX + OH) or the CN product from the reaction of vibrationally excited hydrogen cyanide (X + HCN(lmn) → HX + CN). The notation we use for the vibrationally excited molecule differs for water and hydrogen cyanide. The designation for water, H_2O(|ab>$^{\pm}$|c>), is local mode [14] notation in which a and b designate the number of O-H stretching quanta in each bond and c designates the number of bending quanta in each bond. The superscript indicates the symmetric (+) or antisymmetric (–) combination of a quanta and b quanta in the two bonds. The designation for hydrogen cyanide, HCN(lmn), is conventional normal mode notation with l denoting C-N stretching quanta, m denoting bending quanta, and n denoting C-H stretching quanta.

3 Vibrational Level Dependence in Reactions of Water

Vibrationally exciting water accelerates its reaction with H or Cl atoms substantially. For example, the 60 kJ/mol endothermic reaction has a negligible thermal rate but occurs at roughly the gas kinetic collision rate for water in the |04>$^-$ state [3]. Figure 1 shows the energetics of the reaction with Cl atoms, which are very similar to those with H and O atoms. The reaction rate for water with either Cl or H atoms depends strongly on the vibrational level for states in the region of three quanta of O-H stretching excitation. The reaction cross section for H + H_2O increases sharply from the threshold near |02>$^-$. It is a factor of 8 larger for |03>$^-$ and doubles again, to near the gas kinetic collision rate, for |04>$^-$, as Fig. 2 illustrates [3].

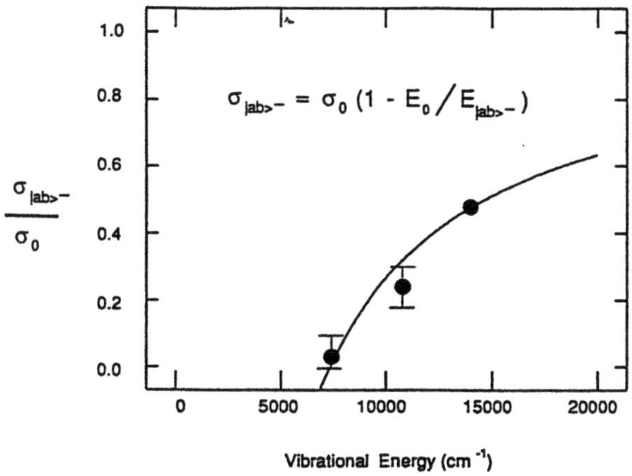

Figure 2: Relative rate for the H_2O + H reaction as a function of vibrational energy. The data are results for the |04>⁻, |03>⁻, and |02>⁻ vibrational states. The experiment determines the rates for the latter two states relative to that for |04>⁻. The solid line is a fit using an arbitrary line-of-centers functional form.

3.1 Mode Dependence of Reaction Rate

Reaction of H or Cl atoms with water also depends strongly on exactly which vibrational mode one excites in the region of the $|03\rangle^-$ state [3]. Three states lying within 1000 cm^{-1} of each other in energy but corresponding to different vibrational motions react at very different rates. The $|12\rangle^-$ state, corresponding to one quantum of stretching excitation in one bond and two in the other reacts at a barely detectable rate, despite having the slightly larger energy shown in Fig. 1. Similarly, the $|02\rangle^-|2\rangle$ state, corresponding to two quanta of stretching excitation and two of bend, has a very small reaction rate. These results are consistent with the non-reacting bond being a spectator. Only the energy in the bond that eventually becomes part of the reaction coordinate contributes to surmounting the energy barrier to the reaction. The energy in the other coordinate, either stretch or bend for the $|12\rangle^-$ state or the $|02\rangle^-|2\rangle$ state, respectively, does not accelerate the reaction. However, the situation is markedly different for the next higher level of excitation. The simple spectator model does not completely describe the vibrational mode dependence of the reaction rate for states in the vicinity of the $|04\rangle^-$ state.

Transitions to the three energy levels, $|04\rangle^-$, $|04\rangle^+$, and $|03\rangle^-|2\rangle$ shown in Fig. 1 appear in a 0.4 nm region of the vibrational overtone spectrum of water in Fig. 3. The upper half of the figure is the photoacoustic spectrum, in which the intensities reflect the vibrational overtone excitation probabilities. The transitions to the symmetric stretching state $|04\rangle^+$ and to the state containing bending excitation $|03\rangle^-|2\rangle$ are inherently much weaker than that to $|04\rangle^-$ because of the different transition dipole moments. They have roughly the same intensity in the spectrum because they come from rotational states with substantially different populations. The lower half of the figure is the action spectrum, obtained by fixing the probe laser on the $Q_1(2)$ transition of the OH(v = 0) product and scanning the wavelength of the vibrational overtone excitation laser. The intensities in the action spectrum reflect not only the vibrational overtone excitation probability but *the*

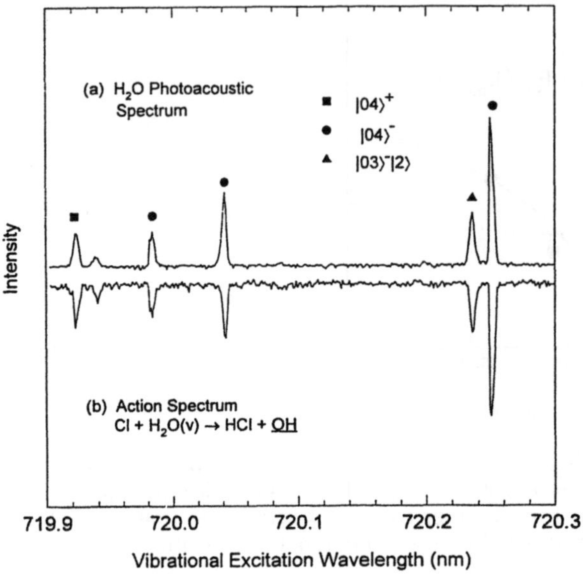

Figure 3: A portion of the (a) photoacoustic and (b) action spectra for the Cl + H_2O reaction. The action spectrum is acquired by monitoring the OH $Q_1(2)$ transition near 313.62 nm.

reaction probability as well. In our study of the |03>⁻ state, this was the key to seeing the difference in reactivity. Transitions to the less reactive states scarcely appeared in the action spectra. Here the situation is different. Transitions to the |04>⁺ and |03>⁻|2> states appear with relative intensities that are comparable to those for transitions to the |04>⁻ state, demonstrating that all three states react at comparable rates.

Table 1 summarizes the results for the reaction of several different states with H or Cl atoms. The results for the states in the region of three quanta of excitation follow the simple spectator model. The first column in the table reports the eight-fold difference in reactivity of the |02>⁻ and the |03>⁻ states with H atoms and also shows that adding two quanta of bend makes the |02>⁻|2> state about twice as reactive as the |02>⁻ state but almost four times *less* reactive than the nearly isoenergetic |03>⁻ state. The reactivities of the Cl atom shown in the second column show a similar pattern for these same states. The state with bending excitation is

Table 1: Reactivities of vibrational states of H_2O with H and Cl relative to the reactivity of the |04>⁻ state. (a) From Ref. 7, (b) from Ref. 3, and (c) from Ref. 5.

Vibrational State, v	H + H_2O	Cl + H_2O
|04>⁻	1.00[a]	1.00[a]
|04>⁺	–	0.95 ± 0.16[a]
|03>⁻|2>	–	0.97 ± 0.16[a]
|03>⁻	0.50[b]	0.33[c]
|02>⁻|2>	0.13 ± 0.03[b]	0.13 ± 0.03[c]
|02>⁻	0.08[b]	–

almost a factor of three less reactive than the state with all of its excitation in stretching excitation. The striking difference comes in the results for the |04>$^-$ and the |03>$^-$|2> states reacting with Cl atoms. *Both states react at comparable rates*, in sharp variance with the spectator model. (The comparable rates for the |04>$^+$ and |04>$^-$ states reflects the essential similarity of the symmetric and antisymmetric local mode wavefunctions with regard to reaction of one bond.)

3.2 Coupling of Bending Vibrations in the Reaction

There are at least two possible origins of the efficacy of the bending state in promoting the reaction. One is that the initially prepared wavefunctions, which we designate by their nominal character, are more highly mixed in the region of four quanta of O-H stretching excitation than at lower levels. For example, if the nominal |04>$^-$ and |03>$^-$|2> states were strongly mixed, such that each contained comparable amounts of stretching and bending excitation, the reactivities would be similar. Detailed calculations by Kellman and co-workers suggest that this is not the case [15]. They calculate the composition of the states using a realistic potential energy function and find very little mixing of the two states [16]. This result is consistent with independent calculations by McCoy and Sibert [17]. Thus, a second possibility, that the bending excitation does in fact project onto the reaction coordinate better in the region of four quanta of O-H stretch than in the region of three, seems likely. This amounts to the reactive collision mixing the states during the encounter that also transfers the atom. Involvement of the bending excitation is consistent with classical trajectory calculations by Schatz and co-workers [18], although the effect

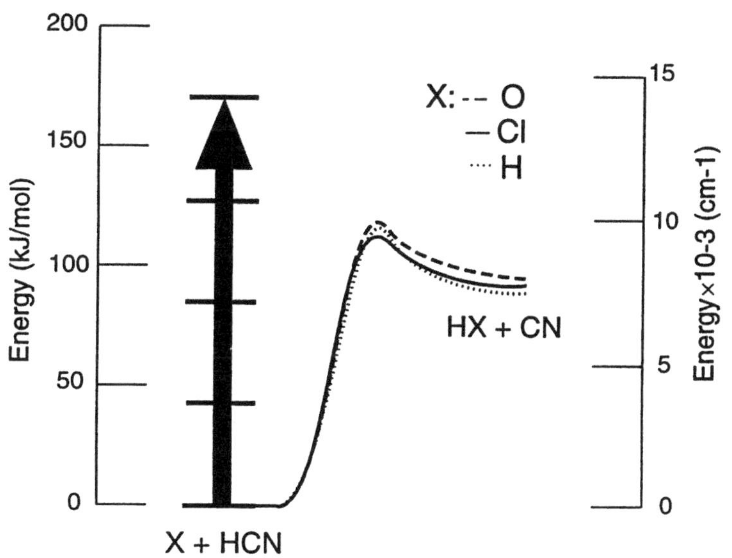

Figure 4: Schematic energy level diagram for the reaction of HCN with H, O, and Cl atoms.

is very sensitive to the details of the potential energy surface and the resulting adiabatic barriers. The participation of bending excitation is also consistent with quantum scattering calculations by Nyman and Clary [19] on HOD with Cl. They find that O-H stretching excitation accelerates the reaction more than bending excitation in the regions of two and three quanta of O-H stretch. They also find that bending excitation is more efficient at promoting reaction in the region of three quanta of O-H stretching excitation than in the region of two quanta of stretching excitation. This generally agrees with our experimental results that bending excitation accelerates the reaction, albeit less effectively than stretching excitation, in the region of three quanta of O-H stretch and is comparably effective in the region of four quanta.

4 Reacting Atom Dependence for Hydrogen Cyanide

Vibrationally exciting hydrogen cyanide accelerates its reactions with H, O, or Cl substantially, qualitatively the same result as with water. The energetics for the hydrogen atom abstraction reaction are very similar to each other and to those for water, as Fig. 4 shows. We use two diagnostics to test the spectator picture in reactions of HCN with the different atoms. One is the reactivity of different states as seen in the action spectrum compared to the photoacoustic spectrum. The other is the product state distributions. We find that the extent to which the nonreacting bond is a spectator depends on the identity of the attacking atom. Reactions with Cl atoms are much further from the simple spectator picture than those with H or O atoms.

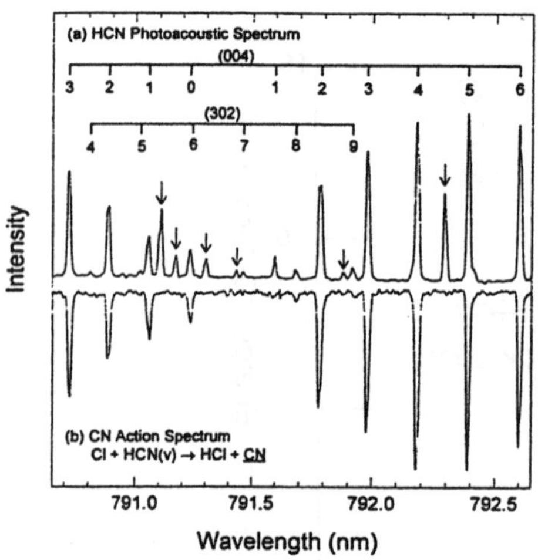

Figure 5: (a) Photoacousitc spectrum of HCN. Combs mark rotational features due to transitions from the vibrational ground state to the (004) and (302) vibrational states. Arrows denote features due to an H_2O impurity. (b) Action spectrum obtained by monitoring the CN (v = 0, J = 7) product of the Cl + HCN reaction.

4.1 Action Spectrum for Cl + HCN → CN + HCl

The photoacoustic spectrum in the upper panel of Fig. 5 shows the relative excitation efficiencies of the state containing four quanta of C-H stretch (004) and of that containing two of stretch and three of bend (302). Both appear in the spectrum although the transition probabilities to the (302) state are much smaller. (Some transitions of water appear in the photoacoustic spectrum as well.) The action spectrum in the lower portion of the figure reflects the relative reactivities since it comes from scanning the vibrational overtone excitation laser while monitoring the v = 0 state of the CN product of the reaction. The water lines do not appear, since reaction with water does not form the CN product, but transitions to both the (004) *and* the (302) state are present in proportion to their strength in the photoacoustic spectrum. Thus, we infer that the two states react at roughly the same rate. Strikingly, this situation is unique to reactions with Cl atoms. We are able to observe little or no reaction of the (302) state of HCN with either O or H atoms. Despite the similarity of the energetics of the H-atom abstraction reactions, the reactivity of the (302) state is different in the case of Cl atoms. While reactivity of the (302) state suggests the C≡N bond is a spectator in reactions with H and O atoms, it seems to be a participant in the reaction with Cl atoms.

4.2 CN State Populations from X + HCN → CN + HX

The distribution of the CN product among its rovibrational states also points to the unique behaviour of Cl atoms in reactions with vibrationally excited HCN. Figure 6 shows the populations of the vibrational states from reaction of HCN(004) with O, H, and Cl atoms and HCN(302) with Cl atoms [9]. Clearly, the reaction with Cl atoms creates CN with significantly more vibrational energy than those with H or O atoms. The reaction with Cl atoms forms more than 40% of the CN radicals in vibrationally excited states, but those with H or O produce less than 25% in vibrationally excited states, differences that are well outside the uncertainties in the measurement. Similarly, the 560 cm^{-1} of rotational energy in the CN product of the Cl-atom reactions exceeds the 140 to 315 cm^{-1} from the O- or H-atom reactions. These results point to the qualitative differences between the reactions with Cl atoms and those with H or O atoms.

4.3 The Influence of Complex Formation

Despite the similarity of the energetics of the H-atom abstraction reaction in all three systems, the energy barriers for other reaction pathways determine their influence on the reactions. The direct abstraction reaction, in which the attacking atom initially encounters the H atom of HCN should be most sensitive to excitation of the C-H stretch since that vibration becomes the reaction coordinate for the direct process. The incoming atom can also attack other portions of the molecule, with the multiple bond being a likely point. Such initial encounters can result in H-atom abstraction only when the attacking atom migrates to the H atom and abstracts it, a process that must involve multiple encounters or complex formation. Thus, a key to the differences in the reactions of the atoms is likely to be the energetics of complex

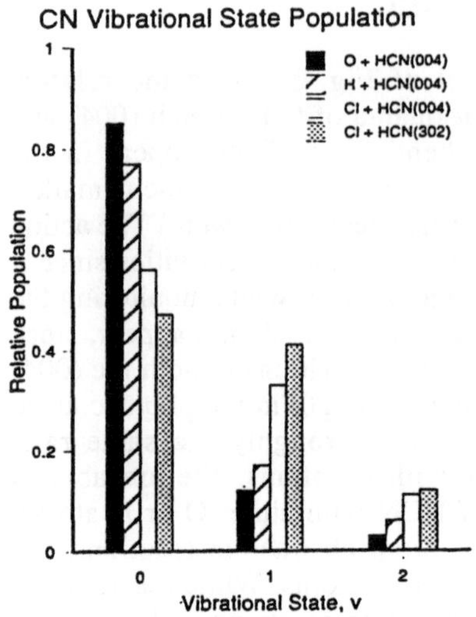

Figure 6: Relative populations of CN (v = 0, 1, 2) from the reactions of O, H, and Cl with HCN(004) and Cl with HCN(302).

decomposition. *Ab initio* calculations provide a means of sorting out the possibilities [9]. Examining calculated transition state energies for the reactions of HCN with O atoms [20], H atoms [21], and Cl atoms [9] shows that there are qualitative differences in the barriers that account for the spectator behaviour of O- and H-atom reactions but the participation of other degrees of freedom in the reactions of Cl atoms.

The crucial point in understanding the differences is the fate of strongly interacting intermediates, if they do form. More detailed statistical calculations support our qualitative conclusions [9]. The complex formed in the reaction of H atoms with HCN almost always decomposes back to reactants, thus making it a nonparticipant in the reaction. The complex formed in the reaction with O atoms almost always proceeds on to products, but *they are not the same products as in the abstraction reaction.* The reaction of O atoms with HCN through a complex preferentially makes H + NCO or NH + CO by decomposing over a much lower barrier than the direct abstraction reaction. Thus, the direct reaction to produce OH + CN proceeds uniquely by the abstraction channel. All indirect reactions lead to different products. The situation differs dramatically for the reaction of Cl atoms with HCN. The barrier to form all products except HCl + CN is too high to permit reaction at the total energies we use. Thus, the CN product comes from *both* direct reaction and indirect reaction. We suspect that the former pathway follows the spectator picture, but it does not account for all of the products. Many come from the indirect pathway in which interaction in a complex couples the C-H stretching and other degrees of freedom, making the CN a full participant that receives a substantial share of the energy. Instead of being a spectator, it becomes an intimate part of the reaction and carries away more vibrational and rotational energy than in reactions with H or O atoms.

5 Summary

Direct abstraction reactions are excellent candidates for vibrational state control of chemical reactivity, and reactions of atoms with both water and hydrogen cyanide show that vibrational excitation can determine the course of the reaction. One conclusion of these studies is that in direct reactions the nonreacting bond, the one that does not become the reaction coordinate, is a spectator in the reaction. Another is that vibrations that do not project directly onto the reaction coordinate do not affect the reactivity. In water, bending vibrations that do not project onto the reaction coordinate for molecules with energies in the range of three quanta of O-H stretch become participants in the range of four quanta, where they couple into the reaction coordinate during the reaction. In hydrogen cyanide, reactions that form the monitored product proceed directly with the nonreacting bond being a spectator, but those that proceed through a complex make the nonreacting bond a full participant.

6 References

[1] J. C. Polanyi and W. H. Wong, J. Chem. Phys. **51**, 1439 (1969); M. H. Mok and J. C. Polanyi, J. Chem. Phys. **51**, 1451 (1969); B. A. Hodgson and J. C. Polanyi, J. Chem. Phys. **55**, 4745 (1971); D. S. Perry, J. C. Polanyi, and C. W. Wilson Jr., Chem. Phys. **3**, 317 (1974).
[2] A. Sinha, M. C. Hsiao, and F. F. Crim, J. Chem. Phys. **92**, 6333 (1990).
[3] A. Sinha, M. C. Hsiao, and F. F. Crim, J. Chem. Phys. **94**, 4928 (1991).
[4] M. C. Hsiao, A. Sinha, and F. F. Crim, J. Phys. Chem. **95**, 8263 (1991).
[5] A. Sinha, J. D. Thoemke, and F. F. Crim, J. Chem. Phys. **96**, 372 (1992).
[6] R. B. Metz, J. D. Thoemke, J. M. Pfeiffer and F. F. Crim, J. Chem. Phys. **99**, 1744 (1993).
[7] J. D. Thoemke, J. M. Pfeiffer, R. B. Metz, and F. F. Crim, J. Phys. Chem., **99**, 13748 (1995).
[8] R. B. Metz, J. M. Pfeiffer, J. D. Thoemke, and F. F. Crim, Chem. Phys. Lett. **221**, 347 (1994).
[9] J. M. Pfeiffer, R. B. Metz, J. D. Thoemke, E. Woods III, and F. F. Crim, J. Chem. Phys., in press (1995).
[10] M. J. Bronikowski, W. R. Simpson, B. Girard, and R. N. Zare, J. Chem. Phys. **95**, 8647 (1991).
[11] M. J. Bronikowski, W. R. Simpson, and R. N. Zare, J. Phys. Chem. **97**, 2194 (1993).
[12] M. J. Bronikowski, W. R. Simpson, and R. N. Zare, J. Phys. Chem. **97**, 2204 (1993).
[13] K.-H. Gericke, J. Chem. Phys., in press (1995).
[14] M. S. Child and R. T. Lawton, Chem. Phys. Lett. **87**, 217 (1982); M. S. Child. Acc. Chem. Res. **18**, 45 (1985); M. S. Child and L. Halonen, Adv. Chem. Phys. **57**, 1 (1984); I. A. Watson, B. R. Henry, and I. G. Ross, Spectrochim. Acta **37A**, 857 (1981).
[15] M. Kellman, private communication.
[16] J. D. Thoemke, M. Kellman, and F. F. Crim, to be published.
[17] A. B. McCoy and E. L. Sibert, J. Chem. Phys. **92**, 1893 (1990).
[18] G. C. Schatz, M. C. Colton, and J. L. Grant, J. Phys. Chem. **88**, 2971 (1984).
[19] G. Nyman and D. C. Clary, J. Chem. Phys. **100**, 3556 (1994).
[20] J. A. Miller, C. Parrish, and N. J. Brown, J. Phys. Chem. **90**, 3339 (1986).
[21] R. A. Bair and T. H. Dunning Jr., J. Chem. Phys. **82**, 2280 (1985).

Reaction Product Imaging: The H + HI Reaction

D. W. Chandler
Combustion Research Facility, Sandia National Laboratories,
Livermore, California 94551-0969, U.S.A.

T. N. Kitsopoulos
Department of Chemistry, University of Crete, and Institute of Electronic Structure and Laser (FORTH), Heraklion, 71110 Crete, Greece

M. A. Buntine
Chemistry Department, University of Adelaide, Adelaide, Australia

D. P. Baldwin
Ames Laboratory, Iowa State University, Ames, Iowa 50011, U.S.A.

R. I. McKay
Department of Chemisty, Baker Lab., Cornell University,
Ithica, NY 14853, U.S.A.

A. J. R. Heck, and R. N. Zare
Chemistry Department, Stanford University,
Stanford California 94305, U.S.A.

Abstract

The technique of reaction product imaging is described and applied to the study of the H + HI reaction in a single molecular beam. Two-dimensional images of quantum-state selected H_2 reaction products from the reaction of photolytically produced H atoms with HI molecules illustrate how reaction product imaging can reveal detailed information about specific reaction channels. The observed product distributions can be explained by a simple physical model of the reaction dynamics.

1 Introduction

In order to understand the dynamics of a reaction, one needs to know as much as possible about the products formed. The most fundamental question concerns the identity of the products; for small systems the answer is often obvious. On these smaller systems it is feasible to investigate the populated quantum states of the products, their velocities, and the correlations between these parameters. For the study of unimolecular dissociation, techniques have been developed to measure the angular distribution [1-3], the recoil velocity [4-6], and the internal energy of fragments produced [7-8]. Doppler profiling techniques have been utilized to study the correlation between the fragments' velocities and internal energies [9-11]. Multiphoton ionization (MPI) time-of-flight (TOF) techniques have been demonstrated that give equivalent information to that obtained by Doppler techniques without the use of sub-Doppler-linewidth lasers [12]. These techniques involve the one-dimensional projection of the three-dimensional velocity distribution of particles. Techniques such as these have rarely been applied to detection of bimolecular

Springer Series in Chemical Physics, Volume 61
Gas Phase Chemical Reaction Systems
Eds.: J. Wolfrum, H.-R. Volpp, R. Rannacher, and J. Warnatz

reaction products. Recently two experiments have used quantum state ionization followed by spatially resolved detection of the ions to learn about the details of bimolecular processes [13,14]. In this paper, we describe a two-dimensional imaging technique wherein a reaction takes place in a single molecular beam and the reaction products are tagged by resonance-enhanced multiphoton ionization (REMPI) and imaged onto a two-dimensional, position-sensitive detector [15,16]. We apply this technique to the study of the reaction of "hot" H atoms with HI to produce $H_2(v, J)$ and an I atom. Analysis of the images allows us, in principle, to determine the branching ratios and energetics of the reaction. We refer to this technique as reaction product imaging (RPI).

As performed here, reaction product imaging involves photolytic preparation of one of the reactants and molecular beam cooling of the other reactant. This photolytic initiation of the reaction (HI photolysis), dictates that the reagents will have precise collision velocities and internal state distributions, and constrains the reaction products, by conservation of energy, to have specific velocities depending upon their internal energy content. State-selective ionization and imaging of the H_2 product molecules allows us to measure the velocity of the ionized fragments and by conservation of energy determine the exothermicity of the reaction and the energy content of the reaction product not ionized, in this case the electronic state of the iodine atom.

The positions at which the ions strike the two-dimensional detector are determined by the velocities with which the neutral reaction products recoil. We place the detector parallel to the symmetry axis of the reaction, in this case the polarization vector of the photolysis laser, and use a numerical inversion technique [17] to reconstruct the three-dimensional velocity distribution of the reaction products. Reaction product imaging has a multiplexing advantage in that the entire distribution of products (in a single quantum state selected by the ionization laser) is sampled every laser shot. The detection system has single ion sensitivity, and signal averaging over a few thousand laser shots produces clear images.

The mass combination of the reactants involved in chemical reactions often influences the partitioning of the available energy among reaction products [18]. The hydrogen-abstraction reaction

$$H + HI \rightarrow H_2 + I \qquad (1)$$

is an extreme example of the $L_{ight} + L'_{ight} H_{eavy} \rightarrow L_{ight} L'_{ight} + H_{eavy}$ mass combination, in which the heavy atom H_{eavy} serves as an effective third body for the combination of the two light atoms L_{ight} and L'_{ight} allowing the reaction to occur over a wide range of impact parameters. The large exothermicity (ΔE_{RN} = 1.424 eV) of the hydrogen-abstraction reaction [19] and the high collision energies available experimentally through photolytic H-atom generation, means that a large number of internal states of the H_2 product are energetically accessible. The resulting broad internal state distributions of the H_2 products have been measured experimentally [20-24] and an extensive comparison with product distributions derived from quasiclassical trajectory (QCT) calculations [22,25-28], has provided a useful test for reaction dynamic models.

Reactions of hydrogen atoms with hydrogen halides have two competing reaction pathways: abstraction and exchange. The homologous series of hydrogen halide reactions H + HX (X = F, Cl, Br, I) has been used to observe this competition and to study systemic trends in the reaction dynamics arising from small changes in the reaction kinematics or potential energy surfaces [20,21,25,29-33]. Early crossed molecular beam studies, conducted in the laboratories of Herschbach [34] and Toennies [35,36], focused on studying the hydrogen atom exchange reaction D + HI → DI + H. Attempts to measure the angular distribution of the molecular product of the abstraction reaction failed because of background problems [34,35]. The cross section for the exchange reaction was predicted by QCT calculations [26,37] to be twice the cross section for abstraction, and was measured experimentally [20,21] to be an order of magnitude larger than the cross section for abstraction.

Two independent experimental studies have measured the H_2 product state distribution for the H + HI abstraction reaction with full quantum state resolution. Valentini and co-workers [21,22,38] used coherent anti-Stokes Raman spectroscopy (CARS) to probe the H_2 yield and determine the state-to-state integral cross section for this reaction at center-of-mass collision energies of 0.6, 1.3, and 1.6 eV. Zare and co-workers [21-24,38] studied this reaction at center-of-mass collision energies of 0.6 and 1.6 eV, using (2+1) REMPI to observe the H_2 product-state distribution. Comparisons between these two sets of results show good agreement. The H_2 product is formed with a high degree of internal excitation; however, not all the energetically accessible levels are populated. The rotational distributions cannot be adequately described by temperatures and a vibrational population inversion is observed. The general appearance of the product-state distributions shows that the coupling between reactant translational energy and product internal excitation is weak, which was interpreted as a strong propensity to conserve translational energy, i.e., the translational energies of the reactants and products are approximately equal and the product internal energies seem to correspond to the reaction exothermicity.

Ab initio calculations for this system are complicated by the large number of electrons present and the consequent possibility of strong relativistic effects. Nevertheless, several approximate HHI potential energy surfaces have been calculated [27-30,39,40]. González and Sayós (GS) [27,28] as well as Aker and Valentini (AV) [22,25,26] have carried out QCT calculations for the H + HI abstraction reaction on these surfaces over the energy ranges studied experimentally by the Valentini and Zare groups (1.3 and 1.6 eV collision energy). Both calculations were performed on London-Eyring-Polanyi-Sato (LEPS) semiempirical surfaces. AV also calculated trajectories on a diatomics-in-molecules potential energy surface including three-center terms (DIM-3C), but obtained results identical to those from a LEPS surface, within statistical uncertainty [25]. After extensive trajectory calculations, an inclusive model was generated by Aker and Valentini [25] that treats the H + HI collision as a simple H + H collision forming H_2 in particular v, J states followed by a scattering of the H_2 from the iodine atom. For low v and high J states the scattering has little effect on the internal states of the H_2 because these molecules are formed with large impact parameters that correspond to geometries where the newly formed H_2 molecule only slightly impacts the iodine atom. Smaller H-H impact parameters b_{H-H} (extrapolation of incoming trajectory to distance of closest approach of the two H atoms involved) lead to lower J states and higher vibrational excitation. Because the opacity function has a peak at impact parameters that would correspond to intermediate J states a vibrational inversion is expected reflecting the shape of the opacity function. This model correctly predicts the rotational cut off observed in H + (HI, HBr or HCl) collisions [25]. We apply this model to our experiments as well. Comparison of the QCT and experimental results shows reasonable agreement in average energy disposal, total cross section, and rotational state distributions.

A comparison of the $I(^2P_{3/2,1/2})$, spin-orbit splitting (0.94 eV) and the reaction exothermicity (−1.42 eV), reveals that a nonadiabatic reaction channel that produces spin-orbit excited-state $I(^2P_{1/2})$ iodine atoms (in the following denoted as I*) is energetically possible for all collision energies. The lack of observed infrared chemiluminescence from nascent I* formed from the abstraction reaction at thermal energies led Polanyi and co-workers [41] to the conclusion that a negligible fraction (< 2%) of the reaction occurred via this nonadiabatic channel. Thereafter QCT calculations and previous experiments have assumed that only ground-state iodine atoms, $I(^2P_{3/2})$, are produced as a product of the H + HI abstraction reaction. Measurements of the total $H_2(v, J)$ quantum yields by both CARS [22,25] and REMPI [24] are unable to distinguish between $H_2(v, J)$ formed concomitantly with $I(^2P_{3/2})$ or I* products. Ion imaging experiments [16] which distinguish both the velocity of the fragments and their internal state distribution can distinguish between these two channels.

In this paper we report the results of experiments using RPI to measure the speed distribution of state-selected H_2 products for a large range of rovibrational states at nominal center-of-mass collision energies of 2.7 eV. This collision energy corresponds to H-atom photolytic generation using ~ 215 nm light producing ground state iodine atoms. We have observed H_2 reaction product in 45 rotational levels spread over four vibrational states. We compare our distributions to those measured at different energies using other techniques. Procedures for reconstructing and subsequently extracting scattering information from ion images are described in the Appendix.

2 Experimental

The technique of reaction product imaging, detection of a reaction product from a bimolecular reaction event using ion imaging techniques, has been described previously [16]. A brief outline is provided here. A schematic of the experimental apparatus appears in Fig. 1. The reactant molecules of interest are delivered through a pulsed molecular beam, having about 1 bar backing pressure, to a high vacuum chamber. A linearly polarized photolysis-laser pulse intersects the molecular beam creating radical reactants. A second linearly polarized laser beam (or in this case the same laser beam that initiated the reaction) intersects the reaction products after a small time delay (1-20 ns) and state-selectively ionizes them. The ideal source of products is a point source but typically the source is defined by the intersection of the molecular beam and the photolysis laser beam creating a cylindrical source of ~1 mm length. The radius of the cylinder is determined by the focusing of the ionization laser. The finite length contributes to a broadening of the images along one axis.

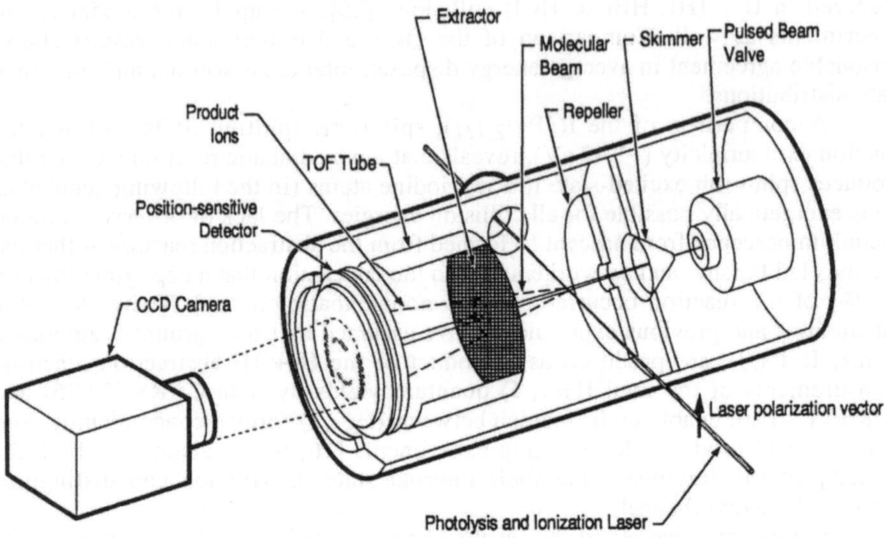

Figure 1: Schematic of the experimental reaction product imaging apparatus. A single laser is used to both photolytically produce H-atom reactants and photoionize the $H_2(v, J)$ products. A single molecular beam of HI is used as a source of both reactant species. The resulting product ions are accelerated along a time-of-flight tube and detected with a position-sensitive detector to obtain $H_2(v, J)$ velocity distributions.

The ions are accelerated along the molecular beam axis and pass into a time-of-flight tube and are detected by a two-dimensional position sensitive detector positioned normal to the flight-tube axis. Since a skimmed and collimated molecular beam is used, the molecules carry very little (~10 K) velocity perpendicular to the detector axis. By choosing the detector axis coincident with the molecular beam axis the initial velocity of the parent molecules does not affect the shape or position of the images. The detector consists of a pair of microchannel plates in front of a fast (~100 ns lifetime) phosphor screen. The front microchannel plate is pulsed in order to record the image from a single mass-to-charge ratio. A background image is typically taken and subtracted so that scattered light and non-resonant ionization contributions to the images are eliminated.

The images that appear on the phosphor screen are recorded with a digital charge-coupled device (CCD) camera (Photometrics 200 thermoelectrically cooled electronic camera) and stored on a microcomputer. A digital recording system allows for signal averaging over long periods of time, background subtraction, signal processing, and a variety of analysis and data reduction techniques. A pixel by pixel summation along either the rows or the columns of the image produces the equivalent of the Doppler profile of the ions.

For photofragment imaging a large Doppler width is a disadvantage. In order to record the entire velocity distribution the bandwidth of the ionization laser must cover the entire Doppler profile. Ideally one would like a laser with a top-hat profile in the frequency domain that completely encompasses the Doppler profile of the species of interest. The "ideal laser" is approximated in the imaging experiments by scanning the frequency of the ionization laser while the image is being collected.

3 Results

3.1 HI Photolysis

Ultraviolet photodissociation of HI proceeds via two energetically accessible product channels:

$$HI + h\nu \rightarrow H_F + I(^2P_{3/2}) \qquad (2a)$$

$$\rightarrow H_S + I(^2P_{1/2}) \qquad (2b)$$

i.e., ground state hydrogen atoms $H(^2S)$ are formed with two different kinetic energies: fast hydrogen atoms H_F associated with formation of iodine atoms in the ground state (I) and slow hydrogen atoms H_S associated with the formation of spin-orbit-excited iodine atoms (I*). From the symmetry of the dissociative surfaces contributing to these two channels, it can be shown that formation of H_F and H_S atoms arises from a perpendicular transition and a parallel transition, respectively [42-47]. The branching ratio between these two channels changes with photolysis wavelength, and must be taken into account in the analysis of our experimental results. HI photodissociation has been studied by a number of different groups at a range of photolysis wavelengths (193, 222, 248, and 266 nm). Branching ratios and angular distributions corresponding to the two photodissociation channels have been reported. A theoretical prediction for these quantities over the entire photolysis wavelength range has been made, based on an empirical fitting of some of these reported observations [48].

To investigate the behaviour of these photolytically produced H atoms in the wavelength range used in this experiment, we measured their velocity distribution using a photolysis wavelength of 205 nm. This wavelength was chosen to be resonant with the two-photon-allowed 1s → 3s transition of the H atom, allowing us to use a single laser to both photolyze HI and ionize the H-atom photofragments using (2+1) REMPI. Figure 2A shows a raw data image of the resulting H^+ ions.

The polarization of the photolysis laser is parallel to the imaging plane (detector face). The angular distribution of photofragments is azimuthally symmetric with respect to the laser polarization [49]. A slice through the center of the original 3D distribution can be calculated [50,51] from the 2D projection of a cylindrically symmetric object onto a plane parallel to its symmetry axis by means of an *inverse Abel transform* (see Appendix). Using this technique, we are able to extract the *2D intensity profile*, i.e., a 2D slice through the original 3D distribution of H atoms, from the raw data image [15]. Figure 2B shows the extracted 2D intensity profile. We refer to such an image as a *reconstructed* image. We clearly see the expected bimodal distribution of H-atom products; slow hydrogen atoms H_S closer to the center of the image and the faster hydrogen atoms H_F observed further from the image center. Angle and speed distributions and relative branching ratios can be determined directly from a reconstructed image (see Appendix). The speed distribution is shown

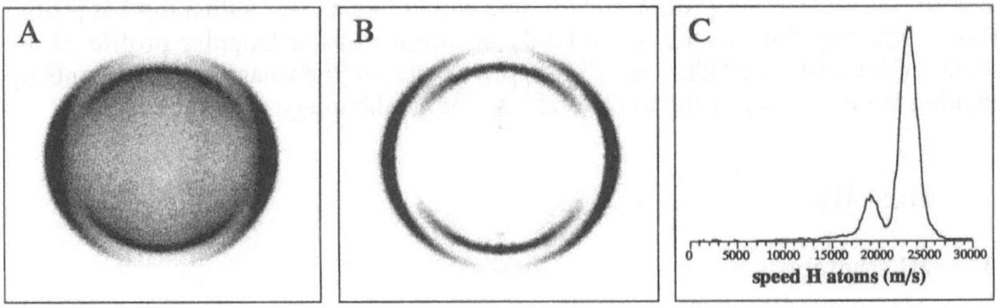

Figure 2: (A) Raw data image of the H atoms produced by the photolysis of HI at 205 nm. The polarization of the photolysis laser is vertical with respect to this image and the position of the HI beam is at the center of the image. (B) Reconstructed image obtained from the raw data image in Fig. 2A, showing the spatial distribution of fast and slow H atoms resulting from the photolysis of HI at 205 nm. (C) Speed distribution obtained of H atoms produced by HI photolysis, determined by integrating reconstructed image 2b around its center point. Each peak in the speed distribution corresponds to a ring in the reconstructed image.

in Fig. 2C. By comparing the intensities of the two speeds seen in the distribution, we determine the branching ratio between the slow and fast H atoms, produced by the 205 nm photolysis of HI, to be 1:5 which is in reasonable agreement with the theoretical predictions of Levy and Shapiro ($I^*:I \cong 0.225$) [48]. We use their calculated $I^*:I$ ratios, i.e., $H_S:H_F$ ratios, for the range of photolysis wavelengths used in this experiment, in the analysis of our H_2 product-state distributions.

3.2 Images of the $H_2(v, J)$ Product

When collecting the images of H_2 formed from the reaction of photolytically produced H atoms with HI reactants in a molecular beam of neat HI a 5 ns long laser pulse is used. This laser pulse both creates the H atom reactant via photolysis of HI and state-selectively ionizes the H_2 (v, J) reaction products via REMPI. Therefore the collision energy of the reaction E_{rel} varies with the laser photon energy $E(h\nu)$ depending on the $H_2(v, J)$ product probed [52,53]. The collision energy varied by

0.6 eV over the entire H_2 product state distribution reported here. For a particular vibrational state of H_2 associated with a single spin-orbit channel, the spread in the collision energy is at most 0.3 eV between the lowest and highest J levels observed. This change in the reactant translational energy is included in our analysis, but does not significantly affect the product state distributions. For the spectral region of interest, the mean collision energy ($\langle E_{rel} \rangle$) for the reaction is 2.7 eV. Note that although two velocities of H atoms are formed (1.8 eV and 2.7 eV) in the initial photolysis event, we concern ourselves only with the faster H atoms formed for reasons we will discuss later.

The $H_2(v, J)$ molecules, resulting from the reaction of photolytically produced H atoms with the remaining HI molecules, are ionized via (2+1) REMPI and their ion images are recorded. Because a linearly polarized photolysis laser is used, a cylindrically symmetric distribution of ions is produced. Because the polarization axis of the laser is parallel to the face of the detector the cylindrically symmetric distribution can be analyzed using an inverse Abel transform thereby obtaining a slice through the three dimensional distribution of the ions. We have taken ion images corresponding to the following 45 product states: H_2 (v = 0; J = 12-17), H_2 (v = 1; J = 3-15), H_2 (v = 2; J = 1-15), and H_2 (v = 3; J = 1-11). In Fig. 3 typical images are presented corresponding to H_2 product in the following rovibrational states: (v = 0, J = 17), (v = 1, J = 11), and (v = 2, J = 15).

A number of corrections need to be made to the ion image intensity during analysis of these images. A correction for solid angle, nuclear indistinguishability (the even (*para*) J levels are multiplied by three [23,54]) and detection sensitivity were all applied to the data. The procedure described by Kliner *et al.* [24,53,55-57] to correct measured signal intensities to allow for differences in ionization efficiency between the various rovibrational states of H_2 arising from the (2+1) REMPI detection scheme was also applied.

The $H + HI \rightarrow H_2(v, J) + I$ reaction is highly exothermic (1.424 eV) and has a low activation energy (0.03 eV) [39,58]. Hence both fast and slow photolytically produced H atoms lead to reaction. For the collision energies in our study four reaction channels are energetically accessible:

$$H_F + HI \rightarrow H_2(v, J) + I(^2P_{3/2}) \qquad \langle E_{rel} \rangle = 2.7\ \text{eV} \qquad \text{FI} \qquad (3a)$$

$$H_F + HI \rightarrow H_2(v, J) + I^*(^2P_{1/2}) \qquad \langle E_{rel} \rangle = 2.7\ \text{eV} \qquad \text{FI}^* \qquad (3b)$$

$$H_S + HI \rightarrow H_2(v, J) + I(^2P_{3/2}) \qquad \langle E_{rel} \rangle = 1.8\ \text{eV} \qquad \text{SI} \qquad (3c)$$

$$H_S + HI \rightarrow H_2(v, J) + I^*(^2P_{1/2}) \qquad \langle E_{rel} \rangle = 1.8\ \text{eV} \qquad \text{SI}^* \qquad (3d)$$

We have labelled these four reaction channels, using F and S to refer to the channels involving fast and slow H atoms reactants respectively, and I and I* to refer to the concomitant production of ground-state and spin orbit-excited iodine atoms in the reaction step. Knowing our experimental conditions and the energetics of the reaction, the speed of the H_2 products formed via these different reaction channels can be calculated and compared directly with the product speeds observed in our images to identify the reaction channels present.

The outermost ring observed in the images corresponds to the FI channel. The rotational population distributions for the $H_F + HI \rightarrow H_2 + I$ abstraction reaction can be determined by integrating the reconstructed ion images intensity within

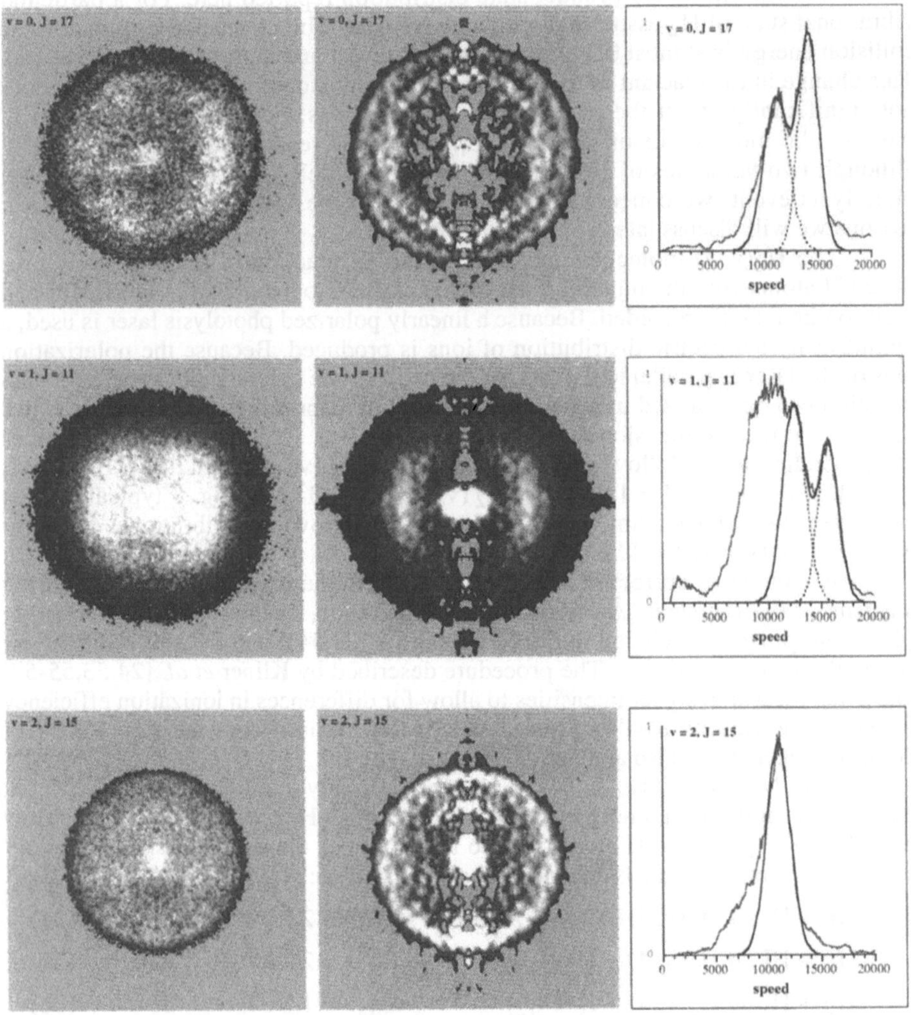

Figure 3: (Left Column) Raw data images of H_2 (v, J = 0, 17 top row, v, J = 1, 11 middle row and v, J = 2, 15 bottom row) produced by the H + HI reaction. The polarization of the HI photolysis laser is vertical with respect to this image and the position of the HI beam is at the center of the images. (Middle column) Reconstructed images obtained from the raw data images, showing the spatial distribution of H_2 product, resulting from the abstraction reaction. The position of the center-of-mass is in the center of the images. (Right column) Speed distributions of H_2 produced by the H + HI reaction, determined by integrating the reconstructed images around their center point. Each peak in the speed distribution corresponds to a ring in the reconstructed image. The curves are a fit to the data as described in the text.

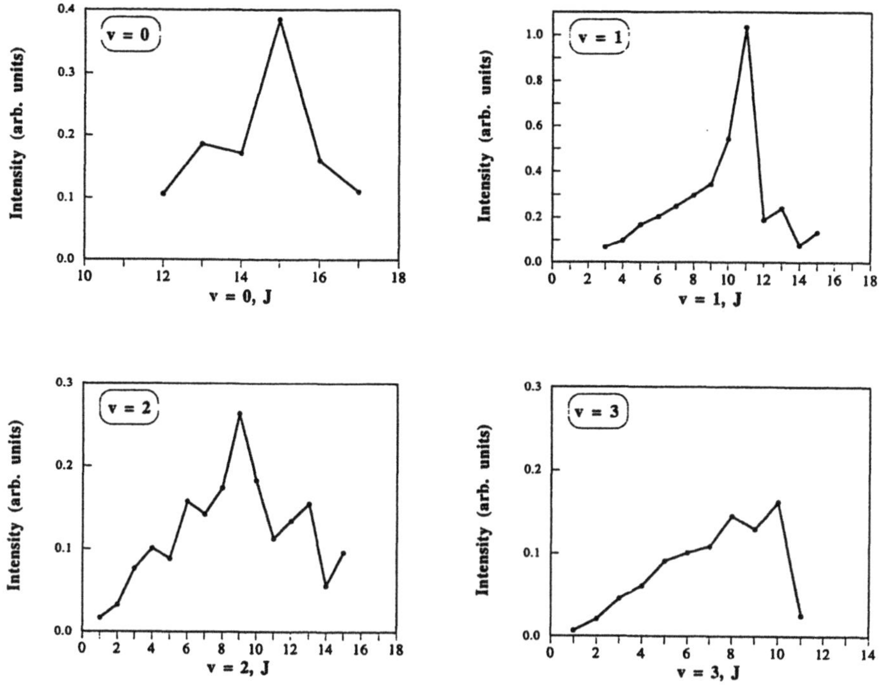

Figure 4: Rotational population distributions of the $H_2(v, J)$ product of the H + HI reaction proceeding via the FI channel (see text) for v = 0, v = 1, v = 2, and v = 3.

this ring. Ion images used in this determination were obtained under identical experimental conditions (e.g., HI beam densities and laser fluencies) and were each averaged for the same number of laser shots. This procedure yields similar information as the experiments of Valentini and co-workers [21,22], and Zare and co-workers [24] but for a single reaction channel. Our resulting rotational population distributions are presented in Fig. 4 and Table 1. By noting the similarities between our rotational population distributions measured at mean collision energy of 2.7 eV, and those measured by Valentini and co-workers [21] and Zare and co-workers [24] at a mixture of collision energies of 0.6 and 1.6 eV, and by Valentini and co-workers [22] at a collision energy of 1.3 eV, we conclude that the internal state distributions appear to be insensitive to the collision energy E_{rel} ,of the reactants.

Most of the ion images we have recorded for this reaction show a peak of variable intensity at the center of the image. This peak is resonant with the H_2 transitions probed, and the spatial width of the peak indicates that its off-axis velocity is small. Rotation of the laser polarization does not affect the shape of this peak, which indicates that it is isotropic. Previously, this central peak was speculated to be caused by the following cluster reaction [16]:

$$(HI)_2 \rightarrow H_2(v, J) + 2I \qquad (4)$$

Table 1. Rotational populations (arb. units) in the H + HI → $H_2(v,J)$ + $I(^2P_{3/2})$ reaction at a nominal collision energy of 2.7 eV.

J	v = 0	v = 1	v = 2	v = 3
0				
1			0.017	0.007
2			0.033	0.022
3		0.067	0.077	0.046
4		0.097	0.101	0.061
5		0.167	0.088	0.091
6		0.203	0.158	0.101
7		0.248	0.143	0.109
8		0.297	0.174	0.145
9		0.344	0.264	0.129
10		0.542	0.183	0.162
11		1.03	0.113	0.025
12	0.106	0.188	0.134	
13	0.186	0.239	0.155	
14	0.171	0.076	0.056	
15	0.385	0.134	0.096	
16	0.159			
17	0.110			
sum	1.12	3.64	1.79	0.90

but a definite assignment was not established. Images were recorded under experimental conditions chosen to discriminate against the production of these low kinetic energy H_2 molecules, i.e., the central peak. Such conditions are achieved by setting the relative timing between the laser and the pulsed nozzle such that the laser intersects the leading edge of the HI gas pulse. It was also found that heating (~40 C°) the nozzle decreased the amount of intensity in the center of the image. Total suppression of the central feature was not achieved. From energetics it is clear that this peak arises from ionized products that have come from a different mechanism than the bimolecular product seen as rings in the images i.e., this peak likely corresponds to H_2^+ formed within dimers or higher order clusters. This peak does not interfere with our determination of rotational state distributions for the FI reaction channel because it is separated in velocity from the bimolecular product formed via this channel. The broad innermost ring shows the same rotational dependence as the center spike and remains centered near 7500 m/sec velocity independent of product rotational state investigated. We assign this ring to H_2^+ formed within large clusters that have become multiply charged due to ionization of several HI monomers within the cluster. These clusters undergo Coulomb explosion – as recently observed by Castleman and co-workers in rare gas clusters [59] – thereby giving the H_2^+ the ~7500 m/sec velocity which overlaps the velocities expected for the SI* and FI*/SI channel. Therefore this ring does interfere with the determination of the intensity of the H_2 products formed in these reaction channels. We see no clear indication that the SI* channel is active, due to this overlap with the "Coulomb exploded" channel, in our experiment but there is a clearly defined ring of H_2^+ at a radius that corresponds to the combined SI and FI* reaction channel. We believe this to be due exclusively to the SI channel but do not attempt here to quantify the rotational distributions due to the overlapping signal from the cluster formed and "Coulomb exploded" H_2^+ signal described above. The FI channel is essentially unperturbed by overlap with the "Coulomb exploded" H_2^+ signal and we therefore concentrate our analysis on this single channel.

3.3 Reaction Kinematics

To interpret our images, we consider the kinematics involved. Provided the spread of velocities within the molecular beam is sufficiently small, we can disregard the beam velocity by defining the laboratory frame to coincide with the beam frame. The photolytically produced H-atom distribution ($\mathbf{v}_H$), and hence the distribution of relative velocities, **g**, is cylindrically symmetric with respect to the polarization of the photolysis laser. For reactive scattering, the product angular distribution is cylindrically symmetric with respect to **g**. Thus the H_2 product angular distribution will be cylindrically symmetric with respect to the laser polarization. This fulfils the condition necessary to use image reconstruction techniques described in the Appendix to obtain 2D intensity profiles of the H_2 product. It can be shown that lab frame velocities are equivalent to center-of-mass velocities for our mass combination, allowing trivial extraction of product translational energies from the images.

Figure 3 (middle column) shows reconstructed images (2D intensity profiles) for the three $H_2(v, J)$ states shown in the left column of Fig. 3. These profiles consist of a series of concentric rings, corresponding to different values of H_2 velocity, and hence different reaction channels. The number of rings varies with the H_2 rovibrational product state probed. This can be seen more clearly from speed distributions obtained by integrating the intensity of each of the reconstructed images around its center and weighting each velocity by the appropriate volume element (see Appendix). The speed distributions corresponding to the three images shown in the middle column of Fig. 3 are shown in the right column of Fig. 3. Each ring in the reconstructed image appears as a peak in the speed distribution.

The Gaussian curves in Fig. 3 (right column) are obtained by fitting the outer two peaks of the experimental speed distributions (corresponding to the FI and SI channels) to Gaussian distributions with the following functional form:

$$P(u) = \sum_{i=1-2} (N_i \exp(-A(x_i - x)^2)) \tag{5}$$

A and N_i are variables, x_i is the position of the predicted speed for each reaction channel, and i = 1 and 2 refers to the two separable reaction channels, SI and FI. Each set of H_2 products having a different speed, i.e., produced by a different reactive channel, is represented by a Gaussian distribution. The width of the Gaussian is chosen to optimize the fit of the data. The widths are found to be approximately the same as the H-atom widths of Fig. 2. By fitting both the SI and FI channels to Gaussian distributions we account to some extent for overlap in intensity between the different channels.

3.4 H_2 Product Internal State Distributions

For the range of photolysis wavelengths used in this experiment, the relative amount of slow and fast reactant H atoms, i.e., the I*:I ratio predicted by Levy and Shapiro, ranges from 0.24 to 0.65 [48]. This ratio is used to scale the state populations associated with the FI reaction channel. The rotational distributions are presented in Table 1.

Presented in Fig. 4 are plots of the derived rotational population distributions for each vibrational level. A number of qualitative observations can be made from these plots. The H_2 product is formed with a high degree of rotational excitation. Note that the rotational constant of H_2 in its electronic ground state is large (~60 cm^{-1}) [19], and a 300K Boltzmann rotational distribution for H_2 has appreciable population only up to J = 3, and a most probable value of J = 1. Despite the high degree of rotational excitation present in the H_2 products, none of the rotational distributions extend to the total energy available. The energetic limits for the FI channels are not indicated in Fig. 4 but extend far beyond the limits of the plots (4.2 eV of available energy which is very close to the dissociation limit of H_2). Not only are the energetically accessible high J states unpopulated, but also only a small amount of products is observed in low J states as well. With the exception of the FI(v = 3) channel, the rotational distributions are sharply peaked with the distributions becoming broader with increasing vibrational excitation. The peaking of each rotational distribution occurs around the same internal energy of the H_2 product, close to the exothermicity of the reaction.

The vibrational distribution for each reaction channel is calculated by summing the rotational distributions for each vibrational state. Normalization between vibrational states is accomplished by scaling the intensities of nearby rotational levels of two adjacent vibrational states measured from data images which were obtained under identical conditions. The populations of rotational levels not probed were taken as zero, an approximation that is reasonable considering the sharply peaked appearance of the rotational distributions. We quote the populations obtained

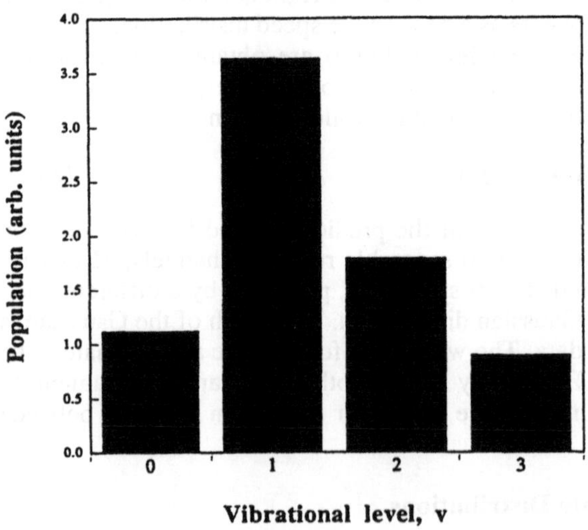

Figure 5: Vibrational distribution of H_2(v, J) produced from the FI channel in the reaction H + HI → H_2(v, all J) + I at a collision energy of 2.7 eV.

using this approximation as lower limits. The $H_2(v = 3)$ rotational distribution corresponding to the FI channel is relatively flat and possibly extends to rotational levels much higher than those probed. Hence our quoted $H_2(v = 3)$ vibrational population for the FI channel constitutes a lower limit. Similarly, because we did not measure the populations of the $H_2(v = 0; J = 0\text{-}11)$ product states, our $H_2(v = 0)$ population also represents a lower limit. Extrapolation of higher rotational levels indicates that little population is expected in these lower rotational states. Our experimentally determined vibrational distribution is shown in Fig. 5. As in the case of rotational excitation, the observed product does not appear to be populated in the highest vibrational level that is energetically allowed. We observe a strong propensity for production of $H_2(v = 1)$, i.e., the vibrational inversion observed in previous experiments on this system.

4 Discussion

The large exothermicity of the hydrogen-abstraction reaction, and the high collision energies used in this experiment, results in a wide range of energetically accessible internal states of the H_2 product. The resulting broad internal state distribution of the H_2 product allows the collection of an extensive experimental data base for this reaction. Reaction product imaging is a quantitative technique. Using procedures described in the Appendix, we have extracted relative rotational and vibrational distribution from our images. The H_2 product is formed over a range of rovibrational states. Not all energetically allowed product states are observed. The vibrational distribution is inverted with the highest population occurring for $v = 1$. The rotational distribution is sharply peaked at an internal energy close to the reaction exothermicity, with surprisingly little population in both high and low energetically accessible J states. These general observations have been well reproduced by QCT calculations on modified LEPS and DIM surfaces by Aker and Valentini (AV) [22,25,26] for the lower collision energies as well as by González and Sayós [27,28]. The QCT calculations do fit the qualitative nature of the observed product state distributions, and a description of the physical picture of this reaction process arising from examination of the trajectories is useful to consider. A number of our observations are supported by the earlier experimental work of Valentini and co-workers [21,22] and Zare and co-workers [24]. Comparison of our results with this earlier work suggests that the average internal energy in the H_2 molecule is approximately independent of the reaction collision energy.

As may be expected for the high collision energies involved, the structured nature of the H_2 rotational and vibrational product distributions and the absence of many energetically allowed product states suggests a direct reaction, i.e., the H + HI abstraction reaction does not proceed through a long-lived complex, which would tend to create a statistical product state distribution. QCT calculations show that for the high collision energies used here, a small number (6%) of the reactive trajectories do indeed undergo an indirect reaction, forming an orbiting resonance complex before abstraction [25].

The H + HI abstraction reaction is very exothermic and proceeds upon a potential energy surface with an early barrier and low activation energy. In such a system, there is a propensity to convert the translational energy of the reaction into product vibrational energy, resulting in an inversion in the product vibrational distribution [60]. The H_2 vibrational distribution observed for the H + HI abstraction is inverted with the highest population occurring for v = 1, see Fig. 5. Early QCT calculations were unable to reproduce the observed vibrational inversion for the collision energies used in the experiments reported [22,25,27]. This was explained to be due to the high collision energy allowing sampling of the more repulsive regions of the potential energy surface other than the minimum energy path. At lower collision energies, where the reaction is likely to follow the collinear minimum energy path on the potential energy surface, calculations were able to predict an inversion [22,25,27]. QCT calculations by González and Sayós produced a vibrational inversion at experimental collision energies by empirically modifying the potential characteristics, in particular, by reducing the H–H bond length for the transition state [28].

The internal states of the H_2 product are not populated to the energetic limit. Indeed, the internal state distribution peaks at 1.5 eV (close to the reaction exothermicity of 1.4 eV) and does not extend to the total energy available, ~4.2 eV. A possible interpretation of this observation is that the translational energy of the reactants is uncoupled from the internal energy of the products, i.e., all the product internal energy arises from the making of the new H_2 bond and the breaking of the old HI bond. In the context of a simple stripping model [18] this explanation is unsatisfactory because a stripping model would predict an increase in internal energy with increasing reactant velocity. An alternative explanation comes from the trajectory studies of Aker and Valentini [25]. Their calculations clearly show a general trend that the high J states are populated by collisions that have large H-H' impact parameters ($b_{H\text{-}H'}$) and that the low J states are populated by collisions having lower $b_{H\text{-}H'}$. The H-H' impact parameter is defined as the impact parameter between the incoming H atom and the H' atom that is attached to the iodine atom. This observation led to a two-step model for the reaction of H + H'X systems where the incoming H atom reacts directly with the H' forming H_2 in a particular (v, J) state. The H_2(v, J) molecule then interacts with the iodine atom. This interaction is weak and does not substantially change the rotational or vibrational distribution formed from the initial interaction between the two H atoms. If we take the maximum $b_{H\text{-}H'}$ to be the maximum $b_{H\text{-}H'X}$ from the trajectory calculations minus the H'-X bond distance this model predicts the rotational cut-offs of the v = 0 distribution to within one rotational quantum number for H + HI, H + HBr and H + HCl for 1.6 and 0.68 eV collision energies. This model can be stated:

$$|\mathbf{L}_{H\text{-}H'}| = |\mathbf{J}'_{H_2}|. \tag{6}$$

where **L** is the incoming orbital angular momentum of the two H atoms and **J** is the rotational angular momentum of the H_2 molecule. When we apply this simple relationship we find that we need to use a maximum $b_{H\text{-}H'I}$ of 2.8 Å, leading to a value of $b_{H\text{-}H'}$ of 1.3 Å, in order to reproduce the rotational cut-off we observe J = 19 for v = 0. This $b_{H\text{-}H'I}$ of 2.8 Å is smaller than $b_{H\text{-}H'I}$ of 3.1 Å found from calculations [25] at 1.6 eV collision energy but is consistent with the decrease in the cross section with increasing velocity observed in the calculations, from about 1.7 $Å^2$ at 1.3 eV collision energy to about 1.4 $Å^2$ at 3 eV collision energy. We therefore conclude that the reason the rotational cut-off remains essentially constant between the 1.6 eV data of both Zare and co-workers and Valentini and co-workers and our data at 2.7 eV is that the increase in velocity of the collision (thereby increasing the amount of angular momentum available) is offset by a decrease in the maximum impact parameter leading to reaction (thereby decreasing the amount of angular momentum available to the reaction) and that Eq. 6 adequately describes the angular momentum constraint that is limiting the rotational distributions. If we allow for angular momentum transfer during the initial step of the reaction into orbital angular momentum of the H_2 around the iodine atom then J values on the order of 70 $\hbar$ are possible.

Note that this model, using Eq. 6, predicts that J_{max} is independent of the vibrational state formed. Typically, this is what has been observed. Rotational distributions extend to about the same limit for v = 0, 1 and 2, but they peak at lower J values for higher v values, see Fig. 4. High values of $b_{H\text{-}H'}$ correspond to distant and highly tangential interactions of the H atom centers. This corresponds to high angular momentum for the system and, by Eq. 6, leads to significant rotational energy in the products. Lower $b_{H\text{-}H'}$ values (more head-on H-H' collisions) lead directly to lower rotational excitation and apparently to more vibrational excitation as well, thereby explaining the shift of the rotational distributions to lower values of J with increasing vibrational level. The shape of our observed vibrational distribution, Fig. 5, can be somewhat rationalized by this simple two step model and the assumption that more head-on H-H' collisions leads to more vibrational excitation of the H_2 product. If we assume that the impact with the iodine does not scramble the vibrational or rotational distributions initially formed, then summing all population in a particular rotational state, independent of vibrational level, and associating, through Eq. 6, this rotational state with an incoming impact parameter we should recover the opacity function of the reaction. Because the opacity function peaks at intermediate $b_{H\text{-}H'I}$ values this peak is reflected in the vibrational distribution peaking. This model also helps explain the lack of very low J states populated because these can only be formed by direct impact $b_{H\text{-}H'} = 0$ on the H atom, an extremely low probability event. This model predicts that if D + DI were studied that the maximum rotational level observed would increase by a factor of $\sqrt{2}$ which is what is observed in new results from our laboratory [61]. The maximum J state observed in the product D_2 for v = 0 is found to be about J = 25.

5 Concluding Remarks

We have used RPI to obtain rotational and vibrational state population distributions for a single reactive channel of the H + HI $\rightarrow$ H_2 + $I(^2P_{3/2})$ reaction at average center-of-mass collision energy of ~2.7 eV. We determine that the rotational distributions do not change significantly between the 2.7 eV collisions energy at which we perform our experiment and the 1.3 eV and 0.6/1.6 eV collision energy experiments others have performed. We speculate that this is due to a balancing between the increased orbital angular momentum available and the changing shape and magnitude of the opacity function. We find that the appearance of the product rotational distributions, including the maximum J state observed in the rotational distributions, can be qualitatively described by a simple model, suggested by Aker and Valentini, that assumes two distinct interactions. The first interaction is between the two H atoms and a second one between the formed H_2 and the iodine atom. This model predicts the maximum rotational states observed and the shift in the maximum upon isotopic substitution. We observe a vibrational inversion in the formed H_2 quantum states.

6 Appendix

This appendix describes the reconstruction technique, and general procedures that can be used in conjunction with the reaction product imaging technique to extract scattering information from ion images.

6.1 Abel Transform of a Two-Dimensional Function

We follow the treatment of Bracewell [62] by considering a two-dimensional function f(x,y) that has cylindrical symmetry, i.e. f(x,y) = f(r). Let an arbitrary line AB lie at a distance R from the origin. Then the projection of f(x,y) on that line can be expressed as

$$f_L(R,\theta) = \int_{-\infty}^{\infty} \int_{-\infty}^{\infty} f(x,y)\delta(x\cos\theta + y\sin\theta - R)\,dxdy \tag{A1}$$

We are free to choose θ because the projection on any line in the xy-plane is the same for a cylindrically symmetric function. Choosing $\theta = 0$, we obtain R = x and thus

$$f_L(x) = \int_{-\infty}^{\infty} \int_{-\infty}^{\infty} f(r)\delta(x-R)\,dxdy = 2\int_{r=x}^{\infty} f(r)\,dy \tag{A2}$$

or

$$f_L(x) = 2\int_{x}^{\infty} \frac{f(r)\,dr}{\sqrt{r^2 - x^2}} \tag{A3}$$

The line integral $f_L(x)$ can be recognized as a form of the Abel integral equation [63], and is referred to as the *Abel transform* of f(r). Of more interest to us is f(r) which can be obtained from $f_L(x)$ by the *inverse* Abel transform

$$f(r) = -\frac{1}{\pi}\frac{\partial}{\partial r}\int_{r}^{\infty} \frac{rf_L(x)\,dx}{x\sqrt{x^2 - r^2}} \tag{A4}$$

Equation A4 relates the experimentally measured projection $f_L(x)$ to the original distribution function f(r). This unique relationship between cylindrically symmetric 2D functions and their line integrals enables us to reconstruct angular distributions for both photofragmentation and bimolecular reaction scattering processes, using the procedures described below.

6.2 Reconstruction of a Three-Dimensional Distribution

At any time following a photofragmentation process at a point source, photofragments with the same speed can be found on the surface of a sphere, centered on the center-of-mass of the process, which we call the *scattering sphere.* The radius of the scattering sphere is proportional to the fragment speed, and thus photofragments of different speeds will create a series of concentric spherical shells. For a one-photon photodissociation process the angular distribution of photofragments forming each spherical shell is described by [64]

$$I(\theta,\phi) = \frac{1}{4\pi}\{1 + \beta P_2(\cos\theta)\} \qquad \text{(A5)}$$

where θ and ϕ are the polar and azimuthal scattering angles with respect to the laser polarization, β is the anisotropy parameter, and $P_2(\cos\theta)$ is the second Legendre polynomial. By inspection of Eq. A5, we see that $I(\theta, \phi)$ is independent of ϕ, and thus the photofragment angular distribution on each spherical shell is cylindrically symmetric with respect to the polarization vector of the photolysis laser. The data image recorded by our ion imaging technique is a two-dimensional *projection* of this three-dimensional spatial distribution of photofragments, i.e., the 3D distribution is crushed flat onto the 2D detector surface. The experiment is configured so that the polarization of the photolysis laser is parallel to the imaging plane (detector face), and so the 2D projection of the cylindrically symmetric 3D distribution is made parallel to its symmetry axis.

Planar cuts through the scattering spheres, made perpendicular to the azimuthal symmetry axis, will yield a series of concentric uniform circular rings. The total photofragment 3D distribution can be thought of as a sum of 2D ring distributions. We have just seen that a unique mathematical relationship exists between cylindrically symmetric 2D functions and their line integrals. Thus one 2D projection of a cylindrically symmetric 3D object, taken parallel to its symmetry axis, is sufficient to completely determine the 3D object using an *inverse Abel transform* (Eq. A4). Experimentally, the photofragment distribution is recorded on a rectangular CCD array whose orientation is such that the laser polarization vector is parallel to the direction of the pixel columns. Thus, each row of pixels corresponds to the 1D projection of a 2D photofragment ring distribution. Consequently we can reconstruct the entire 3D photofragment distribution by reconstructing one row at a time of the CCD array. The results of such a calculation can be displayed as a slice through the center of the original 3D distribution, that we call a *2D intensity profile* or *reconstructed* image.

The Abel inversion described by Eq. A4 is commonly used in many areas of science and engineering [65-67] and several methods for performing this task are available [68-71] One method involves least squares fitting a functional form to each line of data, or pixel row. Choosing the appropriate functional form allows both the integral and the subsequent differentiation in Eq. A4 to be performed analytically. This method is extremely useful for reconstructing data with a low signal-to-noise ratio. It is however a time consuming process particularly when the image sizes are large (typically ≥ 350 pixels). An alternative method makes use of the following principle: the Fourier transform of a two-dimensional (or one-dimensional) projection of a three-dimensional (two-dimensional) object is exactly equal to the central section of the Fourier transform of the object [72]. Our reconstruction algorithm is based on this principle [73]. The inverse Abel transform is obtained by taking the inverse Hankel transform of the 1D fast Fourier transform of each row of the projection data [74,75]. Although this procedure is much faster than the analytical method (30 sec on an 25 MHz RS3000 MIPS computer), it is sensitive to the amount of noise in the data image. Any noise in the projection data appears amplified in the final reconstruction, with an inverse proportionality to the distance from the symmetry axis, i.e., most pronounced along the vertical axis of the reconstruction. Note that only half an ion image (about the axis of cylindrical symmetry) is required to compute the reconstruction, and so each reconstructed image we present seems perfectly symmetrical as the two sides are mirror images.

To demonstrate both the physical significance and the accuracy of our Abel inversion algorithm, consider the three concentric rings shown in Fig. 6a. The intensity per pixel of each ring is such that the total intensity per ring is equal. The projection of the circles onto a row is shown in Fig. 6b. The spikiness, or noise, in the projection is caused by the discrete nature of the circle which was created on a square grid. Reconstruction of this projection is shown in Fig. 6c. We observe that although the cross sectional intensity profile of the three rings is resolved, the amount of artificial noise introduced by the algorithm is substantial. This data was presmoothed using a small Gaussian convolution mask (typically smoothing 5 pixels with a 3 pixels FWHM mask, *c.f.* typical image sizes with a 350 pixel diameter) before transforming. Although the convolution will caused a slight decrease in the resolution, manifested by peak broadening, the noise is effectively suppressed. The agreement between the peak heights in the original image, shown as stars in Fig 6c, and the intensity in the reconstructed image Fig. 6c is excellent, showing that our transformation algorithm used successfully reconstructs even the central portion of the original object, and does indeed generate the correct intensity profile. As this example demonstrates, caution must be taken when performing the reconstruction so that artificial structure caused by noisy data does not appear in the reconstructed image. Convolution, or *filtering*, of the data using smoothing functions is recommended, although data smoothing will reduce resolution. Deblurring routines described elsewhere [73] and other image restoration processes should be used with caution as they can introduce structural artefacts.

6.3 Ion Images

A reconstructed ion image is a 2D intensity profile, i.e., a 2D slice through a 3D distribution, and represents a direct observation, a simultaneous measurement, of an entire 3D spatial distribution of products resulting from the molecular interaction under study, e.g., a unimolecular photofragmentation, or a bimolecular reaction. Extracting scattering information from reconstructed images is straightforward. The

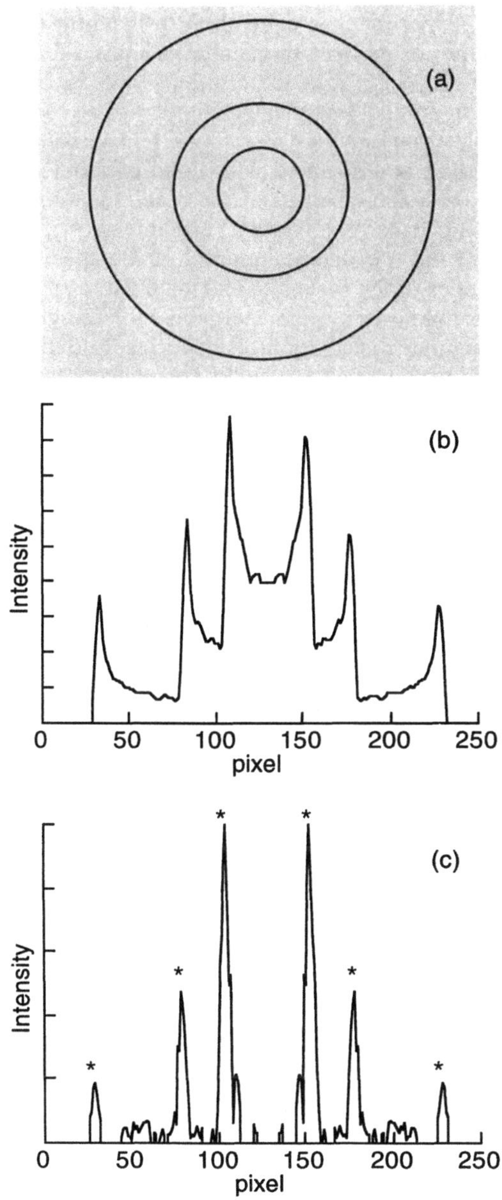

Figure 6: (a) Three concentric rings whose intensity per pixel is such that the total intensity per ring is constant. (b) Intensity profile of the projection of the rings onto an arbitrary row, convoluted with a Gaussian function 5 pixels wide 3 pixels FWHM. (c) Inverse Abel transform of the projection. The intensity profile along a diameter of the rings is indicated by the stars.

projection of a sphere onto a plane is a circle, and so finding the center of mass of the scattering process in an ion imaging experiment is equivalent to locating the center of the image. This is substantially easier than the center-of-mass to lab transformation needed in conventional time-of-flight experiments and is one significant advantage offered by the reaction product imaging technique.

A large amount of data is available from analysis of a single reconstructed image. Any point in the 2D intensity profile is associated with three parameters: radius (distance from the image center, or center-of-mass of the reaction), angle (relative to the cylindrical symmetry axis, or laser polarization) and intensity. The radial position is a direct measurement of the translational energy of the species, which can be coupled to the known energetics of the system to elucidate the reaction channels, or conversely, to derive hitherto unknown system energetics. Ion imaging measures simultaneously the complete angular distribution of a species, allowing determination of reaction β parameters, and in certain cases, the differential cross section of the reaction. Image intensities give a direct measure of the population, or number density, for each product energy and angle. The spatial distributions measured by imaging techniques make it possible to distinguish different reaction channels contributing to the same product, and so relative branching ratios between these different channels can be measured. The resonant nature of laser-based detection allows quantum state resolution. Comparison of ion images corresponding to different product quantum states, allows the measurement of state-resolved relative reaction cross-sections.

6.4 Speed Distributions

The speed of the photofragments is proportional to their distance ρ from the center of mass, which is related to the distance r from the symmetry axis by

$$r = \rho \sin\theta \tag{A6}$$

For each row of data, we have shown that the inverse Abel inversion yields the intensity profile along a diameter of the 2D ring distribution. Hence the total intensity per ring R for a given distance from the symmetry axis r, $P_R(r)$, is given by

$$P_R(r) = 2\pi r\, I(r,\theta) \tag{A7}$$

where $I(r,\theta)$ is the intensity for the pixel position (r,θ). Using Eq. A6 to change variables, Eq. A7 becomes

$$P_R(\rho) = 2\pi\rho \sin\theta\, I(\rho,\theta) \tag{A8}$$

The speed of the photofragments is directly proportional to ρ, thus summing $P_R(\rho)$ over all line projections in the data image yields the speed distribution P(u) for the reaction, i.e.,

$$P(u) \propto \sum_R P_R(\rho) \tag{A9}$$

6.5 Differential Cross Sections

The differential cross section (or angular distribution) is defined as

$$\frac{d\sigma}{d\Omega} = \frac{\text{Number of fragments scattered into } d\Omega \text{ per unit time}}{\text{Intensity of the incident beam}} \tag{A10}$$

where $d\Omega$ is a solid angle. This definition is an empirical one and arises from the fact that conventional TOF scattering experiments were done by measuring the numerator on the right hand side of Eq. A10. The solid angle is defined as

$$d\Omega = \frac{dA}{\rho^2} = \sin\theta\, d\theta\, d\phi \tag{A11}$$

where we have used the definition for the area element in spherical coordinates $dA = (\rho\, d\theta)(\rho \sin\theta\, d\phi)$. For any plane that contains the symmetry axis, let the number of particles scattering into some angle $d\theta$ be denoted as $N(\theta)$. Then the number of the particles $N(\Omega)$ scattering into a solid angle $d\Omega$ is given by

$$N(\Omega) = N(\theta)\sin\theta\, d\phi \tag{A12}$$

because of the azimuthal symmetry of the problem in hand. Thus Eq. A10 can be rewritten as

$$\frac{d\sigma}{d\Omega} = \frac{N(\Omega)}{d\Omega dt} \tag{A13}$$

which upon substitution becomes

$$\frac{d\sigma}{d\Omega} = \frac{N(\theta)}{d\theta\, dt} \tag{A14}$$

where we have assumed that the incident beam intensity has a value of one. dt is a constant for each ion image, and thus Eq. A14 indicates that the differential cross section for the scattering process can be obtained by integrating the reconstructed image along radial sectors centered at the center of mass of the scattering process.

6.6 Bimolecular Scattering

For bimolecular scattering the azimuthal symmetry axis lies along the direction of the relative velocity vector of the collision, and the analysis of the data images to obtain scattering data is identical to the procedure just described for photofragmentation. In the case of a single beam experiment, such as described in this work, the photodissociation process does not define uniquely the direction of the relative velocity (Eq. 5). The angular distribution of the reactants means that an angular distribution of products can yield only the second moment of the differential cross section [76,77].

7 Acknowledgements

We thank Mark Jaska for his expert technical assistance. R.I.McK., A.J.R.H., M.A.B. and R.N.Z acknowledge support from the National Science Foundation. This work is supported by the U.S. Department of Energy, Office of Basic Energy Sciences, Division of Chemical Sciences.

8 References

[1] J. Solomon, J. Chem. Phys. **47**, 889 (1967).
[2] C. Jonah, P. Chandra, and R. Bersohn, J. Chem. Phys. **55**, 1903 (1971).
[3] J. Solomon, C. Jonah, P. Chandra, and R. Bersohn, J. Chem. Phys. **55**, 1908 (1971).
[4] G. E. Busch, R. T. Mahoney, R. I. Morse, and K. R. Wilson, J. Chem. Phys. **51,** 449 (1969).
[5] R. W. Diesen, J. C. Wahr, and S. E. Adler, J. Chem. Phys. **50,** 3635 (1969).
[6] M. Dzvonik, S. Yang, and R. Bersohn, J. Chem. Phys. **61**, 4408 (1974).
[7] W. M. Jackson and R. J. Cody, J. Chem. Phys. **61**, 4183 (1974).
[8] A. P. Baronavski and J. R. McDonald, Chem. Phys. Lett. **45,** 172 (1977).
[9] J. L. Kinsey, J. Chem. Phys. **66,** 2560 (1977).
[10] R. Schmiedl, H. Dugan, W. Meier, and K. H. Welge, Z. Phys. A **304,** 137 (1982).
[11] I. Nadler, D. Mahgerefteh, H. Reisler, and C. Wittig, J. Chem. Phys. **82**, 3385 (1985).
[12] R. O. Loo, H.-P. Haerri, G. E. Hall, and P. L. Houston, J. Chem. Phys. **90,** 4222 (1989).
[13] T. N. Kitsopoulos, M. A. Buntine, D. P. Baldwin, R. N. Zare, D. W. Chandler, Science **260,** 1605 (1993).
[14] A. G. Suits, L. S. Bontuyan, P. L. Houston, B. J. Whitaker, J. Chem. Phys. **96,** 8618 (1992).
[15] D. W. Chandler and P. L. Houston, J. Chem. Phys. **87,** 1445 (1987).
[16] M. A. Buntine, D. P. Baldwin, R. N. Zare, and D. W. Chandler, J. Chem. Phys. **94,** 4672 (1991).
[17] K. R. Castleman, ***Digital Image Processing*** (Prentice-Hall, Englewood Cliffs, 1979).
[18] R. D. Levine and R. B. Bernstein, ***Molecular Reaction Dynamics and Chemical Reactivity*** (Oxford University Press, 1987).
[19] K. P. Huber and G. Herzberg, ***Molecular Spectra and Molecular Structure IV, Constants of Diatomic Molecules*** (Van Nostrand Reinhold, New York, 1979).
[20] P. M. Aker, G. J. Germann, K. D. Tabor, and J. J. Valentini, J. Chem. Phys. **90,** 4809 (1989).

[21] P. M. Aker, G. J. Germann, and J. J. Valentini, J. Chem. Phys. **90,** 4795 (1989).
[22] P. M. Aker, G. J. Germann, and J. J. Valentini, J. Chem. Phys. **96,** 2756 (1992).
[23] D. A. V. Kliner, K.-D. Rinnen, and R. N. Zare, J. Chem. Phys. **90,** 4625 (1989).
[24] D. A. V. Kliner, K.-D. Rinnen, M. A. Buntine, D. E. Adelman, and R. N. Zare, J. Chem. Phys. **95,** 1663 (1991).
[25] P. M. Aker and J. J. Valentini, Isr. J. Chem. **30,** 157 (1990).
[26] P. M. Aker and J. J. Valentini, J. Phys. Chem. **97,** 2078 (1993).
[27] M. González and R. Sayós, Chem. Phys. Lett. **164,** 643 (1989).
[28] M. González, Faraday Disc. Chem. Soc. **91,** 339 (1991).
[29] T. H. Dunning, Jr., J. Phys. Chem. **88,** 2469 (1984).
[30] T. H. Dunning, Jr. and L. B. Harding, in ***Theory of Chemical Reaction Dynamics***, Vol. 1, edited by M. Baer (CRC Press, Boca Raton, 1985), pp. 1.
[31] M. R. Levy, Progress in Reaction Kinetics 10, 1 (1979).
[32] C. A. Parr and D. G. Truhlar, J. Phys. Chem. **75,** 1844 (1971).
[33] R. N. Porter, L. B. Sims, D. L. Thompson, and L. M. Raff, J. Chem. Phys. **58,** 2855 (1973).
[34] J. D. McDonald and D. R. Herschbach, J. Chem. Phys. **62,** 4740 (1975).
[35] W. Bauer, L. Y. Rusin, and J. P. Toennies, J. Chem. Phys. **68,** 4490 (1978).
[36] W. H. Beck, R. Götting, J. P. Toennies, and K. Winkelmann, J. Chem. Phys. **72,** 2896 (1980).
[37] L. M. Raff, H. H. Suzukawa, Jr., and D. L. Thompson, J. Chem. Phys. **62,** 3743 (1975).
[38] G. J. Germann and J. J. Valentini, J. Phys. Chem. **92,** 3792 (1988).
[39] M. Baer and I. Last, in ***Potential Energy Surfaces and Dynamics Calculations for Chemical Reactions and Molecular Energy Transfer***, edited by D. G. Truhlar (Plenum Press, New York, 1981), pp. 519.
[40] L. M. Raff, L. Stivers, R. N. Porter, D. L. Thompson, and L. B. Sims, J. Chem. Phys. **52,** 3449 (1970).
[41] P. Cadman and J. C. Polanyi, J. Phys. Chem. **72,** 3715 (1968).
[42] D. C. Clary, Chem. Phys. **71,** 117 (1982).
[43] P. Brewer, P. Das, G. Ondrey, and R. Bersohn, J. Chem. Phys. **79** (1983).
[44] R. Schmiedl, H. Dugan, W. Meier, and K. H. Welge, Z. Phys. A **304,** 137 (1982).
[45] G. N. A. Van Veen, K. A. Mohamed, T. Baller, and A. E. De Vries, Chem. Phys. **80,** 113 (1983).
[46] Z. Xu, B. Koplitz, and C. Wittig, J. Phys. Chem. **92,** 5518 (1988).
[47] Z. Xu, B. Koplitz, and C. Wittig, J. Chem. Phys. **90,** 2692 (1989).
[48] I. Levy and M. Shapiro, J. Chem. Phys. **89,** 2900 (1988).
[49] R. N. Zare, ***Molecular Photochemistry*** **4,** 1 (1972).
[50] R. C. Gonzalez and P. Wintz, ***Digital Image Processing*** (Addison-Wesley, Reading, 1977).
[51] E. L. Hall, ***Computer Image Processing and Recognition*** (Academic Press, New York, 1979).
[52] K.-D. Rinnen, D. A. V. Kliner, R. S. Blake, and R. N. Zare, Chem. Phys. Lett. **153,** 371 (1988).

[53] K.-D. Rinnen, D. A. V. Kliner, R. N. Zare, and W. M. Huo, Isr. J. Chem. **29,** 369 (1989).
[54] K.-D. Rinnen, D. A. V. Kliner, M. A. Buntine, and R. N. Zare, Chem. Phys. Lett. **169,** 365 (1990).
[55] W. M. Huo, K.-D. Rinnen, and R. N. Zare, J. Chem. Phys. **95,** 205 (1991).
[56] K.-D. Rinnen, M. A. Buntine, D. A. V. Kliner, R. N. Zare, and W. M. Huo, J. Chem. Phys. **95,** 214 (1991).
[57] D. E. Adelman, N. E. Shafer, D. A. V. Kliner, and R. N. Zare, J. Chem. Phys. **97,** 7323 (1992).
[58] H. Umemoto, S. Nakagawa, S. Tsunashima, and S. Sato, Chem. Phys. **124,** 259 (1988).
[59] J. Purnell, E. M. Snyder, S. Wei, and A. W. Castleman, Chem. Phys. Lett. **229,** 333 (1994).
[60] J. C. Polanyi, Accounts of Chemical Research **5,** 161 (1972).
[61] A. J. R. Heck, R. N. Zare, and D. W. Chandler, to be published (1995).
[62] R. N. Bracewell, Aust. J. of Phys. **9,** 198 (1956).
[63] R. N. Bracewell, ***The Fourier Transform and its Applications*** (McGraw-Hill, New York, 1986).
[64] R. N. Zare and D. R. Herschbach, (Proc. IEEE) **173**, 1963.
[65] I. J. D. Craig, Astron. Astrophysics **79,** 121 (1979).
[66] C. J. Cremers and R. C. Birkebak, Appl. Optics **5,** 1057 (1966).
[67] M. Sato, J. of Phys. D **11,** 1739 (1978).
[68] L. M. Smith, D. R. Keefer, and S. I. Sudharsanan, J. of Quant. Spectr. and Rad. Transf. **39,** 367 (1988).
[69] E. W. Hansen and P.-L. Law, J. of the Opt. Soc. Amer. **A2,** 510 (1985).
[70] R. S. Anderson, J. Inst. Math. and its Appl. **17,** 329 (1976).
[71] W. L. Barr, J. Opt. Soc. Amer. **52,** 885 (1962).
[72] M. P. Ekstrom, ***Digital Image Processing Techniques*** (Academic Press, Orlando, Florida, 1984).
[73] R. N. Strickland and D. W. Chandler, Appl. Optics **30,** 1811 (1991).
[74] S. M. Candel, IEEE Transactions of Accoustic Speech Signal Processing **ASSP 29,** 963 (1981).
[75] E. W. Hansen, IEEE Transactions of Accoustic and Speech Signal Processing **ASSR 33,** 666 (1985).
[76] F. Green, G. Hancock, A. J. Orr-Ewing, M. Brouard, S. P. Duxon, P. A. Enriquez, R. Sayos, and J. P. Simons, Chem. Phys. Lett. **182,** 568 (1991).
[77] G. W. Johnson, S. Satyapal, R. Bersohn, and B. Katz, J. Chem. Phys. **92,** 206 (1990).

Dynamics at Unimolecular Transition States

S.K. Kim,[a] E.R. Lovejoy,[b] and C.B. Moore[c]
[a]Division of Chemistry and Chemical Engineering, California Institute of Technology, Pasadena, California 91125, U.S.A.
[b]NOAA/ERL/Aeronomy, E/E/AL-2,325 Broadway, Boulder, Colorado 80303, U.S.A.
[c]Department of Chemistry, University of California, and Chemical Sciences Division, Lawrence Berkeley Laboratory, Berkeley, California 94720, U.S.A.

Abstract

Ketene undergoes three unimolecular reactions with thresholds near 29,000 cm^{-1}, dissociation to either the singlet or triplet state of methylene and carbon monoxide and isomerization through exchange of its methylene and carbonyl carbons atoms. Energy-resolved rate constant measurements near the threshold for triplet methylene formation reveal a stepwise increase in rate at the energy corresponding to each quantized energy level for vibration orthogonal to the reaction coordinate in the region of the transition state. Thus the dynamics in the immediate region of the transition state are approximately adiabatic. The isomerization reaction has a high barrier with a broad shallow well at the top surrounding the oxirene configuration. This well gives rise to quantum resonances for motion along the reaction coordinate and a reaction rate which exhibits narrow peaks as the total energy increases above the reaction threshold. These results provide a clear and detailed confirmation of the fundamental hypothesis of the statistical transition state theory (RRKM) that reaction rates are controlled by the number of energetically accessible vibrational levels at the transition state and a striking demonstration of the importance of the detailed shape of the potential energy surface in the region of the transition state.

1 Introduction

Transition state theory (TST) has dominated our understanding of chemical reaction rates for the last two-thirds of the century since Bodenstein first considered the microscopic dynamics of chemical reactions and their macroscopic consequences. The RRKM theory developed by Rice, Ramsperger, Kassel and Marcus has been widely applied to unimolecular reactions [1-3]. It is a microcanonical ensemble version of transition state theory. It incorporates the fundamental assumptions of TST that there is a local equilibrium between reactants and molecules crossing the transition state toward products along the reaction coordinate and that the molecule crossing through the transition state proceeds to products without recrossing. RRKM theory is based on the additional assumptions that all vibrational states in the excited molecule are equally probable and that vibrational energy flows freely among the different degrees of freedom at a rate much faster than the reaction rate. The RRKM

Springer Series in Chemical Physics, Volume 61
Gas Phase Chemical Reaction Systems
Eds.: J. Wolfrum, H.-R. Volpp, R. Rannacher, and J. Warnatz

rate constant for total energy E and total angular momentum J is given by

$$k(E,J) = \frac{W(E,J)}{h\rho(E,J)} \quad (1)$$

where W(E,J) is the number of energetically accessible states for vibration orthogonal to the reaction coordinate at the transition state, ρ(E,J) is the density of vibrational states of the reactant, and h is Planck's constant. RRKM theory presumes that motion along the reaction coordinate is classical and is decoupled from the bound vibrational motions at the transition state. The passage through the transition state is then vibrationally adiabatic, and the vibrational levels are defined over a sufficiently broad region near the transition state to give well-defined reaction thresholds or quantized channels connecting reactant to products. The rate constant is thus predicted to increase stepwise as the total energy increases through each vibrational level at the transition state.

This step structure in W(E,J) is apparently not an artefact of treating motion along the reaction coordinate classically since the quantum treatments of cumulative reaction probabilities (CRP) [4-6] exhibit quite similar structures. In his elegant treatment of the exact quantum theory of chemical reaction rates, Miller [4] has expressed the CRP as a sum of eigenvalues corresponding to the transition state theory sum of states, N(E) in his work or W(E,J) here. His paper shows that the stepped structure of the TST theory sum of states is found in the exact quantum CRP as a function of total energy for a single value of the total angular momentum. For bimolecular reactions this structure is smoothed out when summation over total angular momentum (integration over impact parameter in classical terms) is carried out [7]. In practicable experiments further smoothing is caused by averaging over the spread in relative translational energy of the reactants. However, for unimolecular reactions with reactant total energy (E) and total angular momentum (J) defined, the rate in Eq. (1) is directly proportional to the exact quantum CRP or more approximately to W(E,J) corrected for one-dimensional tunnelling through the barrier. Thus the influence of transition state structure and dynamics on reaction rates and product energy state distributions can be resolved most clearly for unimolecular reactions. With the resolution of modern laser experiments in cooled, pulsed jets it is not necessary to average over total angular momentum and energy. This suggests a form of spectroscopy in which the energies of the vibrational levels of the transition state are revealed by steps in the rate constant as a function of energy.

The transition state gives the initial conditions for the dynamics of energy release in the repulsive exit valley of the potential energy surface [8,9]. The peaks of product energy state distributions are determined by the geometry of the transition state and the shape of the potential energy surface in the exit valley. Distribution widths are determined by the distributions of atomic positions and momenta at the transition state, that is by the vibrational wavefunction of the molecule as it passes through the transition state. Consequently, if one looks in the wings of the distribution of a particular product quantum number, one expects to see increases in product yield as the energy passes through the threshold for excitation of a transition state vibration which contributes strongly to that product degree of freedom. This can provide a dynamically biased spectroscopy useful for picking out energy levels for specific transition state vibrations [10,11].

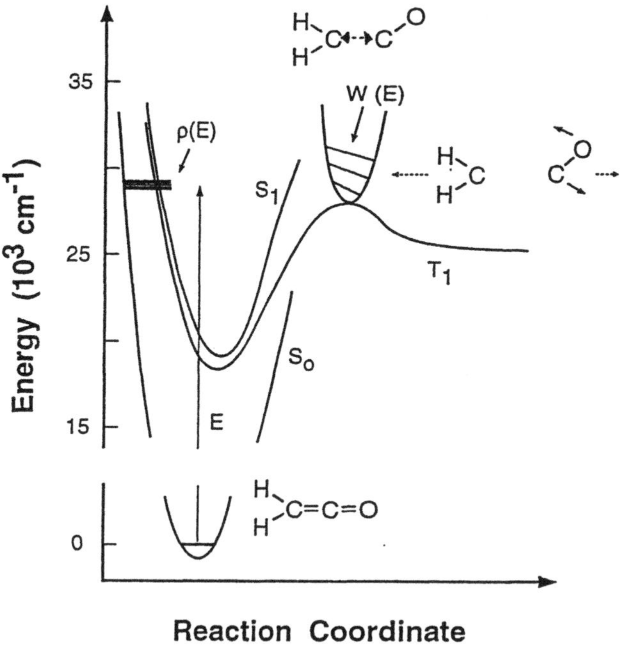

Figure 1: Dissociation of triplet ketene. The reaction coordinate is the distance between the carbon atoms. The solid lines represent the PES for the ground singlet (S_0), the first excited singlet (S_1), and the first triplet (T_1) electronic states. The transition state, which separates the highly vibrationally excited reactant from the products, is depicted as a single potential well perpendicular to the reaction coordinate. This well is meant to represent the eight bound vibrational coordinates at the transition state [11].

2 Vibrational Level Thresholds at the Transition State for Bond Breaking over a Barrier

The stepwise increase in rate constant with energy implied by Equation (1) has been observed for the dissociation of ketene over a barrier on its triplet PES to the triplet ground state of methylene and CO [10,11].

$$CH_2CO \rightarrow {}^3CH_2 + CO \qquad (2)$$

This process [11] is diagrammed in Fig. 1 along with a cartoon of the *ab initio* geometry of Allen and Schaefer [12] of ketene at this transition state. The *ab initio* calculations predict that the lowest frequency is for the torsion about the C-C bond and that the CCO bend and CH_2 out-of-plane wag are the next lowest. Hence, these modes are expected to dominate the structure of the rate constant *vs.* energy in the first few hundred wavenumbers above threshold. Ketene absorbs uv radiation in this energy region due to the $S_0 \rightarrow S_1$ transition dipole moment. Electronic-vibration coupling, internal conversion and intersystem crossing, are sufficiently strong that the wavefunction of the excited molecule may be considered to be a statistical mixture of S_0, T_1 and S_1 basis states of comparable total energy. The density of vibrational states, the sum of that for all three electronic states, is essentially that of the ground state. The uv spectrum is smooth and flat [13, 14]. Thus as a UV exciting laser is tuned, it produces a constant number of hot ketene molecules ready to react along any accessible channel.

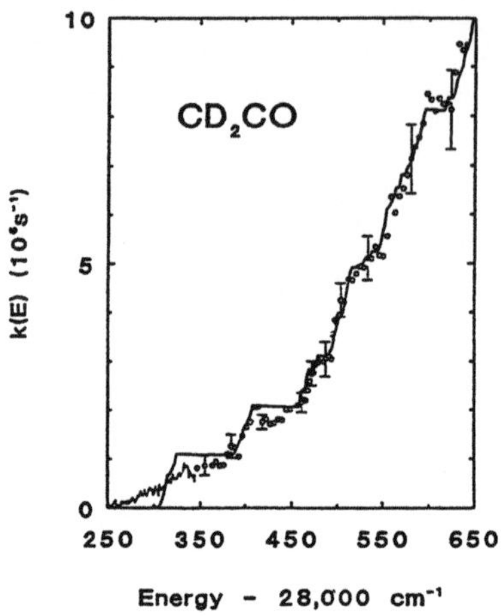

Figure 2: Stepped structure in the rate constant for CD_2CO dissociation as a function of the photolysis energy. The CO (v = 0, J = 12) PHOFEX curve (Δt = 1.7μs) is shown in the reaction threshold region; its intensity is arbitrarily scaled. The solid line is a RRKM fit using the parameters in Table 1 [11].

The rate constant data of Fig. 2 were obtained by uv excitation of ketene cooled to 4K in a pulsed jet [11]. A vuv pulse probed the CO product over a range of delay times to measure the product rise time. To confirm the observed structure in k(E), a photofragment excitation spectrum (PHOFEX) was measured using a short delay time so that the signal was proportional to the reaction rate constant as well as to the fraction of product formed in the CO(J) state probed. For the first 200 cm^{-1} above threshold, in which the CO(J) distribution does not change, the PHOFEX spectra show steps in the identical places as k(E) (Fig. 3). The observed thresholds, Table 1, give barrier heights above product zero-point energies of 1281 ± 15 cm^{-1} for CH_2CO and 1071 ± 40 cm^{-1} for CD_2CO. The implied difference in zero-point energies at the transition state is 1096 ± 20 cm^{-1} compared to the theoretical prediction of 1091 cm^{-1}. The barrier height is some 40% less than the *ab initio* calculation [10,11]. The torsion about the C-C bond has a barrier height within the energy range of these experiments and thus must be treated carefully as a hindered rotor coupled to the free rotation about the C-C bond. Figure 4 shows the potential curve and corresponding energy levels which give the best fit of RRKM theory to the observed k(E), such as in Fig. 2. The positions of many steps for CH_2CO and CD_2CO are fit with a two-parameter potential function. The torsional barrier of 240 cm^{-1} is 35% less than that calculated *ab initio*.

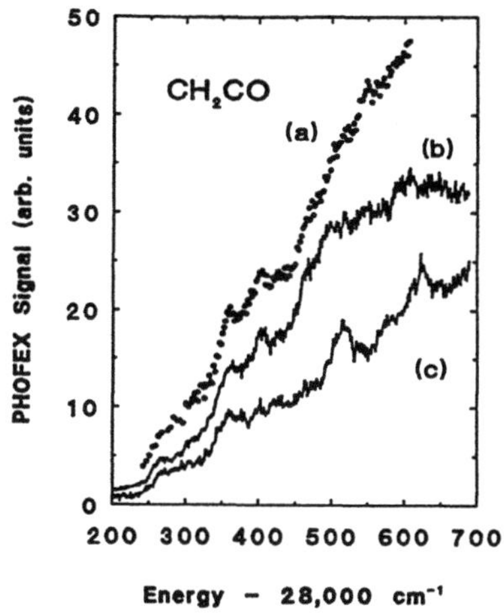

Figure 3: Dynamically biased spectroscopy of a transition state for dissociation. (a) PHOFEX curve calculated from the measured rates for $CH_2CO \rightarrow {}^3CH_2 + CO$ assuming that the distribution of CO(J) states is independent of energy. (b) PHOFEX spectrum at the peak of the CO(J) distribution, CO(J = 12). Evidently, the fraction of CO(J = 12) begins to decrease at about 28,500 cm^{-1} where this curve flattens out. (c) PHOFEX spectrum for the wing of the CO(J) distribution, CO(J = 2). Peaks show the energies of levels with one quantum of CCO bending excitation at the transition state. The delay time of 50 ns for these curves is short compared to the time for dissociation to be complete [11].

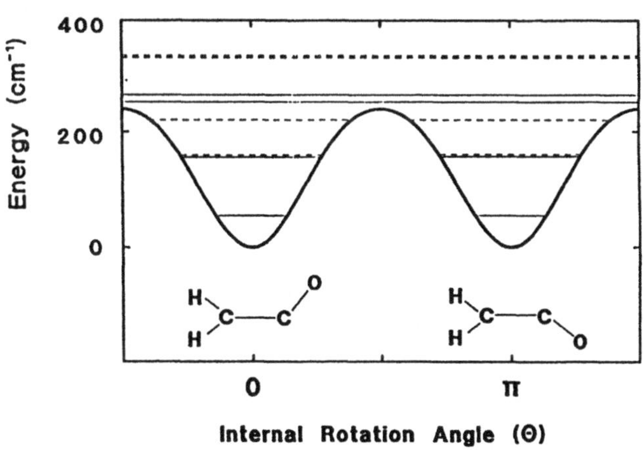

Figure 4: The potential function for hindered internal rotation of the transition state for triplet ketene, $V(\theta) = (1/2)V_0(1-\cos 2\theta) + (1/2)V_1(1-\cos 4\theta)$, where $V_0 = 240$ cm^{-1} and $V_1 = 20$ cm^{-1}. The *ab initio* transition-state geometry is drawn. At the maximum, the O is out of the plane with the CCO plane bisecting the HCH angle. The calculated energy levels are shown for K = 0 (— , ortho; ---, para). The zero-point level splitting is less than 1 cm^{-1} [11].

Table 1: Transition state information from the RRKM fits to experimental results. The values in parentheses are uncertainties obtained from the fits. [a]The imaginary frequency (ν_{im}) is used only for the fit to the first step in k(E) for CH_2CO dissociation. [b]$\rho(E) = \rho_{WR}(E) \times f$. The $\rho_{WR}(E)$ is the density of vibrational states of reactant calculated from the Whitten-Rabinovitch approximation; $\rho_{WR}(E) = 1.36 \times 10^4$ (/cm^{-1}) for $CH_2CO(S_0)$ and 4.78×10^4 (/cm^{-1}) for $CD_2CO(S_0)$ at E = 28,500 cm^{-1}. [c]Assignments are tentative.

	CH_2CO	CD_2CO
E_{th}	28,250 (10) cm^{-1}	28,310 (15) cm^{-1}
ν_{im} [a]	100 (40) i cm^{-1}	–
g_{tf} [b]	3.34 (0.34)	3.56 (0.20)
V_0	240 (30) cm^{-1}	240 (30) cm^{-1}
V_1	20 (20) cm^{-1}	20 (20) cm^{-1}
$\nu_7^‡$ (C-C-O bend)	250 (15) cm^{-1}	[c] 185 (20) cm^{-1}
$\nu_8^‡$ ($CH(D)_2$ wag)	290 (15) cm^{-1}	[c] 240 (20) cm^{-1}

The CCO bending vibration was identified by dynamically biased PHOFEX spectroscopy. Figure 3 shows PHOFEX spectra at the center of the CO(J) distribution, J = 12, and in the wing at J = 2. Molecules coming through the transition state with one quantum of CCO bend excited have a CO(J) distribution with two maxima symmetrically displaced from J = 12 [11]. Such molecules are detected with strong preference at J = 2. The distinct peak observed at 250 cm^{-1} for CH_2CO PHOFEX of CO(J = 2) matches the *ab initio* frequency for the CCO bend of 252 cm^{-1} and is so assigned. The second peak occurs at the combination of the first excited torsional level with the CCO bend. The CH_2 wag was identified with an otherwise unassigned step at 290 cm^{-1}; this is significantly lower than the *ab initio* value of 366 cm^{-1}. These bending frequencies were identified at 185 and 240 cm^{-1} in CD_2CO and may be significantly mixed and perturbed.

In the application of RRKM theory it is usually assumed not only that all vibrational degrees of freedom of a molecule are strongly mixed but also that Coriolis coupling mixes in the K_a degree of freedom [3]. To test this hypothesis for ketene, measurements were made as a function of jet temperature. For the cold ground state at 4K, several J levels are populated but only the lowest K_a levels of each nuclear spin symmetry, K_a = 0, 1, are populated. When the molecule is excited, strong K_a mixing causes all K_a values up to the limit of ±J to be equally populated. The associated energy locked up in rotation as the molecule crosses the transition state is $A^‡K_a^2$, where $A^‡$ is the a-axis rotational constant at the transition state. At a jet temperature of 30 K this requires zero to many tens of cm^{-1} of additional energy to cross through any particular vibrational threshold. The result of K-mixing would be a significant decrease in rate constant with increasing jet temperature. The observed rates were almost identical at 4 and 30K, while the RRKM model predicts differences of up to 20% for K_a mixed. Thus for values of $J \leq 6$, K_a appears to be a good quantum number for highly excited ketene [11].

The step heights in Fig. 2 give the density of states for ketene, or more precisely the density divided by the effective triplet degeneracy at the transition state, $0 \leq g_t \leq 3$. It is not currently known how many of the three symmetries of triplet spin sublevels are coupled strongly to the singlet levels by intersystem crossing. The data for CH_2CO are accurately fit by $\rho = 3.34\rho_{WR}/g_t$ and for CD_2CO by $\rho = 3.56\rho_{WR}/g_t$, Table 1 [11]. These and other parameters derived for the transition state are shown in Table 1. There is ample precedent for the density of levels of molecules of comparable size to be a factor of three to ten larger than ρ_{WR} calculated using the Whitten-Rabinovitch [15] approximation. It will be necessary to understand and to be able to predict these unexpectedly large densities of states in order for RRKM theory to be completely quantitative and predictive.

There are significant discrepancies in detail apparent between Equation (1) and the data of Fig.s 2 and 3. The slope of the step at threshold is rather gradual compared to those at higher energy. The k(E) curves actually show peaks, whereas W(E,J) corrected for 1-dimensional tunnelling can only increase monotonically with energy. Clearly motion along the reaction coordinate and its coupling to other degrees of freedom needs to be treated quantum mechanically and a detailed knowledge of the potential function in the region of the transition state obtained. The reaction discussed in the following section provides a dramatic illustration of the importance of quantized motion along the reaction coordinate.

Other limitations to the application of the classic form of RRKM theory to unimolecular reactions can arise when IVR is not complete. For bonds which break at relatively low energies, partially good vibrational quantum numbers may exist and lead to rate constants which depend on quantum state as well as on total energy. Striking dependencies of reaction rate on rotational and vibrational quantum numbers have been observed for HFCO [16]. Even when vibrational mixing is complete and the wavefunctions of individual vibrational energy levels are in the ergodic or chaotic limit, rate constants must vary from state to state in the limit where initial quantum states are fully resolved [17]. For formaldehyde, average dissociation rate constants agree with RRKM theory, but rates fluctuate by well over an order of magnitude from one individual level to the next [18-20]. The observed distributions of rates quantitatively matched the predicted quantum statistical fluctuations. Completely *ab initio* calculations show such a statistical behaviour for the dissociation of HO_2 [15]. Similar calculations for HCO with a much lower barrier to dissociation give rate constants which depend systematically on clearly assignable vibrational quantum numbers [20]. This provides an example of an RRKM molecule in the quantum-state-resolved limit. The rate constant for statistical molecules with overlapping resonances has been treated theoretically [23].

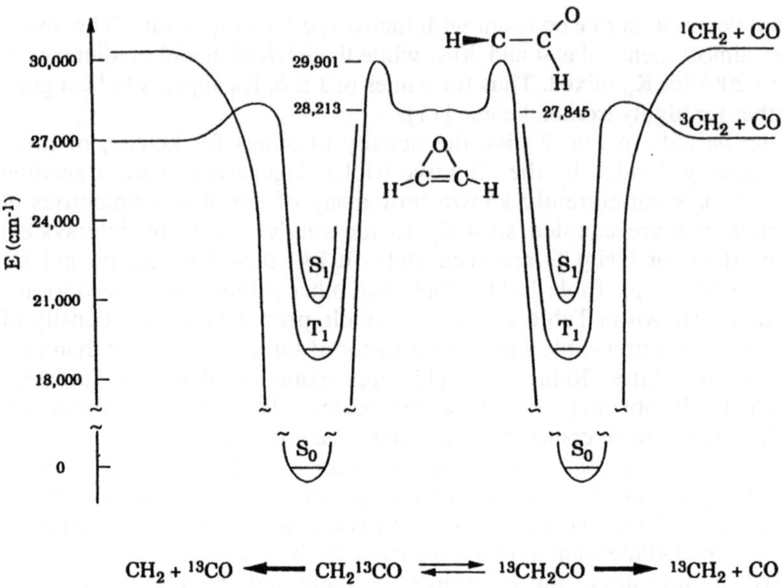

Figure 5: Potential energy diagram for all three unimolecular reactions of ketene. The structures and energies for the C-atom exchange isomerization are from Ref. 26.

3 Quantized Motion Along the Reaction Coordinate in a Well at the Top of an Isomerization Barrier

The thresholds for three different unimolecular reactions of ketene lie within an energy range of 2000 cm^{-1} (Fig. 5) [24]. These reactions occur on potential energy surfaces with qualitatively different shapes and exhibit distinctly different dynamics. The carbon atom exchange reaction passes through the highly strained three membered ring geometry of oxirene in which the carbon atoms are equivalent [25,26]. At energies well above threshold, the measured reaction rates are fit accurately by RRKM theory with the transition state frequencies equal to the *ab initio* frequencies of oxirene [27]. The in-plane motion of O parallel to the C=C bond is taken to be the reaction coordinate. Near the threshold, the rate constant exhibits a series of distinct peaks (Fig. 6) [24]. The coarse structure shows peaks separated by some 80 cm^{-1} with rates decreasing by as much as a factor of 2.3 for an energy increase of less than 20 cm^{-1}. Even the finest structures apparent in Fig. 6 are completely reproducible.

This is the first example of such peaks in a rate constant. These peaks may be understood in terms of tunnelling through a barrier with a minimum at the top (Fig. 5) [24]. Quasi-stable resonances or vibrational levels exist inside the well for motion along the reaction coordinate. The tunnelling rate is greatly enhanced at the energies of these resonances. Resonances can also be expected above the top of the barrier since the reaction coordinate changes from mostly O-atom motion in the central region of the barrier to mostly H-atom motion at the transition states. At the energies of these resonances, motion along the reaction coordinate tends to be reflected at transitions between H motion and O motion. *Ab initio* calculations have established that there is a broad well at the top of the barrier which is about 2000 cm^{-1} deep [25,26]. It has been difficult to establish whether oxirene is a minimum or a saddle point on the surface at the center of this well. For dynamical calculations

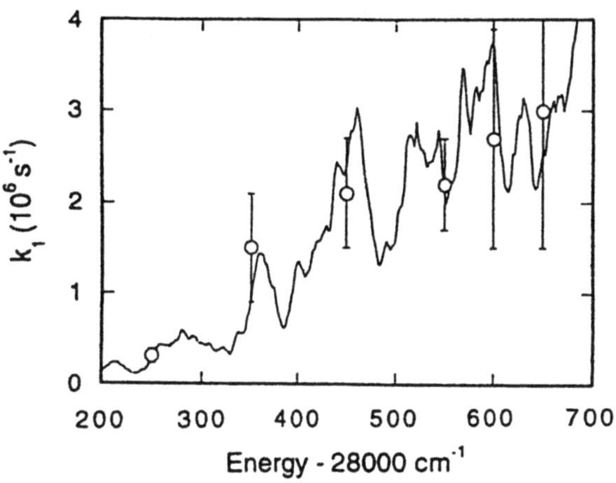

Figure 6: Rate constant for the isomerization $^{13}CH_2CO \rightarrow CH_2{}^{13}CO$ in the threshold region. The solid curve is from PHOFEX spectra of ^{12}CO and ^{13}CO products. All peaks of $0.2 \times 10^{-6} s^{-1}$ and larger are completely reproducible. The points are direct time evolution measurements with 2σ error bars [24].

the important point is that oxirene is at the center of the barrier and very nearly at the bottom of the well within it. Quantum dynamical calculations in which the coordinates most closely coupled with the reaction coordinate are treated together in the region of the barrier reproduce the qualitative features of the observed rate data [28]. The structures in Fig. 6 serve as a dramatic reminder that quantum mechanical resonances in the motion along the reaction coordinate and couplings between the reaction coordinate and other coordinates can be responsible for dramatic deviations from the classical motion along the reaction coordinate represented in Eq. (1).

4 The Completely Loose Transition State

Singlet methylene is so reactive that it recombines with CO to form ketene without any barrier (Fig. 5). Thus $^1CH_2$ is formed precisely at the threshold for ground state product formation, 30,116 cm^{-1} [13]. The reaction competes with dissociation to $^3CH_2$ and so the yield of product, PHOFEX signal in Fig. 7, is directly proportional to the rate constant [13,29,26]. In the first few cm^{-1} there is only one state of singlet methylene energetically possible for each nuclear spin state (ortho and para states of the two H nuclei are conserved throughout the dissociation process) [13]. Thus the curve in Fig. 7 is rigorously proportional to the rate constant for the first 20 cm^{-1}. The PST calculation and the experiment are in precise agreement even to the summation over the thermally populated rotational states of ground-state ketene at 4K in the jet. Thus the positions of the steps in W(E,J) are exceptionally clear and within a fraction of a cm^{-1} of the energy levels of the isolated product fragments. This is undoubtedly the sharpest spectrum showing the location of the energy levels of a transition state. More significantly, this provides an example of the limiting case of a completely loose transition state and the opportunity to study the transition from loose to tight as energy increases above a few tens of cm^{-1} [14,29].

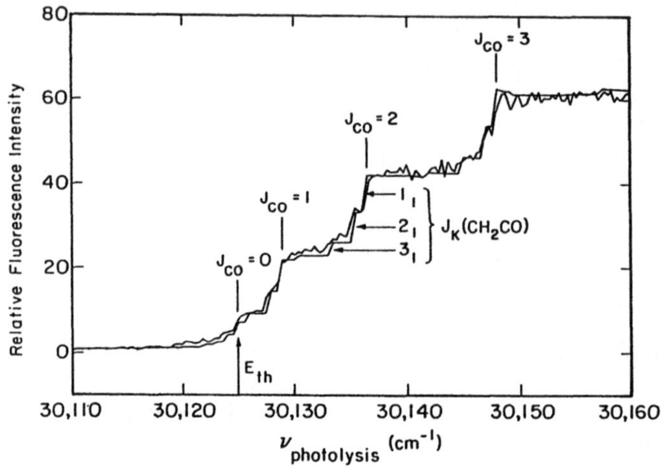

Figure 7: PHOFEX spectrum of the lowest rotational state of *ortho* singlet methylene near the threshold for $CH_2CO \rightarrow {}^1CH_2 + CO$. The smoother line is the phase-space theory rate constant. The step positions match the rotational energy levels for free CO.

5 Conclusion

Understanding of unimolecular reactions has advanced rapidly in recent years. The use of lasers with jet-cooled samples has improved energy and angular momentum resolution for the reactant and time resolution for the rate constant by orders of magnitude. The resolution of product quantum states has added a new dimension to unimolecular dynamics. The concept of the transition state has sharpened greatly. In the past, the geometry, barrier height and vibrational frequencies of the transition state in RRKM theory were adjusted to fit thermal unimolecular reaction rate data. Since rate constants as a function of temperature can usually be described with three independent parameters, the geometry and vibrational frequencies of the transition state can not be derived from thermal rate data. Thus chemists formed a comfortable but largely qualitative view of the transition state as an estimate of the phase space available for reaction to various products. Now we understand that the concept of quantized vibrational level thresholds at the transition state is quantitatively meaningful. There have been successful quantitative tests of the ability of *ab initio* theory to calculate transition state geometries accurately and barrier heights to a few kJ/mol for simple molecules. Predicted frequencies tend to be somewhat too high for the softest modes which are of most importance in determining rates; however, the basic normal modes and sequence of frequencies seem to be correctly predicted. RRKM theory can be used with *ab initio* results to predict rate constants to within a factor of two or three and may be used for quantitative extrapolation to conditions not accessible in the laboratory but important in practical situations. Experiments on single molecular eigenstates have revealed quantum-statistical fluctuations in rates which are predicted quantitatively in the appropriate extension of RRKM theory. Many experiments seeking to demonstrate non-statistical or non-RRKM dynamics have demonstrated the very wide range of applicability of the RRKM model. A few such experiments have demonstrated a lack of complete vibrational energy randomization in a reactant molecule. Dynamical theory has provided an exact quantum analogue to RRKM theory which will combine with future experiments to define the extent to which quantized motion along the reaction coordinate and coupling between the reaction coordinate and vibrational degrees of freedom at the transition state are important.

6 Acknowledgements

I am pleased to acknowledge many stimulating discussions with and advance copies of results from Professors W.H. Miller and H.F. Schaefer, III. The work in my own laboratory at Berkeley has been supported by the National Science Foundation under grant CHE9316640 and by the Chemical Sciences Division of the U.S. Department of Energy under contract DE-AC03-76SF0098.

7 References

[1] R. A. Marcus and O. K. Rice, J. Phys. Colloid Chem., **55**, 894 (1951); R. A. Marcus, J. Chem. Phys., **20**, 359 (1952).
[2] W. Forst, ***Theory of Unimolecular Reactions*** (Academic, New York, 1973).
[3] R. G. Gilbert and S. C. Smith, ***Theory of Unimolecular and Recombination Reactions*** (Basil Blackwell, Oxford, 1990).
[4] W. H. Miller, *Proceedings of The Robert A. Welch Foundation 38th Conference on Chemical Research* (Welch Foundation, Houston, 1994).
[5] D. C. Chatfield, R. S. Friedman, D. G. Truhlar, B. C. Garrett, and D. W. Schwenke, J. Am. Chem. Soc., **113**, 486 (1991).
[6] T. Seideman and W. H. Miller, J. Chem. Phys., **97**, 2499 (1992).
[7] S. L. Mielke, G. C. Lynch, D. G. Truhlar, and D. W. Schwenke, J. Phys. Chem., **98**, 8000 (1994).
[8] I-C. Chen and C. B. Moore, J. Phys. Chem., **94**, 269 (1990).
[9] R. Schinke, ***Photodissociation Dynamics; Cambridge Monographs on Atomic, Molecular and Chemical Physics, Vol. 1*** (Cambridge University, New York, 1993).

[10] E. R. Lovejoy, S. K. Kim, and C. B. Moore, Science, **256**, 1542 (1992).
[11] S. K. Kim, E. R. Lovejoy, and C. B. Moore, J. Chem Phys., **102**, 3202 (1995).
[12] W. D. Allen and H. F. Schaefer III, J. Chem. Phys., **89**, 329 (1988); **84**, 2212 (1986).
[13] W. H. Green, I-C. Chen, and C. B. Moore, Ber. Bunsenges. Phys. Chem., **92**, 389 (1988); I-C. Chen, W. H. Green, and C. B. Moore, J. Chem. Phys., **89**, 314 (1988).
[14] I. Garcia-Moreno, E. R. Lovejoy, and C. B. Moore, J. Chem. Phys., **100**, 8890 (1994); J. Chem. Phys., **100** 8902 (1994).
[15] G. Z. Whitten and B. S. Rabinovitch, J. Chem. Phys., **38**, 2466 (1963).
[16] Y. S. Choi and C. B. Moore, J. Chem. Phys., **97**, 1010 (1992).
[17] W. F. Polik, C. B. Moore, and W. H. Miller, J. Chem. Phys., **89**, 3584 (1988).
[18] W. F. Polik, D. R. Guyer, and C. B. Moore, J. Chem. Phys., **92**, 3453 (1990).
[19] W. F. Polik, D. R. Guyer, C. B. Moore, and W. H. Miller, J. Chem. Phys., **92**, 3471 (1990)
[20] R. Hernandez, W. H. Miller, C. B. Moore, and W. F. Polik, J. Chem. Phys., **99**, 950 (1993).
[21] H.-J. Werner, C. Bauer, P. Rosmus, H.-M. Keller, M. Stumpf, and R. Schinke, J. Chem. Phys., **102**, 3593 (1995).
[22] A. J. Dobbyn, M. Stumpf, H.-M. Keller, W. L. Hase, and R. Schinke, J. Chem. Phys., **102**, 7070 (1995).
[23] U. Peskin, H. Reisler and W. H. Miller, J. Chem. Phys., **101**, 9672 (1994); U. Peskin, W. H. Miller and H. Reisler, J. Chem. Phys., **102**, 4084 (1995).
[24] E. R. Lovejoy and C. B. Moore, J. Chem. Phys., **98**, 7846 (1993).
[25] G. Vacek, J. M. Galbraith, Y. Yamaguchi, H. F. Schaefer, III, R. H. Nobes, A. P. Scott, and L. Radom, J. Phys. Chem., **98**, 8660 (1994).
[26] A. P. Scott, R. H. Nobes, H. F. Schaefer, III, and L. Radom, J. Am. Chem. Soc., **116**, 10159 (1994).
[27] E. R. Lovejoy, S. K. Kim, R. A. Alvarez and C. B. Moore, J. Chem. Phys., **95**, 4081 (1991).
[28] D. Gezelter and W. H. Miller, J. Chem. Phys., in press.
[29] S. K. Kim, Y. S. Choi, C. D. Pibel, Q.-K. Zheng, and C. B. Moore, J. Chem. Phys., **94**, 1954 (1991).
[30] W. H. Green, A. J. Mahoney, Q.-K. Zheng, and C. B. Moore, J. Chem. Phys., **94**, 1961 (1991).

The Reactions of Na_2 with O_2

H. Hou, K-T. Lu, V. Sadchenko, A.G. Suits and Y.T. Lee
Department of Chemistry, University of California at Berkeley, and Chemical Sciences Division, Lawrence Berkeley National Laboratory Berkeley, CA 94720

Abstract

The reactions of Na_2 with O_2 were studied in a crossed-beam experiment at collision energies (E_c) of 8 and 23 kcal/mol. The formation of NaO_2 + Na was observed at both collision energies, with the angular distributions of NaO_2 in the center of mass coordinates peaking strongly forward with respect to the direction of the O_2 beam, suggesting that the reaction is completed in a time scale that is shorter than one rotational period of the molecular system. From the velocity distribution of the products, we found that the newly formed NaO_2 molecules are internally excited, with less than 20% of the available energy appearing in the translational motion of the separating products. These results indicate a "spectator stripping" mechanism, with the O_2 stripping one Na off the Na_2 molecule. At E_c = 23 kcal/mol, the cross section for this reaction channel, σ_{NaO2}, is estimated to be 0.8 Å^2. Another reaction channel which produces NaO + NaO was seen at E_c = 23 kcal/mol. The angular distribution for NaO is broad and forward-backward symmetric in the center-of-mass frame. A substantial fraction of the available energy is released into the relative motion of the products. This reaction is likely to proceed on an excited potential energy surface since a charge transfer to the excited O_2^- orbitals seems necessary for breaking the O-O bond. The measurement yields a bond energy of 60 kcal/mol for the Na-O molecule, and a total cross section for this reaction channel at E_c = 23 kcal/mol, σ_{NaO} = 2 Å^2.

1 Introduction

The reactions of alkali metals (M) with oxygen molecules have been studied for decades. In combustion processes, alkali atoms can rapidly form superoxides, MO_2, in an oxygen rich flame through recombination reactions such as $M + O_2 + N \rightarrow MO_2 + N$ (where N is a third body) [1-3]. The knowledge of the rate constants for these reactions and that for the unimolecular decomposition of MO_2 is essential for modelling the kinetics of the combustion processes and calculating the composition of the free radicals in the flame. However, the alkali dimer reactions, which also produce MO_2, were often neglected in kinetics modelling because there was less information for these reactions. Although the concentration of the dimers in a flame is much lower than that of the monomers, it is still appreciable owing to the covalent bonding in the alkali dimer molecules, which is nearly 1 eV in strength. Furthermore, reactions of the dimers could be significant also because they produce MO_2 via direct bimolecular reactions, which is expected to be faster than the three body recombination between M and O_2. Therefore, taking the dimer reactions into consideration could make substantial improvements over the existing kinetics models.

Springer Series in Chemical Physics, Volume 61
Gas Phase Chemical Reaction Systems
Eds.: J. Wolfrum, H.-R. Volpp, R. Rannacher, and J. Warnatz

The various collisional processes of alkali atoms with oxygen molecules have been investigated extensively during the past fifteen years, especially the electronic excitations of the alkali atoms $M + O_2 \rightarrow M^* + O_2$ at collision energies of 2~10 eV [4]. In the Na + O_2 collision, Na D-line fluorescence was observed as the collision energy was increased above the 3P ← 3S excitation energy of 2.1 eV, with its intensity becoming stronger as the collision energy was increased. When the collision energy was above 8 eV, fluorescence from higher electronic states appeared. The cross section for this process was found to be 26 $Å^2$ at a collision energy of 5 eV. The inelastic scattering experiment of Na(4D) + O_2 carried out in our laboratory [5] showed that the translational energy of the reactants could effectively promote Na to long-lived Rydberg states ($\tau \geq 320$ μs). Kleyn and co-workers carried out landmark investigations of the charge transfer processes $M + O_2 \rightarrow M^+ + O_2^-$ at collision energies ranging from 4 to 2000 eV [6, 7]. In their experiment, the relative velocity of the reactants was so large that the "collision time" was comparable with the vibrational period of the O_2^- ($\omega_e = 1089$ cm^{-1}) [8]. Because the electron transfer probability and the critical distance vary strongly as a function of the internuclear distance of O_2^-, these authors observed oscillations in the differential cross sections. These results clearly demonstrate that the measurements of angular distributions of the scattered molecules reveal not only the average lifetime of the collision complexes using the picosecond rotational period as an inherent clock, but also, in the cases of charge transfer, the vibrational motions of the reaction intermediates on the time scale of ~100 femtoseconds. In all these processes mentioned above, charge transfer from alkali metal to O_2 molecules was undoubtedly the essential first step to account for the change of the electronic states of the alkali atoms in these experiments; and the behaviour of the ionic intermediate $M^+O_2^-$ determines the outcome of the scattering. The chemical reaction $M + O_2 \rightarrow MO + O$ however, was not seen in the previous studies.

Alkali dimers have many interesting properties [8-10] that make their chemistry unique. The ionization potentials of the dimer molecules are lower than those of the corresponding monomers such that when dimers interact with electron accepting molecules, the ionic and covalent potential energy surfaces cross at larger intermolecular distances, which could strongly affect the charge transfer probabilities and the overall reaction cross sections. The bond lengths of M_2 and M_2^+ are extraordinarily long (3.0 Å for Na_2, 3.4 Å for Na_2^+) and although the bond length of M_2^+ is longer, the bond dissociation energy of M_2^+ ion is ~50% higher than those of M_2 neutrals. The covalent bonds of the alkali dimers are abnormally weak, with the bond dissociation energies lower than 1 eV, making many chemical reactions involving the cleavage of the alkali dimer bonds exoergic.

Figger and co-workers [11] observed chemiluminescence in their crossed-beam experiment of alkali metal dimers with oxygen molecules. In the reactions of Cs_2, Rb_2, K_2 and Li_2, D-line emission as well as continuous chemiluminescence from the collision zone were recorded. They attributed this emission to the formation of electronically excited products from the reactions $M_2 + O_2 \rightarrow MO_2 + M^*$ and $M_2 + O_2 \rightarrow MO_2^* + M$. No emission was observed for the Na_2 reactions due to their lower exothermicities.

Very recently, Goerke and co-workers [12] reported the results of their study on the reactions of sodium clusters with oxygen molecules using photoionization and ion time-of-flight as detection methods. $Na_nO(2 \leq n \leq 4)$ and $Na_mO_2(2 \leq m \leq 6)$ products were detected for the reactions of $Na_x(3 \leq x \leq 8)$ with O_2. Angular distributions of $Na_nO(2 \leq n \leq 4)$ and $Na_mO_2(m = 2, 3, 5)$ showed strong forward scattering with respect to the sodium cluster beam in the center-of-mass frame. Although the energy distributions of the products were not measured, their experimental data suggested that most of the exothermicity remained in the internal

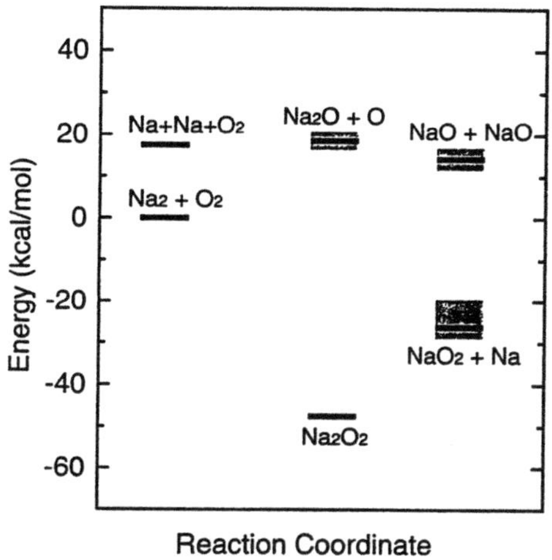

Figure 1: The energy level diagram for Na_2 + O_2. Shaded regions indicate the uncertainties in the energetics. The observed reaction paths are indicated by lines connecting the reactants to the products. Calculations are based on changes in the bond dissociation energies ΔD_0 for each reactions. D_0(Na-O), D_0(O-O) and D_0(Na-Na) are from standard handbooks [8-10]. D_0(Na-ONa) is from M. Steinberg and K. Schofield [23]. D_0(Na-O_2) is based on the experiment of H. Figger, W. Schrepp, and X. Zhu [11]. The energy of Na_2O_2 is from our *ab initio* calculation using the Gaussian 92 program package [34].

degrees of freedom of the products. The cross sections for the reactions were determined to be 50-80 $Å^2$, which led to their conclusion that electron harpooning [13-16] was the first step in the course of the reactions. However, these authors did not include Na_2 reactions in their considerations. Due to the high ionization potential of NaO and NaO_2 they were not able to detect these molecules with the photon energy used in their experiments, although the presence of these molecules was almost certain.

This paper presents our results from crossed-beam studies of the reactions of sodium dimers with oxygen molecules. For this seemingly simple chemical system, there are at least four distinct reaction pathways that were energetically accessible under our experimental conditions:

$$Na_2 + O_2 \rightarrow NaO_2 + Na \qquad (i)$$

$$Na_2 + O_2 \rightarrow Na_2O + O \qquad (ii)$$

$$Na_2 + O_2 \rightarrow NaO + NaO \qquad (iii)$$

$$Na_2 + O_2 \rightarrow Na + Na + O_2 \qquad (iv)$$

The energetics [8-10] for these reactions are plotted in Fig. 1. Reaction (i) is the only exothermic reaction and it becomes Reaction (iv) in the case when the internal excitation of NaO_2 along the Na-O_2 bond exceeds its dissociation limit. In our experiment, we measured the angular and velocity distribution of the products at two collision energies, which enabled us to identify the important reaction channels and elucidate the dynamics underlying these molecule-molecule collision processes.

2 Experimental

The details of the crossed-beam apparatus used in our experiment can be found in many earlier publications [17,18]. Briefly, the alkali dimer source consisted of a resistively heated molybdenum oven and nozzle assembly, with the temperatures of the nozzle and the oven being controlled independently by different heating elements. Sodium vapour carried by an inert gas, which was either He or Ne, expanded out of the 0.2 mm diameter nozzle to form a supersonic beam of Na/Na_2/inert gas mixture. The Na_2 concentration was about 5% molar fraction of the total sodium in the beam when He was used as carrier gas. The beam quality dropped severely when we seeded Na_2 in Ne so the dimer intensity became much weaker. No substantial amount of trimers or larger clusters was detected under our experimental conditions. The Na_2 beam was crossed at 90° by a neat oxygen supersonic beam in the main collision chamber under single collision conditions. The O_2 source nozzle was heated to 473 K to prevent cluster formation. Both sources were doubly differentially pumped. The beams were skimmed and collimated to 2° FWHM in the collision chamber. Under these conditions, the collision energies for the reaction could be varied from 8 kcal/mol to 23 kcal/mol.

A rotatable detector combining a quadrupole mass spectrometer and a Daly-type ion counter was used to measure the angular and velocity distributions of the products in the molecular beam-defined plane. A fraction of the neutral products entering the detector were ionized by 200 eV electron bombardment. Ions formed in the ionization region were extracted into a quadrupole mass filter which was set to have unit mass resolution. The products were modulated at the entrance of the detector with a double-sequence pseudo-random mechanical chopping wheel [18]. The number density of a certain mass was recorded as a function of the flight time that the molecules spent to travel through the distance from the chopping wheel to

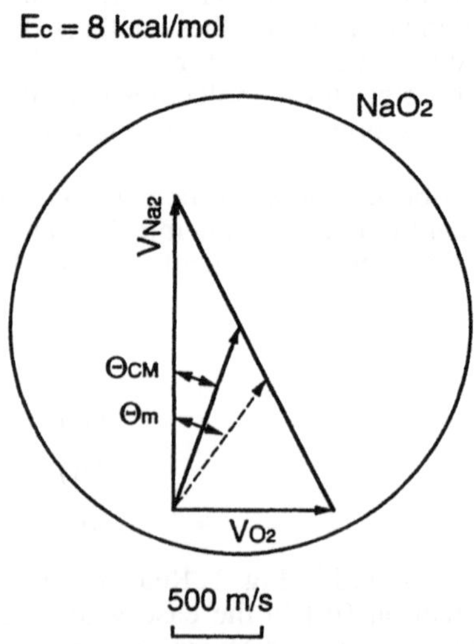

Figure 2: Newton Diagram for $Na_2 + O_2$ at a nominal collision energy $E_c = 8$ kcal/mol. Circles indicate the limiting velocities obtained using Eq. (1) for each ground state product. Θ_{CM} is the Na_2/O_2 center of mass. Θ_m indicates the position of Na/O_2 center of mass, where we observed maximum signal for the NaO_2 channel.

the ion counter. Neutral time-of-flight spectra were obtained by deconvoluting the recorded signal over the chopper gating function and subtracting the ion flight time from ionizer to the ion counter. Under our experimental conditions, we were unable to measure the signal close to the Na_2 beam. At these angles, the number density of elastically scattered Na was very high such that a certain fraction of scattered Na could be excited to long-lived Rydberg states by electron bombardment and subsequently field-ionized in the ion counting region. The mass spectrometer was "transparent" to these Rydberg atoms so they produced strong interference at all the masses of interest. We also recorded the spectra for O and O_2 but the effort to extract any information related to these reactions was not successful because of the high background of oxygen in the mass spectrometer and the intense nonreactive scattering of O_2.

3 Results and Kinematic Analysis

In our experiment, we observed an NaO^+ (m/e = 39) signal which was unambiguously from the reactive scattering since it contains both Na and O. The monomer reaction $Na + O_2 \rightarrow NaO + O$ has very large endothermicity ($\Delta D_0 = -58$ kcal/mol [9]). Under our experimental conditions, even the highest collision energy for $Na + O_2$ was not sufficient for this reaction. Thus the observed NaO^+ signal must be from the reactions of sodium dimers with oxygen molecules. This result is in contradiction with that reported by Goerke et al.[12] who explicitly ruled out the possibility of the dimer reactions in their similar experiment. Neither NaO_2^+ nor Na_2O^+ signal was seen under our experimental conditions. However this does not

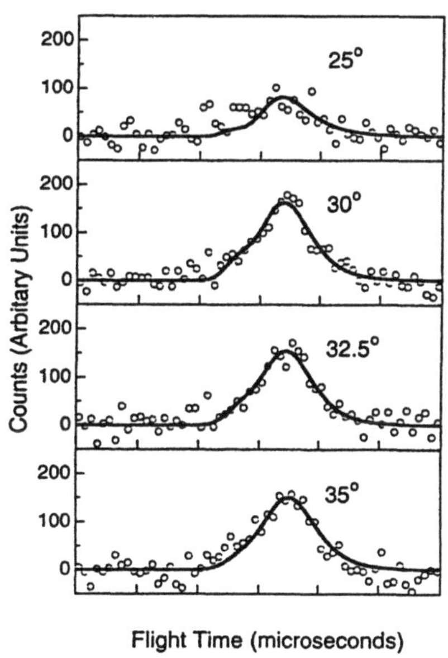

Figure 3: Time-of-flight spectra measured for m/e = 39 (NaO^+) at the collision energy 8 kcal/mol. NaO^+ is from the electron impact fragmentation of NaO_2. Solid lines indicate the best fit using P(E) and T(θ) as shown in Fig. 5.

necessarily mean that only reaction (iii) took place. In the ionization region, electron impact fragmentation is severe, especially for the internally excited molecules. Both NaO_2 from reaction (i) and Na_2O from reaction (ii) could fragment completely and both molecules could form NaO^+. Additional analysis of the kinematics and dynamics of the reactions has to be made in order to identify the actual reaction pathway(s).

3.1. Collision Energy at 8 kcal/mol

The Newton diagram [19-21] for $Na_2 + O_2$ at a nominal collision energy of 8 kcal/mol is shown in Fig. 2. Θ_{CM} denotes the laboratory angle of the center-of-mass velocity vector for $Na_2 + O_2$. For future discussion, we also plot the center of mass velocity (dotted line) for $Na + O_2$, which extends an angle of Θ_m from the Na_2 beam. The direction of the Na_2 velocity in the center-of-mass frame is defined as 0° hereafter. At this collision energy, only reaction (i) could take place, so the observed NaO^+ signal was entirely due to the fragmentation of NaO_2 products. The solid circle in Fig. 2. encloses the zone within which NaO_2 might appear. Figure 3. shows the NaO^+ time-of-flight spectra recorded at several laboratory angles. Beyond the range of these angles, the signal became too weak to observe the time-of-flight peaks. It was observed that most NaO_2 products appeared around Θ_m. Signal intensities obtained by integrating the time-of-flight spectra are plotted in Fig. 4. to give the laboratory angular distribution for NaO_2. Compared with the range of the NaO_2 circle in the Newton diagram, the NaO_2 angular distribution was very narrow.

In Fig. 3. and Fig. 4., the solid lines are the computer simulation of the data obtained by assuming the center of mass translational energy distribution (P(E))

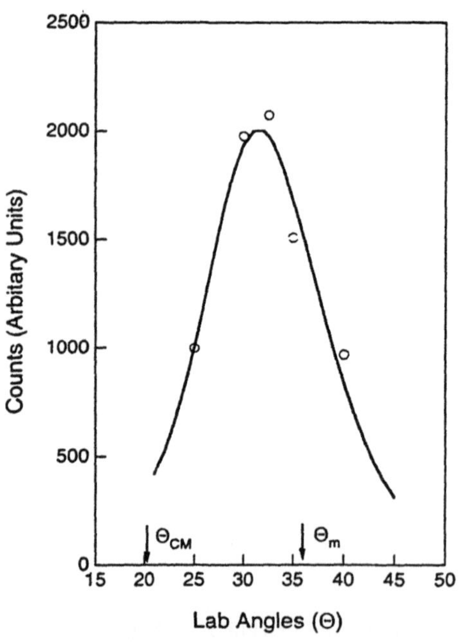

Figure 4: Laboratory angular distribution of NaO^+ signal at E_c = 8 kcal/mol. Solid line is the fit using P(E) and T(θ) as shown in Fig. 5.

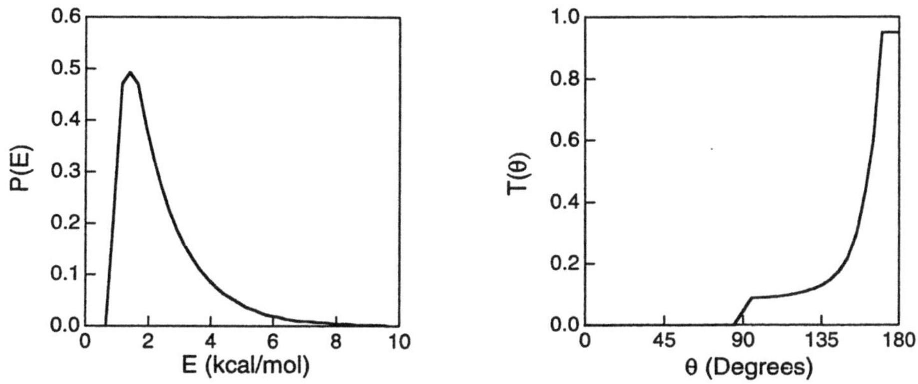

Figure 5: Product center-of-mass translational energy distribution P(E) and angular distribution T(θ) that best fit the experimental data.

and angular distribution (T(θ)) shown in Fig. 5. Peaking in the direction of the O_2 beam (180°), the angular distribution shows remarkable anisotropy, with all the NaO_2 products scattered into the hemisphere toward the O_2 beam. This anisotropy indicates that the reaction takes place during a time that is shorter than the rotational period of the molecular system so that the products are separated before they lose the sense of the well-defined approach direction of the reactants [21]. At this collision energy, the total available energy for the reaction ($E_c + \Delta D_0$) is 34 kcal/mol, of which only an average of 2.6 kcal/mol appears in the translational motion of the products (cf. P(E) in Fig. 5). The rest of the energy has to become the internal excitation of the NaO_2 molecules. When NaO_2 is ionized to form NaO_2^+, the binding energy between Na^+ and O_2 becomes weaker and the internal excitation initially stored in NaO_2 is likely to remain in NaO_2^+ and causes the complete dissociation of NaO_2^+ into Na^+ and O_2. This is probably the reason why the NaO_2^+ parent ion was not detected. However when NaO_2 is excited to higher electronic states during electron impact ionization, the dissociative ionization can produce a stable NaO^+ ion.

3.2 Collision Energy at 23 kcal/mol

All the dimer reactions (i)-(iv) became energetically accessible at a collision energy of 23 kcal/mol. The Newton diagram for this collision energy is shown in Fig. 6. For reaction (ii), the available energy was only 5 kcal/mol. The heavier products Na_2O (compared with O) have to be limited in the small circle. On the other hand, NaO molecules produced by reaction (iii) are much less confined because there is slightly more available energy (9 kcal/mol) and no disparity in the masses of the products. The time-of-flight spectra of NaO^+ recorded at different laboratory angles are plotted in Fig. 7., and the laboratory angular distribution of the NaO^+ intensity is shown in Fig. 8. There are clearly two features in the time-of-flight spectra. Both features exist well out of the Na_2O limit which indicates clearly that the observed signal was not from reaction (ii). The slow peaks which appear at ~120 μs have a very narrow angular distribution, with the intensity peaking sharply around Θ_m. The angular and velocity distributions of this slow signal bear such strong

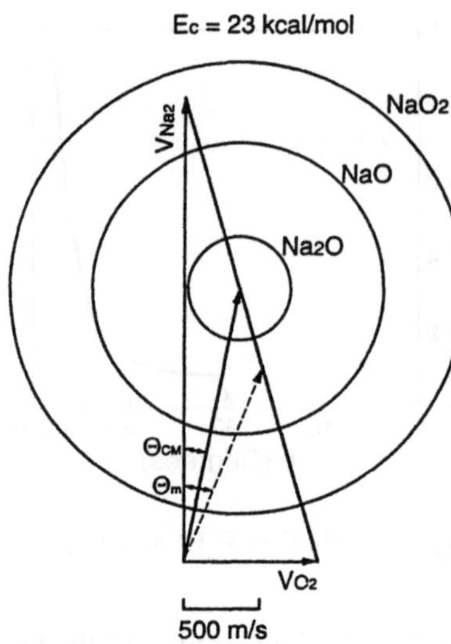

Figure 6: Newton Diagram for $Na_2 + O_2$ at a nominal collision energy $E_c = 23$ kcal/mol.

resemblance to those of the NaO_2 products we saw at low collision energy that we can with little doubt assign this signal to the same reaction channel. The "fast peaks", arriving at ~70 μs, have very different charateristics than the NaO_2 products. Appearing weak at all angles and becoming only slightly stronger at small angles, these fast peaks have much less variation in intensity, an indication of products with larger recoil velocities and broader laboratory angular distributions. This fast signal is, without any doubt, the NaO product from reaction (iii).

The simulation of the time-of-flight and angular distribution data was done by assuming an independent set of P(E) and T(θ) for each reaction channel. An additional parameter γ was assigned to each reaction channel to weight its contribution to the observed signal. The total fits are plotted as solid lines in Fig. 7. and Fig. 8. The dotted lines and the dashed lines show the contributions from reaction (i) and Reaction (iii), respectively. Fig. 9. shows the P(E) and T(θ) for reaction (i). As mentioned above, these distributions are similar to those at low collision energy. T(θ) again peaks sharply in the direction of the O_2 beam with all the NaO_2 scattered into the O_2 hemisphere, showing a strong anisotropy. Of the 49 kcal/mol available energy, only an average of ~8 kcal/mol was released into the translational energy of the separating products. The P(E) had a sharp rising edge at ~7 kcal/mol, which indicates that the internal energy of NaO_2 could not exceed 42 kcal/mol, as otherwise the NaO_2 molecule could not be held together. The subsequent result is the collisional dissociation making $Na + Na + O_2$. Therefore this sharp rising edge in P(E) marks the onset of reaction (iv), which requires at least the energy to break the Na-Na bond. The collision energy of 23 kcal/mol less the cut off energy of 7 kcal/mol in the P(E) is then our experimental measurement of the Na-Na bond dissociation energy. This result of 16 kcal/mol is in agreement with the well known value in the literature [9].

Figure 10. shows the P(E) and T(θ) used in the simulation of reaction (iii). The angular distribution is quite broad. The forward-backward symmetry in the

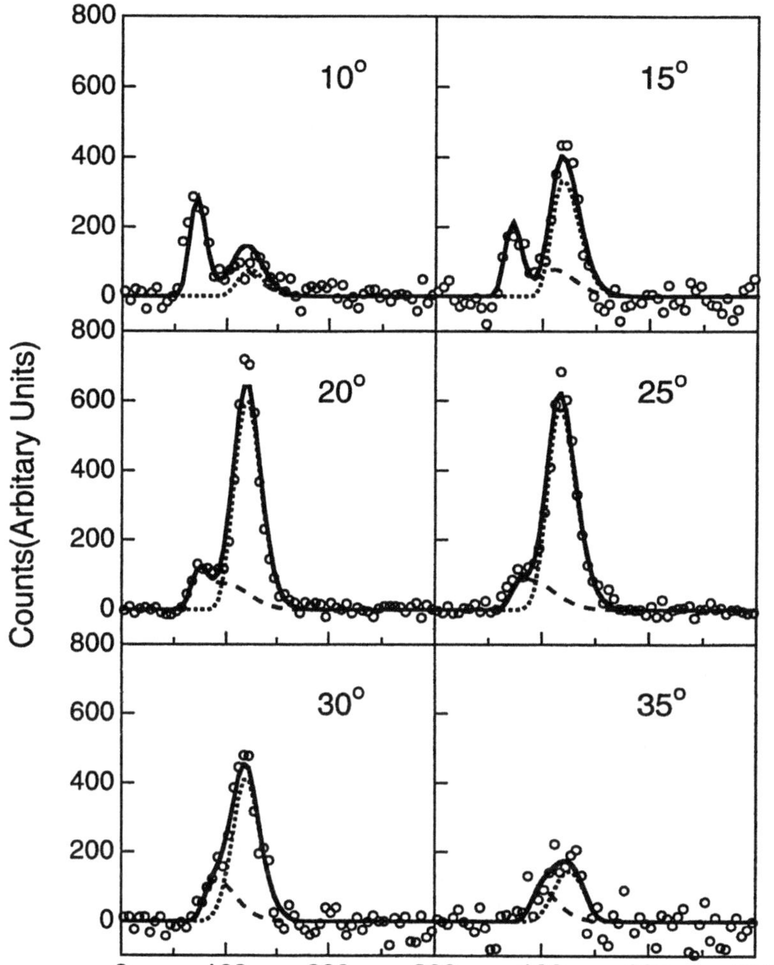

Figure 7: Time-of-flight spectra measured for $m/e = 39$ (NaO^+) at collision energy of 23 kcal/mol. NaO^+ is from the both NaO_2 and NaO channel. Solid lines indicate the total fit. Dotted lines indicate the contribution from each reaction channel.

product angular distribution is a prerequisite rather than an experimental result because in this reaction two identical molecules are always produced with exactly opposite center-of-mass velocities. Under such conditions the angular distribution does not tell us the time duration of the reaction, which is the important information for dynamics studies. The P(E) distribution shows that an average of 3 kcal/mol is released in the translational motion of the products. The high energy cut off at 5.5 kcal/mol in P(E) allows us to estimate Na-O bond strength. From conservation of energy, there is the following relation between the energies of the reactants and the products:

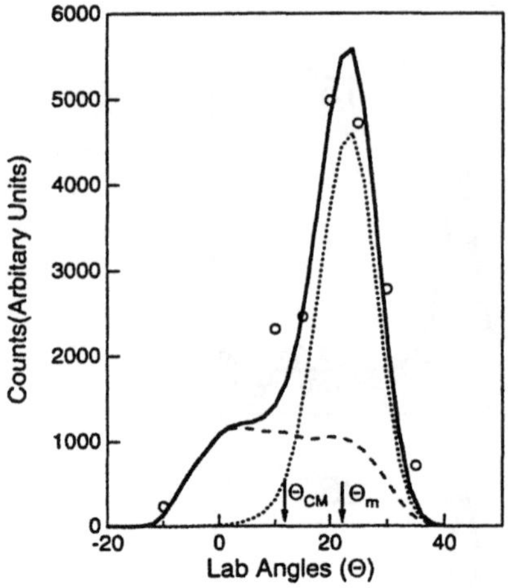

Figure 8: Laboratory angular distribution of NaO^+ signal at $E_c = 23$ kcal/mol. Solid line indicates the total fit. The dashed line indicates the contribution from NaO_2 + Na reaction channel, and the dash-dotted line is for the NaO + NaO channel.

$$E'_{trans} + E'_{int} = E_{trans} + E_{int} + \Delta D_0, \qquad (1)$$

where E_{trans} and E'_{trans} are the center of mass translational energies of the reactants and products, respectively. Similarly, E_{int} and E'_{int} are the internal energies, i.e. the total electronic, vibrational and rotational energies, in the reactants and the products, respectively. ΔD_0 is the change of the bond dissociation energy for the reaction. E_{int} is assumed to be negligible for the supersonic beams. E'_{trans} reaches its maximum value when E'_{int} is a minimum, which will be assumed to be zero. Namely at the highest translational energy release, both NaO products are assumed to be in the ground state. From Eq. (1) and the maximum translational energy release in P(E), we calculated the change of bond dissociation energy ΔD_0 for reaction (iii) to be -17 kcal/mol. Using the accurately known values of D_0(O-O), 119 kcal/mol, and D_0(Na-Na), 18 kcal/mol [9], we obtained D_0(Na-O) = 60 kcal/mol, which can be compared with the literature value of 61.2 ± 4.0 kcal/mol [9].

From the best fit, the relative contributions of these two reactions to the observed signal at *m/e* = 39 are:

$$\gamma_i : \gamma_{iii} = 1 : 1.8\,. \qquad (2)$$

In Eq. (2), we use the subscripts to denote the reaction channels.

In order to estimate cross sections for the reactions, we also measured time of flight spectra for the Na^+ signal under the same experimental conditions. It is known that both NaO and NaO_2 dissociate heavily to Na^+ under electron impact [22,23]. The collisional dissociation reaction (iv), if present, would also generate Na^+. But most Na^+ detected came from elastic/inelastic scattering of the Na monomer with O_2 because the number density of the monomers at the collision zone was roughly 20 times higher than that of the dimers. Na^+ time-of-flight spectra at 25° and 30° are shown in Fig. 11. At these angles, nonreactive scattering of Na_2 was negligible. For comparison, the scattering signal of Na from N_2 at corresponding angles was also measured and is plotted in the same figure. The differences due to the reactions are obvious. We can fit the Na^+ spectra for Na/Na_2 + O_2 scattering with four different channels: the elastic scattering from Na + O_2, the inelastic scattering from Na + O_2,

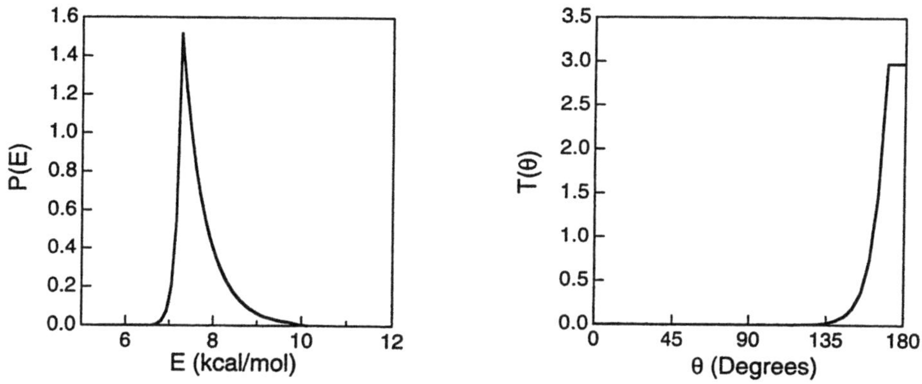

Figure 9: Translational energy distribution P(E) and angular distribution T(θ) that best fit the NaO_2 channel.

reaction (i) from $Na_2 + O_2$ and reaction (iii) from $Na_2 + O_2$. We use the same translational energy and angular distributions as those for the NaO^+ signals to fit the reactive signals in the Na^+ spectra although we have to use different weights (γ') to account for the different fragmentation patterns of NaO and NaO_2 in the ionization region. The results are also plotted in Fig. 11, where the solid lines show the overall fit and the broken lines indicate the contribution from individual channels, as described in detail in the figure captions. The discrepancies between the fit and the data are possibly due to the collisional dissociation reaction (iv), which is difficult to account for because it is a three body event. Nevertheless, leaving

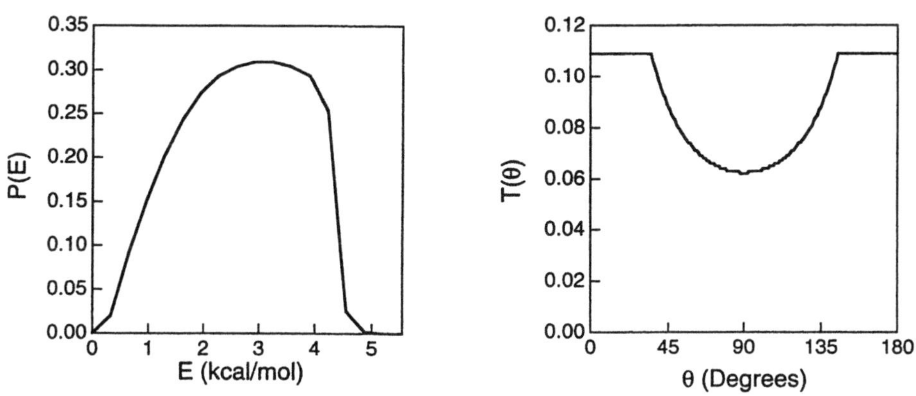

Figure 10: Translational energy distribution P(E) and angular distribution T(θ) that best fit the NaO channel.

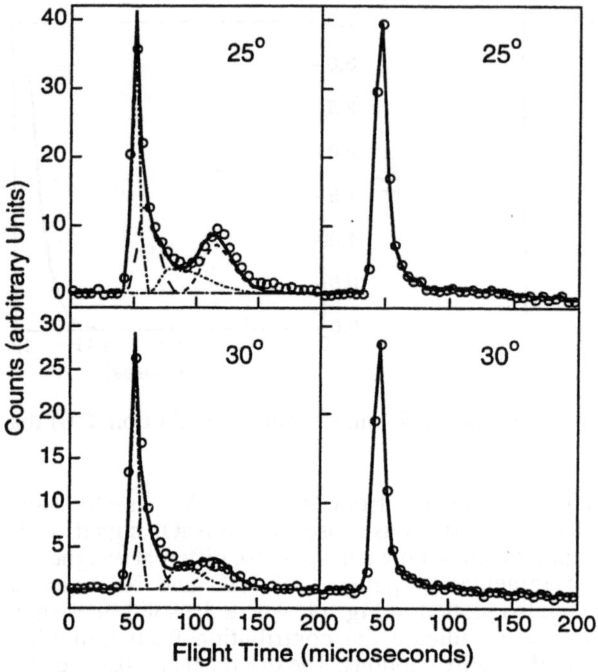

Figure 11: Na^+ time-of-flight spectra from $Na/Na_2 + O_2$ (left) and $Na/Na_2 + N_2$ (right). Theoretical fit is given for the $Na/Na_2 + O_2$ data. The total fit (solid lines) is composed of four channels including elastic scattering (dash-dot-dot), inelastic scattering (long dash), NaO_2 channel (dash-dot) and NaO channel (short dash).

Reaction (iv) out will not affect our final results of obtaining the cross sections for Reaction (i) and Reaction (iii) because we are comparing these reaction channels with the nonreactive signal only. From the fit, the relative weight for each channel is:

$$\gamma'_{elastic} : \gamma'_{inelastic} : \gamma'_{i} : \gamma'_{iii} = 105 : 171 : 0.5 : 3.0. \qquad (3)$$

Because we did not observe any NaO_2 signal, it is safe to assume that NaO_2 molecules, if ionized, dissociated completely to either NaO^+ or Na^+. Figure 12 illustrates the relations between the parent molecules directly from the scattering and the measured signal, which will serve as a road map for our discussions in the following sections. Here σ_i and σ_{iii} are the reaction cross sections for reaction (i) and reaction (iii), respectively. $\sigma_{elastic}$ and $\sigma_{inelastic}$ are the cross sections for the elastic scattering and the inelastic scattering of Na + O_2. $\sigma_{ion}(X)$ denotes the ionization cross section of molecule X by electron bombardment. The ions so formed undergo further dissociation to give the observed signal. We define ξ as the fraction of NaO^+ among the total Na-containing positive ions produced by electron bombardment. According to the assumption that NaO_2 dissociates completely upon electron bombardment, the fraction of Na^+ from NaO_2 is (1-ξ). In a similar way, we define the fraction of Na^+ among the total Na-containing positive ions produced from ionization of NaO products to be η. Accordingly the branching ratio between reaction (i) and reaction (iii) can be obtained. From Eq. (2), we have:

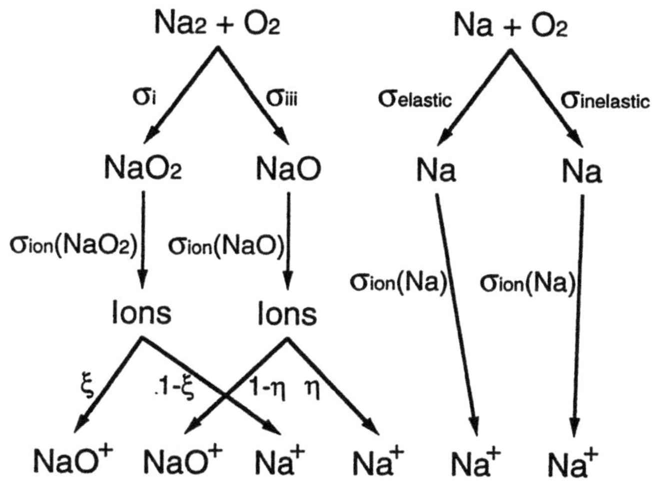

Figure 12: Schematic drawing that illustrates the relations between the observed signal and various products. σ_i and σ_{iii} are the reaction cross sections for reaction (i) and Reaction (iii), respectively. $\sigma_{elastic}$ and $\sigma_{inelastic}$ are the cross sections for the elastic scattering and the inelastic scattering of Na + O_2. σ_{ion}'s are the ionization cross sections by electron bombardment. The ions so formed undergo further dissociation to give observed signal. ξ is defined as the fraction of NaO^+ among the total Na-containing positive ions produced by electron bombardment. According to the assumption that NaO_2 dissociates completely upon electron bombardment, the fraction of Na^+ from NaO_2 is (1-ξ). η is defined as the fraction of Na^+ among the total Na-containing positive ions from the ionization of NaO.

$$\gamma_i : \gamma_{iii} = [\sigma_{ion}(NaO_2) \times \sigma_i \times \xi] : [2 \times \sigma_{ion}(NaO) \times \sigma_{iii} \times (1 - \eta)] = 1 : 1.8. \quad (4)$$

The factor "2" accounts for the fact that two NaO were produced in Reaction (iii). Similarly, from Eq. (3), we derive another relation between σ_i and σ_{iii}:

$$\gamma'_i : \gamma'_{iii} = [\sigma_{ion}(NaO_2) \times \sigma_i \times (1 - \xi)] : [2 \times \sigma_{ion}(NaO) \times \sigma_{iii} \times \eta] = 0.5 : 3.0. \quad (5)$$

In order to compare the cross section for each channel, we need the ionization cross sections for Na, NaO and NaO_2 by 200 eV electron bombardment. The ionization cross section of Na is known to be $\sigma_{ion}(Na) = 2.46$ Å^2 [24]. The ionization cross section of NaO can be estimated [25] from the ionization cross section of O_2, and the ratio of the polarizabilities (α's) of O_2 [9] and NaO using the following equation:

$$\sigma_{ion}(O_2) : \sigma_{ion}(NaO) = \alpha_{O_2} : \alpha_{NaO} = 1.58 : 3.05, \quad (6)$$

where α_{NaO} was obtained by adding the polarizability of Na^+, $\alpha_{Na^+} = 0.155$ Å^3, and the polarizability of O^-, $\alpha_{O^-} = 2.85$ Å^3. The result is $\sigma_{ion}(NaO) = 5.2$ Å^2. Similarly, we can calculate the ionization cross section for NaO_2, which gives $\sigma_{ion}(NaO_2) = 6.5$ Å^2.

Covinsky [22] measured η as a function of internal excitation of NaO in the Na + O_3 experiment. η is roughly 0.8 for internally cold NaO (which was the case in this experiment because there is very little excess energy for reaction (iii) and a considerable amount of this excess energy was released into the translational motion of the products). Using this value, we obtained $\xi = 0.45$ and $\sigma_i : \sigma_{iii} = 1 : 2.5$.

The next step is to compare the reactive signal with the total nonreactive scattering. From Eq. (3), we have the following relations:

$$\frac{\gamma_{i'}}{\gamma_{elastic'} + \gamma_{inelastic'}} = \frac{\sigma_{ion}(NaO_2) \times \sigma_i \times (1-\xi)}{20 \times \sigma_{ion}(Na) \times \sigma_{nonreactive}} = \frac{0.5}{276}, \qquad (7)$$

and

$$\frac{\gamma_{iii'}}{\gamma_{elastic'} + \gamma_{inelastic'}} = \frac{2 \times \sigma_{ion}(NaO) \times \sigma_{iii} \times \eta}{20 \times \sigma_{ion}(Na) \times \sigma_{nonreactive}} = \frac{3.0}{276}, \qquad (8)$$

where $\sigma_{nonreactive}$ is the sum of the cross sections of elastic scattering and inelastic scattering of Na + O_2. We added factors "20" in Eq. (7) and Eq. (8) because the number density of the monomers in the collision zone was roughly 20 times higher than that of the dimers. The interaction between a Na monomer and O_2 becomes strong at the crossing of the covalent and ionic surfaces because of the charge transfer. If we neglect the long range van der Waals interaction, the internuclear distance at the surface crossing can be estimated using the following equation which is based on the simple "harpoon mechanism" [13-16]:

$$R_c = \frac{14.4}{IPNa(eV) - EAO_2(eV)} \quad (\text{Å}), \qquad (9)$$

where IP is the ionization potential of Na [9] and EA is the electron affinity of O_2 [9], both in electron volts. Eq. (9) gives a value of 3.1 Å for R_c, which is larger than the averaged hard sphere radius between Na and O_2. Since there is no reaction between a Na monomer and O_2, we can assume R_c to be the nonreactive collision radius. Therefore we obtain the total nonreactive scattering cross section to be: $\sigma_{nonreactive} = \pi R_c^2 = 30$ Å^2. Using these numbers in Eq. (7) and Eq. (8), we obtain the cross sections for reaction (i), $\sigma_i = 0.8$ Å^2, and reaction (iii), $\sigma_{iii} = 2$ Å^2.

4 Discussion

The MO_2 molecules have been studied quite extensively in the past. Alexander generated semiempirical potential energy surfaces for LiO_2 and NaO_2.[26] He showed that the ground state NaO_2, which was 1.6 eV more stable than the separated ground state Na and O_2, has C_{2v} geometry with Na at the apex of an isosceles triangle. The interaction within the molecule could be best described as $Na^+ ... O_2^-$ singly charged ionic bonding. This is in accord with the infrared and Raman [27,28] or ESR [29] spectroscopy results in low temperature matrices. Calculations of the NaO molecule also indicate that it is singly ionic, $Na^+ ... O^-$ [30]. These results suggest that during the course of reaction (i) and reaction (iii), the $Na_2 + O_2$ system has to undergo a change from a covalent to ionic character. This resembles the well known M/M_2 + halogen molecules (XY) reactions and was commonly characterized by the "harpoon mechanism". Nevertheless, the $Na_2 + O_2$ reaction shows very different results than the M/M_2 + XY system. In the case of M_2 + XY [31], for example, the main pathways involve both M_2 and XY dissociation of the form MX + M + Y or MX + MY with the total cross sections being larger than 100 Å^2. For the MX + M + Y channel, the MX product peaked strongly towards the M_2 direction in the center-of-mass velocity space.

M_2 + XY reactions are model systems that proceed via a long range electron transfer mechanism. Due to the low ionization potential of M_2 and high electron affinity of XY, the potential energy surfaces of the ionic state (M_2^+...XY^-) and the covalent state (M_2...XY) intersect at intermolecular distance R_c = 7-8 Å (Eq. (9)) which is much larger than the "hard sphere" radius. At this internuclear distance, there is a fair chance for the valence electron to "jump" from M_2 to the XY. A vertical ionization from the outer turning point of ground state M_2 puts M_2^+ in the attractive well and vertical ionization from the inner turning point sets M_2^+ on the slightly repulsive (1 eV/Å) surface. Because M_2^+ is almost 50% more strongly bound than M_2 [10], along with the fact that both M_2 and M_2^+ have very low vibrational frequencies, the two alkali nuclei will separate only very slowly after electron transfer. On the other hand the XY^- bonding is much weaker than that of XY neutral molecule. A vertical Frank-Condon transition puts XY^- above its dissociation limit on a very strong repulsive wall (6 eV/Å), primarily due to the strong antibonding σ_{pz}* orbital. As a result, X^- and Y separate promptly after the vertical electron transfer while the M and M^+ will linger together for a longer time. The reaction proceeds as if the M_2^+ "strips" the X^- off to form the M_2X molecules, which peak preferentially towards the M_2 direction, leaving the other halogen atom Y as a "spectator". The highly vibrationally excited M_2X nascent molecule then boil off a M atom to form the final MX product. The initial repulsive force between the two halogens is offset by their attraction at longer distance as the two halogen atoms separate. Consequently, the spectator Y does not carry away much of the excess energy.

For the $Na_2 + O_2$ system, the scenario is quite different. O_2 has the much lower electron affinity of 0.44 eV [9] so that the nominal crossing distance R_c calculated from Eq. (9) is only 3.2 Å and only collisions with impact parameter smaller than R_c will experience the electron transfer. This limits the total cross section for charge transfer to 32 $Å^2$ at most. Nevertheless, at such a short intermolecular distance, the interaction between the ionic surface and the covalent surface is very strong. Once the system arrives at this region, it will have almost unit probability to switch adiabatically from covalent surface to ionic surface. The final outcome of the reaction is determined by the interaction between Na_2^+ and O_2^- after electron transfer. Unlike the case of XY, the O_2^- bond is only slightly longer than that of O_2 and the binding energy is almost as large. A vertical transition from O_2 to the ground state of O_2^- would put O_2^- in a deep potential well. It takes some extra 90 kcal/mol in order to break O_2^- to form O + O^-. On the other hand, the weakly bound Na_2^+ is much more vulnerable to dissociate into Na + Na^+ under the influence of the strong Coulomb attraction. In the case when the impact parameter is large, there is little chance to cleave the O_2^- molecule during the time when two molecules glance off each other. The strong Coulomb interaction between Na^+ and O_2^- leads to the formation of NaO_2 molecules that are highly excited along the newly formed bonds. The departing Na atom should hardly be affected during the collision because the interaction between Na^+ and Na is relatively weak compared with the Coulomb interaction between Na^+ and O_2^-. Therefore it acts similarly as a "spectator" in the case of the halogen reactions. The most important role for this Na atom is to carry away excess energy partly through breaking of the Na-Na bond, much like a "third body" as in the recombination reactions, which is the reason why the NaO_2 products are most likely to appear at Θ_m – the result is just like a simple recombination of Na and O_2. Using this "spectator stripping" model, we calculate the translational energy releases for E_c = 8 and 23 kcal/mol to be 2.4 and 6.6 kcal/mol, respectively, matching well with the peaks in the corresponding P(E)'s.

It is worth pointing out that the formation of NaO_2 results from the transfer of only one electron from Na_2 to O_2. If both valence electrons are transferred to O_2, the molecular system is likely to sample the deep well on the potential energy surface, forming the Na_2O_2 complex (cf. Fig. 1). Since the Na_2O_2 complex dissociates mainly through the low energy channels, forming NaO_2 + Na, we would expect to see for the NaO_2 products a forward-backward symmetric angular distribution, which is a characteristic of the complex formation. However, in the experiments we observe a very anisotropic angular distribution. Therefore complex formation mechanism can be excluded.

Contrary to the spectator mechanism of Reaction (i), Reaction (iii) requires considerable rearrangement of the electronic configuration of the system. First of all, both valence electrons have to be transferred because each NaO molecule has $Na^+ ... O^-$ singly ionic character. This requires a close collision and a favourable approach geometry between the reactants. This was not directly observed for this reaction because the symmetric angular distribution determined by the kinematics does not provide us with much information on the scattering. But evidence could be found from the chemi-ionization reaction $Ba + Cl_2 \rightarrow BaCl^+ + Cl^-$ in which $BaCl^+$ is backward scattered with respect to the Ba beam, demonstrating that the favoured geometry is a collinear head-on collision [33]. Second, the strong O-O bond has to be broken. The exact pathways leading to the breaking of the O-O bond could not be deduced from our data, but some insights could be gained by looking at early results of dissociative attachment of O_2, and the results of our recent *ab initio* calculations [34] using the Gaussian 92. The formation of $O + O^-$ by electron attachment was studied in great detail and a single electronically excited state $^2\Pi_u$ was found to be responsible for this process, although there are many other states in the vicinity. The state specificity is largely due to the symmetry restriction which could be removed in the case of chemical reaction by the presence of the Na atoms. The results do suggest that breaking of the O-O bond is likely to be initiated by electron transfer to the excited O_2^- orbital, which will result in substantial bond stretching in O_2^-. The argument that the excited O_2^- orbital is important to the NaO + NaO reaction is further supported by our recent calculations in which we computed the energetics for the molecular system at various geometries. It is found for all potential energy surfaces with ground state O_2 characteristics, NaO_2 is the dominant product in the exit channel. The details of the calculations will be reported in a forthcoming paper.

5 Acknowledgements

This work was supported by the Director, Office of Energy Research, Office of Basic Energy Science, Chemical Sciences Division of the U.S. Department of Energy under Contract No. DE-AC03-76SF00098 and The Office of Naval Research.

6 References

[1] E. Bulewicz, C. G. James, and T. M. Sugden, Proc. R. Soc. London A **235**, 89 (1956).
[2] M. J. McEwan and L. F. Phillips, Combust. Flame **9**, 420 (1965); *ibid.* **11**, 63 (1967).
[3] C. H. Muller, K. Schofield, and M. Steinerg, J. Chem. Phys. **72**, 6620 (1980).
[4] V. Kempter, W. Mecklenbrauck, M. Menzinger, and Ch. Schlier, Chem. Phys. Lett. **11**, 353 (1971).
[5] H. Hou, K. T. Lu, A. G. Suits, and Y. T. Lee, unpublished results.

[6] A. W. Kleyn, Ph. D. Thesis, *"Stichting voor Fundamenteel Onderzoek der Materie"*, Amsterdam, The Netherlands (1982).
[7] A. W. Kleyn, M. M. Hubers, and J. Los, Chem. Phys. **34**, 55 (1978).
[8] JANAF Thermochemical Tables, 3rd Ed., J. Physical and Chemical Ref. Data, 14 (1985), Supp. No.1.
[9] D. R. Lide, Editor-in-Chief, ***CRC Handbook of Chemistry and Physics,*** 74th edition (1993-1994).
[10] K. P. Huber and G. Herzberg, ***Molecular Spectra and Molecular Structure, IV. Constants of Diatomic Molecules,*** Van Nostrand Reinhold Company (1979).
[11] H. Figger, W. Schrepp, and X. Zhu, J. Chem. Phys. **79**, 1320, (1983).
[12] A. Goerke, G. Leipelt, H. Palm, C. P. Schulz, and I. V. Hertel, Z. für Physik D **32**, 311, (1995).
[13] J. L. Magee, J. Chem. Phys. **8**, 687 (1940).
[14] A. W. Kleyn in ***Alkali Halide Vapors,*** edited by P. Davidovits and D. L. McFadden, Academic Press (1979).
[15] D. R. Herschbach, Appl. Opt. Supp. **2**, 128 (1965).
[16] R. D. Levine and R. B. Bernstein, ***Molecular Reaction Dynamics and Chemical Reactivity,*** Oxford University, New York (1987).
[17] Y. T. Lee, J. D. McDonald, P. R. LeBreton, and D. R. Herschbach, Rev. Sci. Instrum. **40**, 1042 (1969).
[18] H. F. Davis, A. G. Suits, and Y. T. Lee, J. Chem. Phys. **96**, 6710 (1992) and the references therein.
[19] D. R. Herschbach, Disc. Faraday Soc. **33**, 149 (1962).
[20] Y. T. Lee, in ***Atomic and Molecular Beam Methods, Vol. 1,*** edited by G. Scoles, Oxford University Press, New York (1988).
[21] W. B. Miller, S. A. Safron, and D. R. Herschbach, J. Chem. Phys. **56**, 3581 (1972).
[22] M. Covinsky, Ph.D. Thesis, University of California at Berkeley (1990).
[23] M. Steinberg and K. Schofield, J. Chem. Phys. **94**, 3901 (1991).
[24] R. H. McFarland and J. D. Kinney, Phys. Rev. **137**, A1058 (1965).
[25] P. S. Weiss, Ph. D Thesis, Chapter III, University of California at Berkeley (1985).
[26] M. H. Alexander, J. Chem. Phys. **69**, 3502 (1978).
[27] L. Andrews, J. Chem. Phys. **54**, 4935 (1971); R. R. Smardzewski and L. Andrews, J. Phys. Chem. **77**, 801 (1973); L. Andrews, J. T. Hwang, and C. Trindle, *ibid.* **77**, 1065 (1973).
[28] H. Huber and G. A. Ozin, J. Mol. Spectrosc. **41**, 595 (1972).
[29] D. M. Lindsay, D. R. Herschbach and A. L. Kwiram, Chem. Phys. Lett. **25**, 175 (1974).
[30] P. A. G. O'Hare, and, A. C. Wahl, J. Chem. Phys. **56**, 4516 (1972).
[31] W. S. Struve, J. R. Krenos, D. L. McFadden, and D. R. Herschbach, J. Chem. Phys. **62**, 404 (1975).
[32] L. King and D. R. Herschbach, Faraday Disc. Chem. Soc. **55**, 331 (1973).
[33] A. G . Suits, H. Hou, and Y. T. Lee, J. Phys. Chem. **94**, 5672 (1990).
[34] K. T. Lu, H. Hou, A. G. Suits, and Y. T. Lee, unpublished results.

Reaction Dynamics of Three-Atom and Four-Atom Systems

M. Alagia,[a] N. Balucani,[a] L. Cartechini,[a] P. Casavecchia,[a]
D. Stranges[b], and G. G. Volpi[a]
[a]Dipartimento di Chimica, Università di Perugia,
06100 Perugia, Italy
[b]Dipartimento di Chimica, Università di Roma "La Sapienza",
00100 Roma, Italy

Abstract

We highlight some recent work from our laboratory on reactions of atoms and radicals with simple molecules by the crossed molecular beam scattering method with mass-spectrometric detection. Emphasis is on three-atom ($Cl + H_2$) and four-atom ($OH + H_2$ and $OH + CO$) systems for which the interplay between experiment and theory is the strongest and the most detailed. Reactive differential cross sections are presented and compared with the results of quasiclassical and quantum mechanical scattering calculations on *ab initio* potential energy surfaces in an effort to assess the status of theory versus experiment.

1 Introduction

The landmark work of Max Bodenstein on the $H_2 + I_2$ reaction around the turn of the century can be regarded as the birth of gas-phase chemical *kinetics* [1]. About 30 years later, following the advent of quantum-mechanics, it became possible for chemists to pursue an understanding also of the *dynamics* of chemical reactions. Since then, tremendous progress has been made in the field of chemical kinetics and dynamics [2]. In particular, during the last decade, our understanding of atom and small radical gas-phase reaction dynamics has deepened enormously owing to exciting developments in molecular beam and laser spectroscopic techniques as well as in theoretical methodologies and computer capabilities.

The reactions $H + H_2$ and $F + H_2$ (and their isotopic variants) have been the benchmark systems in the field of chemical reaction dynamics. For them, fully converged three-dimensional (3D) quantum scattering calculations of state-to-state differential cross sections (DCS) have been performed and accurate comparisons with very detailed experimental observables carried out [3-9]. To date, only for one other neutral three-atom system have the exact (i.e., fully converged) 3D quantum scattering calculations of state-to-state DCS on a reliable *ab initio* potential energy surface (PES) been carried out, namely, for the prototypical reaction $Cl + H_2$ [10], a system chemical kineticists have been interested in since the time of Max Bodenstein (for a historical overview, see the paper by Truhlar in this volume). However, in contrast to $H + H_2$ and $F + H_2$, no experimental dynamical information is available on $Cl + H_2$. Here we highlight the results of the first dynamical investigation of the $Cl + H_2$ and $Cl + D_2$ reactions by the crossed molecular beam

Springer Series in Chemical Physics, Volume 61
Gas Phase Chemical Reaction Systems
Eds.: J. Wolfrum, H.-R. Volpp, R. Rannacher, and J. Warnatz

(CMB) method, which has been accompanied by exact quantum scattering calculations carried out by Truhlar and co-workers [11] on a new, high-quality, *ab initio* PES at the energies and for the conditions of our experiments.

The progress made in the solution of the reactive scattering problem involving three-atoms has prompted, during the last few years, many theoretical dynamicists to tackle the four-atom problem, which is the natural extension of A + BC towards more complex systems [12,13]. As the reactions $H + H_2$, $F + H_2$ and $Cl + H_2$ are the benchmarks for the three-atom case, $OH + H_2$ and OH + CO have been taking the prototypical role for the four-atom reaction family. Despite extensive work on the kinetics of these reactions [14-16] and on the dynamics of the reverse processes [17-20], very little was known about their dynamics. By exploiting the capability to generate *continuous* supersonic beams of OH radicals, measurements of reactive DCS by the CMB method for these two important radical-molecule reactions have been very recently reported from our laboratory [21-24]. This extends beyond A + BC the possibility of detailed comparison with the results of quasi-classical trajectory (QCT) and quantum mechanical (QM) dynamical calculations on *ab initio* PESs, carried out very recently by Schatz and co-workers [13,25,26] and by Clary and co-workers [12,26,27], and is the second topic to be considered below.

2 Experimental

The principles and practical implementation of crossed molecular beam experiments are well understood [28]. The observables which provide information on the reaction dynamics are the angular and velocity distributions (i.e., the doubly differential cross section) of the reaction products, which are measured in the laboratory (lab) frame of reference. For the physical interpretation of the scattering process it is necessary to transform the data into the center-of-mass (c.m.) system [28]. The lab number density and the c.m. flux are related by $N_{lab}(\Theta,v) = I_{c.m.}(\vartheta,u)v/u^2$, where Θ and v are the lab angle and velocity, respectively, and ϑ and u are the corresponding c.m. quantities. The nature of the lab-c.m. transformation is in general well illustrated by the velocity vector (so called "Newton") diagram [28,29]. Analysis of the lab data is usually carried out by forward convolution procedures over the experimental conditions of trial c.m. distributions (i.e., c.m. angular and translational energy distributions are assumed, averaged over experimental conditions and transformed to the lab for comparison with the data). The final outcome is the generation of velocity flux contour maps, $I_{c.m.}(\vartheta, u)$, of the reaction products, i.e., the plot of intensity as a function of angle and velocity in the c.m. system. The product contour map can be regarded as an *image* of the reaction.

The CMB apparatus used in our laboratory follows the classical design of Lee *et al.* [30] and has been described in detail elsewhere [23,24]. Critical in the experiments discussed here has been the high-pressure radio-frequency discharge source used for generating supersonic beams of Cl atoms and of OH radicals starting from dilute mixtures of Cl_2 and H_2O, respectively, in He or Ne seeding gas [23]. The $Cl(^2P_{3/2,1/2})$ spin-orbit state distribution was determined by Stern-Gerlach magnetic analysis [31]. The OH radicals in the beams are in the ground electronic state ($^2\Pi$) and expected to be rotationally as well as vibrationally cold. Further details on the specific experiments highlighted here can be found in Refs. [21-24].

3 Three-Atom Reactions: Cl + H_2 → HCl + H

In the series of the halogen-hydrogen reactions, which have been playing a central role in fundamental chemical kinetics beginning with the classic studies by Max Bodenstein [1,32], the elementary processes F + H_2 and Cl + H_2 have received the most attention at the kinetic level, both experimentally and theoretically [33]. While F + H_2 has been investigated in very great detail also from the dynamics standpoint [3,4,7], for Cl + H_2 only theoretical work [10,34,35] has been reported on the dynamics, with experimental studies completely lacking. The experimental techniques (IR chemiluminescence and CMB) which were so successful in the investigation of the F + H_2 dynamics could not be applied, for obvious reasons, to the homologous Cl + H_2 system. In fact, Cl + H_2 is nearly thermoneutral and the HCl product can only be formed in the ground vibrational level for energies up to 40 kJ/mol (see Fig. 1). Furthermore, the reaction cross section is quite small because of the large energy barrier, the Cl natural isotopic distribution is a complication and the kinematics are very unfavourable, since the heavy HCl product is confined very near the Cl beam and within a very small Newton sphere (see below and Fig. 2).

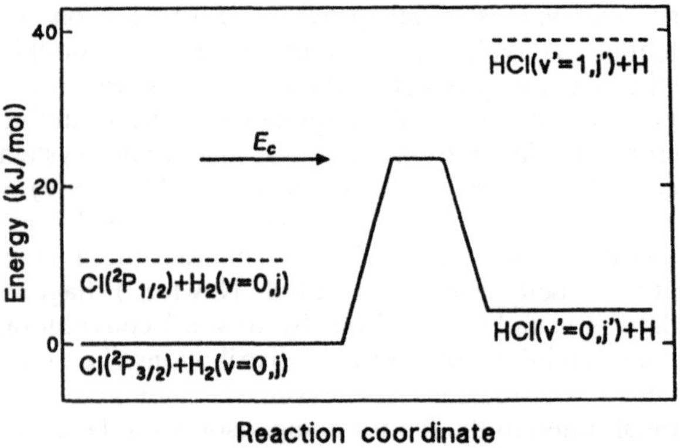

Figure 1: Energy level and correlation diagram for the Cl($^2P_{3/2,1/2}$) + H_2 reaction. The arrow indicates the experimental collision energy.

We have, nevertheless, succeeded in carrying out a CMB study of the reactions Cl + H_2 and D_2 and here we present preliminary data on:

$$Cl + H_2 \rightarrow HCl + H \qquad \Delta H^\circ_0 = 4.31 \text{ kJ/mol}$$

at a collision energy of 24.5 kJ/mol, that is slightly above the barrier to reaction.

In the left-hand-side of Fig. 2 we show the angular distribution of the HCl product (detected at *m/e* = 38) together with the most probable Newton diagram.

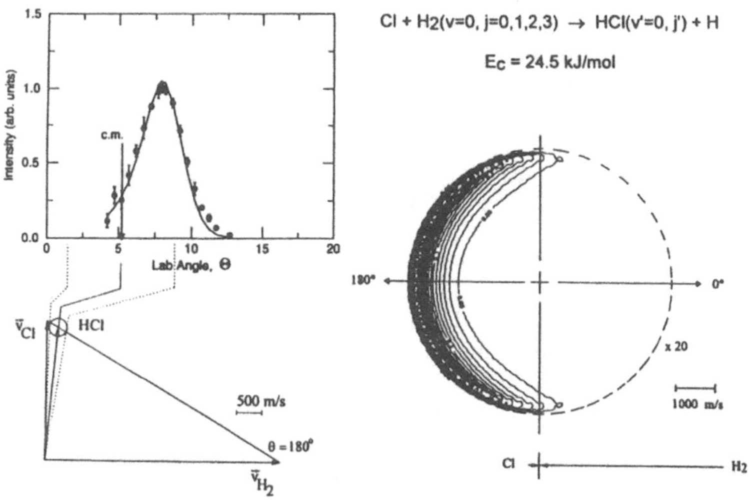

Figure 2: *Left:* Measured (dots) and calculated (solid line – see text) HCl product laboratory angular distribution from the Cl + H_2 reaction at E_c = 24.5 kJ/mol and corresponding Newton diagram. *Right*: HCl product center-of-mass flux (velocity-angle) contour map.

As can be noticed, the product is almost completely confined to the right of the c.m. angle in the lab frame and is sharply peaked at an angle nearly tangential to the maximum Newton circle, within which the product can be scattered on the basis of energy and linear momentum conservation. These two features immediately suggest that the mechanism of the reaction is direct (of the *rebound* type) and that a very large fraction of the total available energy is disposed into product translation. Velocity spectra at four selected angles were also recorded. From the angular and velocity distribution data in the lab frame, c.m. angular and translational energy distributions were derived; they are shown in the 2D product flux contour map representation in the right-hand-side of Fig. 2. As can be seen, the c.m. angular distribution is completely confined in the backward hemisphere and peaks at $\vartheta = 180°$. This suggests that the favoured orientation for reaction is collinear. About 80% of the total available energy is deposited into translation, which reflects strong repulsive forces between HCl and H after the saddle point is passed. The exceptionally high translational excitation of the products can be well appreciated in the contour map (Fig. 2), which shows the HCl product to be confined almost to the limit (dashed circle) of energy conservation. Since at this collision energy the HCl product can energetically be formed only in v' = 0, energy conservation gives us, by difference, also the amount of energy released into rotation (20%). It appears that the dynamics of Cl + H_2 resembles that of H + H_2, suggesting similarities of the PES topologies [35].

Experiments on Cl + D_2 were carried as a function of collision energy from about 12 up to 26.4 kJ/mol, using a beam containing about 75% of ground state $Cl(^2P_{3/2})$ and 25% of spin-orbit excited $Cl(^2P_{1/2})$. The purpose was to explore the reactivity of $Cl(^2P_{1/2})$ operating at E_c below the threshold for the ground state reaction. The vibrationally adiabatic threshold is predicted to be 26 kJ/mol on the G3 surface, while the dynamic one is considerably lower (about 14 kJ/mol) [11]. Reactive signal was observed starting at about 19 kJ/mol and angular distributions were measured at 19.7 and 20.5 kJ/mol. A preliminary fit of the data could, however, be obtained without the need to invoke the extra energy of $Cl(^2P_{1/2})$ (located 10.5 kJ/mol above the ground state) (see Fig. 1). This is pointing to a negligible (within our sensitivity) reactivity of $Cl(^2P_{1/2})$ to give DCl (v' = 0) at low collision energies.

Very recently, Truhlar and co-workers [11] have computed a new *ab initio* ground-state PES for Cl + H_2, termed G3, and have carried out on it exact quantum scattering calculations to be compared with our experimental results. Scattering calculations were also performed by the QCT method on the same PES by Aoiz and co-workers [36]. The *ab initio* PES indicates that collinear approach of Cl + H_2 is the dominant geometry leading to chemical reaction, in agreement with experiment. The exact quantum results offer the possibility of an unambiguous test of the PES when compared with experimental DCSs. A detailed comparison between experiment and theory, which requires to take into account the internal rotational distribution of the $H_2(D_2)$ reactant beam, is in progress. It can be anticipated that there is a good agreement between measured DCSs and those calculated quantum mechanically on the G3 PES.

4 Four-Atom Reactions: OH + H_2 and OH + CO

The four-atom reactions:

$$OH + H_2 \rightarrow H_2O + H \qquad \Delta H^\circ_0 = -61.9 \text{ kJ/mol} \qquad (1)$$

and

$$OH + CO \rightarrow CO_2 + H \qquad \Delta H^\circ_0 = -102.5 \text{ kJ/mol} \qquad (2)$$

have recently emerged as benchmarks for theoretical studies of systems involving four-atoms. In particular, OH + H_2, comprising three hydrogen atoms and an atom of the first row of the periodic table, is expected to play for the four-atom case the same role played by H + H_2 for the three-atom cases. Large scale *ab initio* electronic structure calculations of the PES [37,38] and quasiclassical and (approximate) quantum scattering calculations of the dynamics [12,13,38] were recently reported both for OH + H_2 and OH + CO. In Fig. 3 the energy level and correlation diagrams for the two reactions are shown comparatively. Because of their practical interest in combustion and atmospheric chemistry [39], their kinetics have been studied very extensively over the last 20 years [14-16]. The importance and prototypical nature of these two reactions have also encouraged a variety of state-of-the-art dynamics experiments in recent years [17-20]. All of them, however, have focused on the reverse endothermic processes. By exploiting the hot H-atom technique, absolute reactive cross sections and OH internal state distributions were measured, using pump-probe laser techniques, by Wolfrum and co-workers [18] and also by others [19]. In pioneering work Zewail and co-workers [17a] and, more recently, also Wittig and co-workers [17b], determined the lifetime of the HOCO complex using real-time probing, on the femtosecond time scale, of the OH product from the photoinitiated reaction in a van der Waals complex. A series of elegant experiments in Crim's and Zare's laboratories [19], involving selective initial excitation of the local OH stretching mode of H_2O (HOD), have explored the effect of vibrational excitation on the reaction H + H_2O.

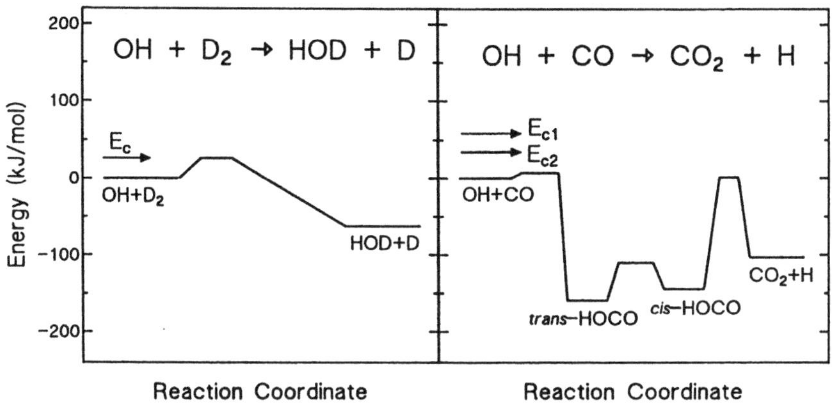

Figure 3: Energy level and correlation diagrams for the OH + D_2 and OH + CO reactions, shown on the same energy scale, according to *ab initio* results. The arrows indicate the experimental collision energies.

Experimental information on the dynamics of the direct processes is much more limited. Early flash photolysis studies [40] coupled to time-resolved infrared detection, did not observe any infrared emission from H_2O or CO_2 product. In a recent flash photolysis study, Smith and co-workers [14] found that a mere fraction of the total energy available to the H + CO_2 products is channelled into vibration at room temperature. Very recently, Wolfrum and co-workers [20] measured absolute reactive cross sections for OH + H_2, OH + D_2 and OH + CO at two different energies.

We have carried out reactive differential cross section measurements in CMB experiments [21,22] and determined the spatial distribution and energy distribution of the products from reaction (1) and (2). The results are compared with those of quasiclassical and quantum mechanical (approximate) dynamical computations on *ab initio* surfaces, which have been carried out by Schatz [13,25,26] and by Clary [12,26,27] at the experimental energies.

Now let us summarize our experimental results. For obvious reasons of simpler detection, we have studied the isotopic variant of reaction (1): OH + D_2. In Fig. 4 we report the laboratory angular distributions of the HOD product at E_c = 26.4 kJ/mol and of the CO_2 product at E_c= 59.0 kJ/mol, together with the corresponding most probable Newton diagrams. The lab angle Θ is measured from the OH beam. In the c.m. coordinate system, $\vartheta = 0°$ is the direction of the OH beam and represents the forward direction with respect to OH.

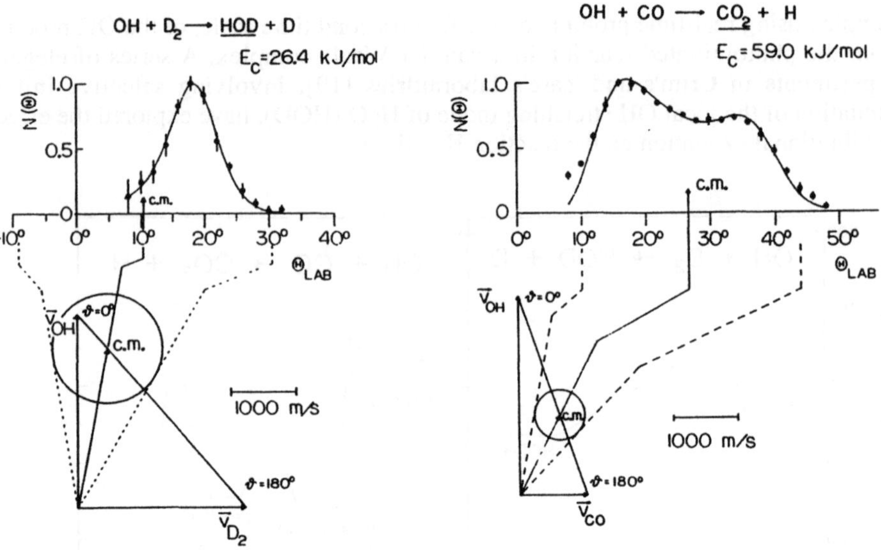

Figure 4: HOD and CO_2 product laboratory angular distributions from the OH + D_2 and OH + CO reactions at E_c = 26.4 kJ/mol and E_c = 59.0 kJ/mol, respectively. The solid line represents the calculated angular distribution with the best-fit c.m. translational energy and angular distributions. The circle in the Newton diagrams delimits the maximum HOD and CO_2 speeds assuming that all the available energy goes into translation.

The experimental results for the two reactions are shown comparatively in the same figure, so that the reader can immediately appreciate the difference in the dynamics. As can be seen, the HOD angular distribution peaks sharply to the right of the c.m. angle, indicating that the product is thoroughly back-scattered. In contrast, the CO_2 lab angular distribution exhibits a quite broad backward-forward structure with a clear preference for more forward scattering. The angular distributions were also accompanied by product velocity distribution measurements at selected angles and, via a forward convolution procedure, the product translational energy and angular distributions in the c.m. system were derived by using a separable form for the c.m. frame product flux distribution $I_{cm}(\vartheta,E') = T(\vartheta) \times P(E')$. These results are well synthesized in the product flux (velocity-angle) representations shown comparatively in Fig. 5. In this figure one can immediately see how the reaction product (HOD and CO_2) is spatially and energetically distributed in the c.m. system. In the case of HOD, it is clearly evident how the product is strongly backward scattered with respect to the OH direction, with the peak well within the limit of energy conservation (the dashed circle), which implies a modest translational energy, and consequently high internal energy, of the product. In the case of CO_2 it is quite clear: (i) the backward-forward structure of the angular distribution with a forward bias, which witnesses the formation of an "osculating complex", (ii) the marked confinement of the peaking of the angular distribution at the limit of energy conservation, which witnesses a "high product translational excitation", (iii) the near isotropy (i.e., mild polarization) of the angular distribution, which witnesses a "high product rotational excitation". While $OH + D_2$ is a direct, simple hydrogen abstraction reaction, in which formation of the new bond occurs simultaneously with the breakage of the old bond, OH + CO goes through a complex intermediate, HOCO, living a time comparable to its rotational period.

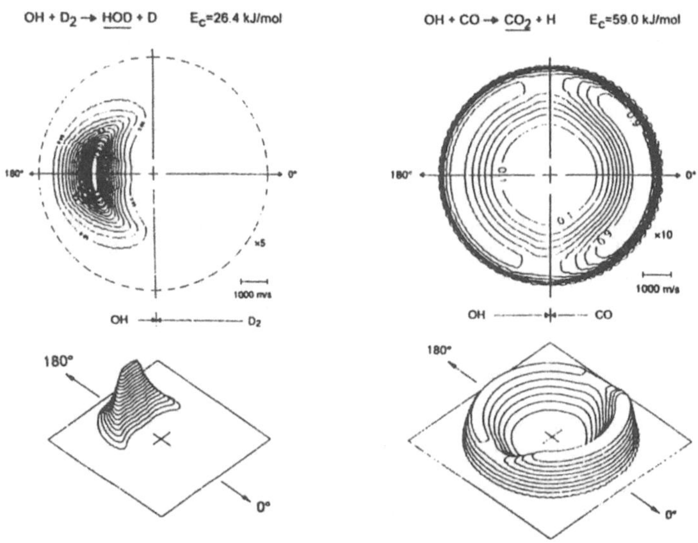

Figure 5: Center-of-mass flux (velocity-angle) contour maps of the HOD and CO_2 products from the $OH + D_2$ and OH + CO reactions with three-dimensional perspective.

4.1 Comparison with theory

$OH + D_2 \rightarrow HOD + D$: To date, among the various theoretical approaches so far devised for the four-atom reactive scattering problem, only the reduced dimensionality treatment of Clary [12] has provided differential cross sections. The angular distribution calculated [27] on the *ab initio* (Walch-Dunning-Schatz-Elgersma) surface is in good agreement with the experimental determination [23]. In contrast, the calculated fraction of energy released in translation is 0.71, which disagrees strongly with respect to the experimental value of 0.34. This disagreement suggests that the PES has some shortcomings and may need improvement. The QCT results of Bradley and Schatz [25] on the same PES are essentially in line with the approximate quantum results of Nyman and Clary [27].

Large scale *ab initio* calculations of a new PES were recently completed by Kliesch and Werner [41] and were tested in dynamical calculations by Clary [42]. The results indicate that the Kliesch-Werner surface, in addition to predicting an angular distribution in good agreement with experiment, predicts also a much larger vibrational excitation of the water product (the energy release into translation is about 44%), which is in closer, although not perfect, agreement with the scattering determination. The *exact* dynamical calculations are, of course, required for an unambiguous test of this new *ab initio* PES.

$OH + CO \rightarrow CO_2 + H$: In Fig. 6 the experimental c.m. angular distribution at E_c = 59.0 kJ/mol is compared with the results of QCT calculations by Kudla and Schatz [13] and of approximate quantum calculations by Clary and Schatz [26] using the same *ab initio* PES. As can be seen, the quantum treatment gives an angular distribution much more polarized at $\vartheta = 0°$ and $\vartheta = 180°$ than the experiment, indicating that a three degrees of freedom model is inadequate to describe the OH + CO reactive scattering. Instead, the angular distribution calculated by the full-dimensional QCT method is in close, although not quantitative, agreement with experiment. Both the rotating-bond-approximation quantum calculations of Clary and Schatz [26] and the QCT results of Kudla and Schatz [13] give a fraction of energy in translation in good agreement with experiment (65%). The still noticeable discrepancy between QCT results and experiment has to be traced back to some inadequacies of the PES. Clearly, an improved PES is needed and full-dimensional quantum scattering calculations are desirable.

The HOCO lifetime, estimated from the asymmetry in $T(\vartheta)$ within the "osculating model" for chemical reactions [43], is about 0.6 ps at E_c = 59 kJ/mol and 1 ps at E_c = 36 kJ/mol, and compares well with the lifetime values measured by Zewail and co-workers [17a] and by Wittig and co-workers [17b] and with theoretical values [13,26].

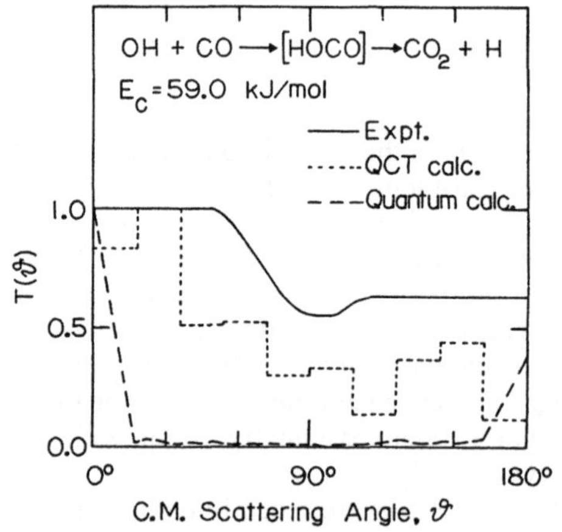

Figure 6: Center-of-mass product angular distributions for the OH + CO reaction compared to the results of QCT and approximate QM dynamical calculations (see text).

5 Acknowledgements

Financial support from the Italian "CNR-Progetto Finalizzato Chimica Fine" and MURST is gratefully acknowledged. Grants from EEC, ENEA, NATO (CRG. 920549) and AFOSR (F617-08-94-C-0013) through EOARD (SPC-94-4042) made possible several aspects of the research reported here.

6 References

[1] M. Bodenstein, Z. phys. Chem. **13**, 56 (1894); *ibid.* **22**, 1 (1897); *ibid.* **29**, 295 (1899).

[2] R.D. Levine and R.B. Bernstein, ***Molecular Reaction Dynamics and Chemical Reactivity***, Oxford University, New York, 1987; I.W.M. Smith, ***Kinetics and Dynamics of Elementary Gas Reactions***, Butterworths, London, 1980; ***Modern Gas Kinetics***, (Ed. M.J. Pilling and I.W.M. Smith) Blackwell, Oxford, 1987.

[3] D.M. Neumark, A.M. Wodtke, G.N. Robinson, C.C. Hayden, and Y.T. Lee, J. Chem. Phys. **82**, 3045 (1985); D.M. Neumark, A.M. Wodtke, G.N. Robinson, C.C. Hayden, K. Shobatake, R.K. Sparks, T.P. Shafer, and Y.T. Lee, J. Chem. Phys. **82**, 3067 (1985) and references therein.

[4] F.J. Aoiz, L. Bañares, V.J. Herrero, V. Sáez Rábanos, K. Stark, H.-J. Werner, Chem. Phys. Lett. **223**, 215 (1994); S.L. Mielke, G.C. Lynch, D.G. Truhlar and D.W. Schwenke, Chem. Phys. Lett. **213**, 11 (1993); M. Baer *et al.* in this volume.

[5] R.E. Continetti, B.A. Balko, and Y.T. Lee, J. Chem. Phys. **93**, 5719 (1990); S.A. Buntin, C.F. Giese, and W.R. Gentry, J. Chem. Phys. **87**, 1443 (1987); Chem. Phys. Lett. **168**, 513 (1990); H. Buchenau, J.P. Toennies, J. Arnold, and J. Wolfrum, Ber. Bunsenges. Phys. Chem. **94**, 1231 (1990); D.P. Gerrity and J.J. Valentini, J. Chem. Phys. **81**, 1298 (1984); J.C. Nieh and J.J. Valentini, J. Chem. Phys. **92**, 1083 (1990).

[6] T.N. Kitsopoulos, M.A. Buntine, D.P. Baldwin, R.N. Zare, and D.W. Chandler, Science **260**, 1605 (1993).

[7] D.E. Manolopoulos, K. Stark, H.-J. Werner, D.W. Arnold, S.E. Bradforth, and D.M. Neumark, Science **262**, 1852 (1993).

[8] W.H. Miller, Annu. Rev. Phys. Chem. **41**, 245 (1990).

[9] J.M. Launay and M. LeDourneuf, Chem. Phys. Lett. **169**, 473 (1990); D. Neuhauser, R.S. Judson, R.L. Jaffe, M. Baer, and D.J. Kouri, Chem. Phys. Lett. **176**, 546 (1991); M. D'Mello, D.E. Manolopoulos, and R.E. Wyatt, J. Chem. Phys. **94**, 5985 (1991); S.L. Mielke, R.S. Friedman, D.G. Truhlar, D.W. Schwenke, Chem. Phys. Lett. **188**, 359 (1992); Y.-S.M. Wu and A. Kuppermann, Chem. Phys. Lett. **201**, 178 (1993); A.A. Lazarides, D. Neuhauser, and H. Rabitz, J. Chem. Phys. **99**, 6653 (1993); R.T Pack and G.A. Parker, J. Chem. Phys. **87**, 3888 (1987); *ibid.* **90**, 3511 (1989).

[10] J.M. Launay and S.B. Padkjaer, Chem. Phys. Lett. **181**, 95 (1991).

[11] D.G. Truhlar, private communication, and in this volume.

[12] D.C. Clary, J. Phys. Chem. **98**, 10678 (1994), and references therein. See also: H. Szichman and M. Baer, Chem. Phys. Lett. **242**, 285 (1995), and references therein; U. Manthe, T. Seideman, and W.H. Miller, J. Chem. Phys. **99**, 10078 (1993); D.H. Zhang and J.Z.H. Zhang, J. Chem. Phys. **101**, 1146

(1994); D. Wang and J.M. Bowman, J. Chem. Phys. **98**, 6235 (1993); D. Neuhauser, J. Chem. Phys. **100**, 9272 (1994); N. Balakrishnan and G.D. Billing, J. Chem. Phys. **101**, 2785 (1994).

[13] K. Kudla and G.C. Schatz, in ***The Chemical Dynamics and Kinetics of Small Radicals*** (Ed. K. Liu and A.F. Wagner) World Scientific, Singapore, 1995, chap. 10.

[14] M.J. Frost, P. Sharkey, and I.W.M. Smith, Faraday Discuss. Chem. Soc. **91**, 305 (1991); M.J Frost, J. Salh, and I.W.M. Smith, J. Chem. Soc. Faraday Trans. **87**, 1037 (1991).

[15] I.W.M. Smith and R. Zellner, J. Chem. Soc. Faraday Trans. 2, **70**, 1045 (1974); A.R. Ravishankara, J.M. Nicovich, R,L, Thompson, and F.P. Tully, J. Phys. Chem. **85**, 2498 (1981), and references therein.

[16] J. Warnatz, *Sandia National Laboratories Report No. SAND83-8606*, Livermore, CA, 1983; T. Dreier and J. Wolfrum, in *18th Symposium (Int.) on Combustion*, The Combustion Institute, Pittsburg, PA, 1981, p. 801; J. Troe, in *22nd Symposium (Int.) on Combustion*, The Combustion Institute, Pittsburg, PA, 1985, p. 559.

[17] (a) N.F. Sherer, C. Sipes, R.B. Bernstein and A.H. Zewail, J. Chem. Phys. **92**, 5239 (1990); L.R. Khundkar and A.H. Zewail, Annu. Rev. Phys. Chem. **41**, 15 (1990); A.H. Zewail, Faraday Discuss. Chem. Soc. **91**, 207 (1991); (b) S.I. Ionov, G.A. Brucker, C. Jaques, L. Valachovic, and C. Wittig, J. Chem. Phys. **97**, 9486 (1992); C. Jaques, L. Valachovic, S. Ionov, E. Böhmer, Y. Wen, J. Segall, and C. Wittig, J. Chem. Soc. Faraday Trans. **89**, 1419 (1993).

[18] A. Jacobs, H.R. Volpp, and J. Wolfrum, *24th Symposium (Int.) on Combustion,* The Combustion Institute, Pittsburgh, PA, 1992, p. 605; J. Wolfrum, Faraday Discuss. Chem. Soc. **84**, 191 (1987); K. Kleinermanns and J. Wolfrum, Chem. Phys. Lett. 4, **104**, 157 (1984); A. Jacobs, M. Wahl, R. Weller, and J. Wolfrum, Chem. Phys. Lett. **158**, 161 (1989).

[19] K. Honda, M. Takayanagi, T. Nishiya, H. Ohoyama, and I. Hanazaki, Chem. Phys. Lett. **180**, 321 (1991); K. Kessler and K. Kleinermanns, Chem. Phys. Lett. **190**, 145 (1992); A. Sinha, M.C. Hsiao, and F.F. Crim, J. Chem. Phys. **92**, 6333 (1990); J. Chem. Phys. **94**, 4928 (1991); M.C. Hsiao, A. Sinha, and F.F. Crim, J. Phys. Chem. **95**, 8263 (1991); M.J. Bronikowski, W.R. Simpson, B. Girard, and R.N. Zare, J. Chem. Phys. **95**, 8647 (1991); M.J. Bronikowski, W.R. Simpson and R.N.Zare, J. Phys. Chem. **97**, 2194 (1993); *ibid.*, **97**, 2204 (1993); (b) D.E. Adelman, S.V. Filseth, and R.N. Zare, J. Chem. Phys. **98**, 4636 (1993); C.R. Quick, Jr. and J.J. Tiee, Chem. Phys. Lett. **100**, 223 (1983); S.K. Shin, Y. Chen, S. Nickolaisen, S.W. Sharpe, R.A. Beaudet, and C. Wittig in ***Advances in Photochemistry***, Vol. 16, (Ed. D. Volman, G. Hammond, and D. Neckers) Wiley, New York, 1991, pp. 249-363, and references therein; J.K. Rice and A.P. Baronavsky, J. Chem. Phys. **94**, 1006 (1991).

[20] S. Koppe, T. Laurent, P.D. Naik, H.-R. Volpp, and J. Wolfrum, Can. J. Chem. **72**, 615 (1994); A. Jacobs, H.-R. Volpp, and J. Wolfrum, J. Chem. Phys. **100**, 1936 (1994); Chem. Phys. Lett. **218**, 51 (1994); H.-R. Volpp, and J. Wolfrum, in this volume.

[21] M. Alagia, N. Balucani, P. Casavecchia, D. Stranges, and G.G. Volpi, J. Chem. Phys. **98**, 2459 (1993).

[22] M. Alagia, N. Balucani, P. Casavecchia, D. Stranges, and G.G. Volpi, J. Chem. Phys. **98**, 8341 (1993).

[23] M. Alagia, N. Balucani, P. Casavecchia, D. Stranges, and G.G. Volpi, J. Chem. Soc. Faraday Trans. **91**, 575 (1995).
[24] P. Casavecchia, N. Balucani, and G.G. Volpi, in ***The Chemical Dynamics and Kinetics of Small Radicals***, (Ed. K. Liu and A.F. Wagner) World Scientific, Singapore, 1995, chap. 9.
[25] K.S. Bradley and G.C. Schatz, J. Phys. Chem. **98**, 3788 (1994).
[26] D.C. Clary and G.C. Schatz, J. Chem. Phys. **99**, 4578 (1993).
[27] G. Nyman and D.C. Clary, J. Chem. Phys. **100**, 3556 (1994).
[28] Y.T. Lee, in ***Atomic and Molecular Beam Methods*** (Ed. G. Scoles) Oxford University, New York, 1987, Vol. 1.
[29] T.T. Warnoch and R.B. Bernstein, J. Chem. Phys. **49**, 1878 (1968).
[30] Y.T. Lee, J.D. McDonald, P.R. Le Breton, and D.R. Herschbach, Rev. Sci. Instr. **40**, 1402 (1969).
[31] M. Alagia, N. Balucani, L. Cartechini, P. Casavecchia, G.G. Volpi, V. Aquilanti, D. Ascenzi, D. Cappelletti, F. Pirani, work in progress.
[32] M. Bodenstein, Z. für Elektrochem. **85**, 329 (1913); *ibid.* **22**, 335 (1916).
[33] J.B. Anderson, Adv. Chem. Phys. **41**, 229 (1980), and references therein; J.T. Muckerman, in ***Theoretical Chemistry-Advances and Perspectives***, Vol. VIA. (Ed. H. Eyring and D. Henderson) Academic, New York, 1981; M. Baer *et al.* in this volume.
[34] M.J. Stern, A. Persky, and F.S. Klein, J. Chem. Phys. **58**, 5697 (1973); S.C. Tucker, D.G. Truhlar, B.C. Garrett, and A.D. Isaacson, J. Chem. Phys. **82**, 4102 (1985) and references therein; A. Persky, J. Chem. Phys. **60**, 2932 (1977); *ibid.* **68**, 2411 (1978); *ibid.* **70**, 3910 (1979).
[35] S. Takada, K. Tsuda, A. Ohsaki, and H. Nakamura, in ***Advances in Molecular Vibrations and Collision Dynamics: Quantum Reactive Scattering***, Vol. IIA (Ed. J.M. Bowman) JAI, Greenwich, 1994, p. 245.
[36] F.J. Aoiz and L. Bañares, private communication.
[37] S.P. Walch and T.H. Dunning, Jr., J. Chem. Phys. **72**, 1303 (1980); M. Aoyagi and S. Kato, J. Chem. Phys. **88**, 6409 (1988).
[38] G.C. Schatz and H. Elgersma, Chem. Phys. Lett. **73**, 21 (1980); G.C. Schatz, M.S. Fitzcharles, and L.B. Harding, Faraday Discuss. Chem. Soc. **84**, 359 (1987); K. Kudla, G.C. Schatz, and A.F. Wagner, J. Chem. Phys. **95**, 1635 (1991); K. Kudla, A.G. Koures, L.B. Harding, and G.C. Schatz, J. Chem. Phys. **96**, 7465 (1992).
[39] J. Warnatz, in ***Combustion Chemistry*** (Ed. W.C. Gardiner, Jr.) Springer, New York, 1984, chap. 5; J.A. Miller and G.A. Fisk, Chem. Eng. News **65**, 22 (1987); R.P. Wayne, ***Chemistry of Atmospheres,*** Clarendon, Oxford, 1985.
[40] D.W. Trainor and C.W. von Rosenberg, Jr., *15th Symposium (Int.) on Combustion,* The Combustion Institute, Pittsburgh, PA, 1974, p. 755; D.W. Trainor and C.W. von Rosenberg, Jr., Chem. Phys. Lett. **29**, 35 (1974).
[41] H.-J. Werner, private communication.
[42] D.C. Clary, private communication, and in this volume.
[43] G.A. Fisk, J.D. McDonald, and D.R. Herschbach, Faraday Disc. Chem. Soc. **44**, 228 (1967).

Part II

Microscopic Dynamics of Elementary Reactions: Theory

Die photochemische Bildung des Chlorwasserstoffs Dynamics of Cl + $H_2 \rightleftharpoons$ HCl + H on a New Potential Energy Surface: The Photosynthesis of Hydrogen Chloride Revisited 100 Years after Max Bodenstein

T.C. Allison,[a] S.L. Mielke,[a] D.W. Schwenke,[b]
G.C. Lynch,[a] M.S. Gordon,[c] and D.G. Truhlar[a]
[a]Department of Chemistry, Chemical Physics Program, and Supercomputer Institute,
University of Minnesota, 207 Pleasant Street SE,
Minneapolis, MN 55455-0431, U.S.A.
[b]NASA Ames Research Center, Mail Stop 230-3, Moffett Field, CA
94035-1000, U.S.A.
[c]Department of Chemistry, Iowa State University, Ames, IA 50011, U.S.A.

Abstract

In this contribution, we present a preliminary account of our recent work on the Cl + H_2 and Cl + D_2 reactions, which includes a new potential energy surface, variational transition state theory and semiclassical tunneling calculations for both reactions and for other isotopomeric cases, and accurate quantum dynamical calculations of rate constants and state-to-state integral and differential cross sections.

1 Introduction

Halogen atom–hydrogen molecule reactions and their reverse hydrogen atom–hydrogen halide reactions have played a major role in the development of chemical kinetics. The work of Bodenstein and Lind [1] on the H_2–Br_2 reaction led to the development of the chain mechanism [2–4] for free radical reactions and ultimately, with the steady-state assumption, to a value for the rate constant of the Br + H_2 reaction [5–10]. The work of Bodenstein and Dux [11–13] on the Cl_2–H_2 reaction also led to a chain mechanism for that case, with Cl eventually established as the chain carrier [7,14]. The BrH_2 and ClH_2 triatomic systems were the subjects of historical papers [15–17] in the development of the London-Eyring-Polanyi-Sato (LEPS) semiempirical valence bond treatment [18] for potential energy surfaces of atom transfer reactions. The F + H_2 reaction also has a long history and has been labelled the "bell-wether" elementary reaction for showing the power and limitations of the methods of chemical dynamics [19]. The I + H_2 reaction is highly endothermic and hence slow; like the other halogen-hydrogen reactions it has been the subject of considerable controversy [10,20].

The Cl + $H_2 \rightleftharpoons$ HCl + H reaction has a particularly long and interesting history, including thermal and photochemical studies of the H_2–Cl_2 system, molecular beam experiments, measurements of perhaps the largest number of kinetic isotope effects known for any system, and a host of theoretical treatments. We will mention only a few particularly relevant studies. Early experimental studies of this reaction were often initiated by photodissociation of Cl_2, and some of the papers are entitled "Photosynthesis of Hydrogen Chloride "(a translation of *die Photochemische*

Springer Series in Chemical Physics, Volume 61
Gas Phase Chemical Reaction Systems
Eds.: J. Wolfrum, H.-R. Volpp, R. Rannacher, and J. Warnatz

Bildung des Chlorwasserstoffs). These experimental studies of the hydrogen-chlorine photochemical reaction were complicated by the existence of an induction period. The induction period was discovered by Draper in 1843 and studied further in 1857 by Bunsen and Roscoe, who believed it to be inherent in the mechanism; however, it was later shown by van't Hoff to be an experimental artifact. In 1906, Burgess and Chapman further clarified the effect as due to impurities. Chapman and McMahon suggested NCl_3 as the culprit a few years later, and the mechanism was only completely established in 1934 [21]. Additional controversy, in this case started by Bodenstein and Dux [11] and Chapman and Underhill [22], concerned the inhibiting effect of hydrogen, and this was resolved in 1933 [23].

At the same time when the mechanistic issues were being sorted out a rough estimate of the rate constant emerged. Bodenstein [24], based on earlier work on chain reactions by himself and others, estimated the probability P of the reaction H + Cl_2 to be of the order of magnitude of 10^{-2}, that for H + HCl to be of the order of 10^{-4}, and the equilibrium constant for Cl + $H_2 \rightleftharpoons$ HCl + H to be of order unity (a more accurate modern value at 298 K is 0.047 [26]); and hence his estimates yielded P(Cl + H_2) to be ~10^{-4}. With a room-T collision rate coefficient of 4×10^{-10} $cm^3molecule^{-1}s^{-1}$, this yields $k = 4 \times 10^{-14}$ $cm^3molecule^{-1}s^{-1}$. In 1933, Ritchie and Norrish obtained additional experimental evidence for P(Cl + H_2)/P(H + Cl_2) = 10^{-2} and repeated the estimate P(Cl + H_2) = 10^{-4}.

The first "direct" measurement of the rate was carried out by Rodebush and Klingenhoefer [27] in 1933. They produced Cl by an electrodeless discharge, passed Cl and H_2 through a flow tube, quenched the reaction, and took samples after about 5–10 minutes (kilosecond chemistry). They titrated the products with KOH and methyl orange indicator and obtained k(Cl + H_2) = 1.5×10^{-14} $cm^3molecule^{-1}s^{-1}$ at room temperature, a factor of 3 lower than Bodenstein's estimate. The best value available at present is 1.50×10^{-14} $cm^3molecule^{-1}s^{-1}$ [28], which is (fortuitously, but amazingly) identical to the 1933 value! (At 273 K, the best modern value is about 30% higher than the 1933 value).

In 1932, Semenoff suggested that the activation energy E_a be estimated by equating the rate constant to the collision rate constant times a Boltzmann factor; he obtained E_a = 5.5–6.6 kcal [29]. In 1933 Ritchie and Norrish [25] estimated E_a more directly from experiment. Using previous work [30–32] on the temperature coefficient (i.e., the amount by which the reaction rate goes up for a 10 deg increase in T) of the overall $H_2 - Cl_2$ chain reaction, which they took as 1.14 (with a large uncertainty), they estimated $E_a \cong$ 2.3 kcal. In contrast, by the less reasonable assumption that two-body recombination of Cl occurs via $Cl(^2P_{1/2}) + Cl(^2P_{3/2}) \rightarrow Cl_2 + h\nu$, they obtained E_a = 3.6 kcal. However a more recent (1932) measurement [33] of the temperature coefficient in an oxygen-free mixture (that they did not use) gave 1.37, which yields E_a = 5.8–6.0 kcal [6,10]. The direct measurement of Rodebush and Klingenhoefer in 1933 yields E_a = 5.6 kcal (obtained by fitting their data to standard Arrhenius form rather than the collision theory form they used). The best modern value is E_a = 4.4 kcal, which would correspond to a temperature coefficient of 1.27 at 298 K [28].

The first kinetic isotope effect (KIE) measurement was reported in 1934, and it yielded $k_{H_2}/k_{D_2} \approx 10$ at room temperature [34]. Bigeleisen *et al.* obtained the same result in 1959 [35], and Chiltz *et al.* obtained 9 in 1963. The most recent values are 9.1 [95] and 7.5 [28]. By 1973, kinetic isotope effects were available for quite a few isotopomeric versions of the reaction [34–38].

The reverse abstraction reactions H + HCl $\rightarrow$ H_2 + Cl, D + DCl $\rightarrow$ D_2 + Cl, H + DCl $\rightarrow$ HD + Cl, and D + HCl $\rightarrow$ HD + Cl as well as the exchange reactions,

H + DCl ↔ HCl + D, which compete with the back reactions, have also been studied experimentally, with the early work being quite controversial (even contradictory), as summarized elsewhere [26,39–42]. Two issues are involved: (i) Does detailed balance hold, i.e., does the forward rate constant divided by the backward abstraction rate constant equal the equilibrium constant? Some evidence pointed to possible deviations as great as a factor of two or three, which would raise interesting questions regarding nonequilibrium internal state distributions [43,44] during reaction. (ii) What is the relative rate of exchange compared to abstraction in the forward reaction? Issue (i) was settled by Miller and Gordon [26], who measured the reaction in both directions, and found that detailed balance holds quite well, as usually assumed. The best results for issue (ii) are also due to Miller and Gordon [41]. They placed an upper limit of 2×10^{-3} on k_{exch}/k_{abs} for D + HCl at 325 K. This implies a barrier height greater than about 7 kcal for exchange, whereas the molecular beam and infrared fluorescence experiments of McDonald and Herschbach [45] and Wight *et al.* [46] yield an upper bound in the range 20–22 kcal.

In 1991, Barclay *et al.* [47] reported single-collision studies of the competitive pathways of the D + HCl reaction for collision energies 27 and 43 kcal. The ratio of exchange to abstraction cross sections, $\sigma_{exch}/\sigma_{abs}$, was found to be about 2–3 at these high energies.

Returning to the forward reaction, we note that, due primarily to its small probability and nearly thermoneutral character, detailed dynamical studies (i.e., state-sensitive results or cross sections rather than thermal rate constants) have become available only very recently, in the crossed molecular beam (CMB) experiments of Casavecchia and coworkers [48]. At a collision energy of 6.4 kcal, they found that the DCl produced in the Cl + D_2 reaction is mainly scattered more than 80° backwards from the incident *D* direction (in the center-of-mass frame) and that about 80% of the total available energy is disposed into relative translation of the products.

Theoretical work, as always, must begin with a potential energy surface. Early work on the potential energy surface was quantitatively and sometimes qualitatively unreliable and is reviewed elsewhere [18]. In 1973, Stern, Persky, and Klein [49] created three semiempirical potential energy surfaces of the extended-LEPS type for comparison to kinetic isotope effects on the Cl + H_2 reaction. Although realistic for Cl–H–H type geometries, these surfaces are not globally satisfactory because they are quite unrealistic for H–Cl–H type geometries. The first globally realistic potential energy surface was due to Baer and Last, whose surface has a barrier height of 8.1 kcal for Cl + H_2, 5.1 kcal for the reverse abstraction reaction, and 12.5 kcal for the exchange reaction [50]. Unfortunately this surface predicts somewhat inaccurate rate constants and KIEs [51]. A more successful surface, based in part on electronic structure calculations with scaled [52] electron correlation for H–Cl–H type geometries, was published in 1989 [53]. This surface, called GQQ, has a barrier height of 7.7 kcal for the forward reaction, 4.7 kcal for the reverse abstraction, and 18.1 kcal for the exchange reaction. Barclay *et al.* [47] concluded from trajectory studies that the GQQ surface is adequate for the abstraction reaction and for the exchange reaction at 27 kcal but not for the exchange reaction at 36 kcal and higher.

It was noted in the concluding remarks of the paper presenting the GQQ surface [53] that this surface could be improved by incorporating *ab initio* calculations on the Cl–H–H bend potential. A surface incorporating this improvement has now been created [54] and is called G3. The barrier heights on the G3 surface are 7.9 kcal for Cl + H_2 and 4.9 and 18.1 kcal for the H + HCl abstraction and exchange reactions.

Returning for a moment to the general theme of halogen-H_2 reactions, we note that the methods used to obtain the GQQ and G3 surfaces for ClH_2 have also been used to obtain the most accurate currently available surfaces for FH_2 [55] and BrH_2 [56].

Dynamics calculations on the ClH_2 reactions also have a long history. Early work was based on conventional transition state theory (TST), without tunneling or with one-dimensional tunneling [16,47,57]. Later work included trajectory calculations [45,58–64], reduced-dimensionality studies [65–67], variational transition state theory (VTST) with multidimensional tunneling (MT) [68,69] or optimized multidimensional tunneling (OMT) [51,53,54], and approximate quantum scattering calculations [70]. In 1991 and 1993, Launay and coworkers presented converged quantum scattering calculations for the Cl + H_2 reaction on the GQQ surface [71,72], and Takada *et al.* [73] calculated accurate quantal reaction probabilities. The present account summarizes some of our recent dynamics calculations based on the G3 surface.

2 G3 Potential Energy Surface

Ab initio electronic structure calculations were carried out for 63 Cl–H–H geometries by Møller-Plesset 4th order perturbation theory (MP4) [74] using valence triple zeta basis sets on Cl [75] and H [76] augmented by a set of five *d* functions on Cl and a set of *p* functions on H. The correlation energies were scaled by the MP4-SAC method [77] with scale factor 1/0.82. The geometries consisted of 21 sets of nearest-neighbor bond distances, each with Cl–H–H bond angle equal to 180, 170, and 160 deg.

The GQQ surface was modified by making the H-Cl triplet interaction in the LEPS function be an explicit function of all three internuclear distances. This allowed us to fit the *ab initio* Cl–H–H bending potentials with an RMS error of 0.08 kcal, while retaining the general shape of the GQQ potential for H–Cl–H and collinear Cl–H–H geometries.

3 Rate constants for Cl + H_2

Quantum mechanical rate constants were calculated for the Cl + H_2 reaction in three steps. First, we calculated converged cumulative reaction probabilities [78–80] for total angular momenta $J = 0$–6. The calculations were performed using the outgoing wave variational principle [81,82] and techniques presented elsewhere [83,84]. Second, we used the separable rotation approximation (SRA) [85,86] to generate from a single J the cumulative reaction probability up to as high a J as is required for convergence. Third, we integrated these cumulative reaction probabilities, weighted by a Boltzmann factor and the appropriate kinematical and electronic partition function factors, to obtain thermal reaction rates.

Before presenting the results we comment on the validation of the SRA and the accuracy attainable with accurate quantum dynamical calculations by considering recent results for the D + H_2 → HD + H reaction [85,86]. The SRA is a shortcut allowing us to generate quantum dynamical rate constants, fully summed over J, from calculations at only one or a few low J. In our initial studies of the method [85,86], we found that it is much more accurate if it is based on $J \geq K_{conv}$ where K_{conv} is the highest value of the vibrational angular momentum quantum number [87] of the

Table 1. Rate constants ($cm^3 molecule^{-1} s^{-1}$) for $Cl + H_2$

T(K)	SRA $J=3$	SRA $J=5$	VTST/OMT	Experiment [28]
200	8.49(−16)	8.66(−16)	8.96(−16)	3.92(−16)
300	2.48(−14)	2.50(−14)	2.31(−14)	1.57(−14)
400	1.50(−13)	1.51(−13)	1.31(−13)	9.89(−14)
600	1.00(−12)	1.01(−12)	8.68(−13)	7.19(−13)
800	2.79(−12)	2.80(−12)	2.52(−12)	2.24(−12)
1000	5.41(−12)	5.46(−12)	4.92(−12)	4.85(−12)

Table 2. Kinetic isotope effects k_{AB}/k_{CD}

AB/CD	T(K)	Experiment '73[49]	Experiment '83[95]	Experiment '94[28]	VTST/OMT anhar	TST anhar	TST har
H_2/D_2	245	14.6			15.0	8.7	12.3
	255		12.9	9.4	13.9	8.3	11.4
	298		9.1	7.5	10.6	6.7	8.7
	345	7.5	6.9	6.1	8.3	5.6	7.0
	500		4.0	3.6	4.9	3.8	4.3
	600			3.0	3.9	3.2	3.5
	1000			2.1	2.3	2.2	2.3
H_2/T_2	275	34.2			50.7	20.4	30.0
	345	18.3			26.5	13.1	17.4
HD/DH	300	1.76			2.1	1.29	1.41
	445	1.37			1.79	1.25	1.31
H_2/(HD+DH)	245	3.4			4.4	2.5	2.9
	345	2.5			3.1	2.1	2.3
H_2/(HT+TH)	245	6.5			10.8	3.4	4.2
	345	4.1			6.0	2.7	3.2
H_2/(DT+TD)	275	20.7			25.7	11.8	16.6
	345	12.1			15.3	8.3	10.7

transition state that makes a significant contribution to the reaction rate. We would expect $K_{conv} < 6$ for reactions like $D + H_2$ and $Cl + H_2$ at temperatures of interest here, and rate constants for $D + H_2$ calculated from either $J = 3$ or $J = 5$ agree with full calculations with an average absolute error of only 3% over the range 200–1000 K [85,86]. Furthermore, full rate calculations agree with experiment [88] with an average absolute deviation of only 8% over this range of T. For $D + H_2$ accurate *ab initio* kinetics agree so well with experiment because the potential energy surfaces

[89,90] are apparently very accurate. For Cl + H_2, then, comparison to experiment provides a check on the potential energy surface.

Table 1 shows SRA results for Cl + H_2 based on $J = 3$ and $J = 5$, and for this reaction the two sets of calculations agree with an average absolute deviation of only 1% over the 200–1000 K range. In contrast the average absolute deviation of the SRA results based on $J = 0$ (not shown) from those based on $J = 3$ or $J = 5$ is 23%.

Table 1 also shows a comparison to the values recommended [28] in the most recent evaluation of experimental data. On average the experimental rate constants are 42% lower than the quantal ones; the deviation could be accounted for by the barrier being too thin and the barrier height being about a half kcal (or more) too low on the G3 surface.

In addition, Tab. 1 also includes VTST/OMT results, in particular calculations carried out by improved canonical variational theory [91] with tunneling contributions included by the least-action ground-state approximation [92]. Anharmonicity is included by the WKB method [93] for stretches and by a semiclassical centrifugal oscillator method [94,95] for the bend. The average absolute deviation of the VTST/OMT calculations from the quantum scattering calculations is only 10%. This excellent agreement for the most quantum mechanical of the isotopes validates the use of the very inexpensive VTST/OMT method. Thus it is meaningful to use the VTST/OMT method for calculating KIEs.

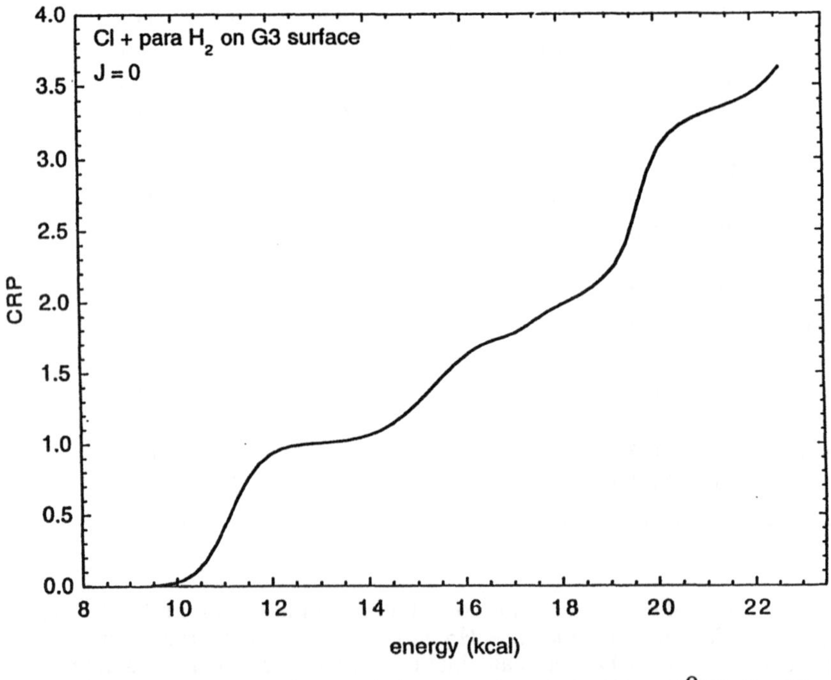

Figure 1: Accurate quantal cumulative reaction probability $N^0(E)$ for Cl + para-H_2 $\rightarrow$ HCl + H on the G3 surface.

Table 2 compares the present results to experiment for the KIEs. In this table, AB denotes Cl + AB $\rightarrow$ ACl + B, and BA denotes Cl + BA $\rightarrow$ BCl + A. Three sets of experimental data are shown: the collection of results from Stern, Persky, and Klein [49] and two sets of k_{H_2}/k_{D_2} ratios: the results of Miller and Gordon [96] and the recommendation of Michael and coworkers [28]. The present results are given in the VTST/OMT column, followed by two sets of calculations based on conventional TST without tunneling. All the calculated results are based on the G3 surface; two of the calculations include anharmonicity as described above, and the final column shows conventional TST with no tunneling and harmonic partition functions, just to provide an old-fashioned point of reference. First of all we see that anharmonicity tends to decrease the predicted KIEs while more accurate inclusion of dynamics tends to increase them. The overall conclusion is that the present results may overestimate the tunneling, especially if we accept the most recent experimental results. This may be caused by the barrier being too thin, which is quite possibly a remnant of using the LEPS form as part of the fitting function, since the LEPS form seems to give barriers that are too thin, as discussed elsewhere [97].

The HD/DH results in Table 2 are especially worth emphasizing since this KIE is a very sensitive test of potential energy surfaces. In a previous survey [51] of eleven potential energy surfaces the calculated values for k_{HD}/k_{DH} at 300 K were, in chronological order of the development of the surfaces, 0.2, 0.2, 2.2, 1.8, 3.6, 2.6, 3.0, 0.3, 4.4, 3.1, and 1.9. Only three of these are within 40% of the experimental 1.8. Thus attaining a value within 40% of experiment can serve as a criterion of good quality.

Figure 1 shows the quantum scattering theory calculations of the cumulative reaction probability (CRP), $N^0(E)$, for Cl + para-H_2 for total angular momentum zero. The CRP is the sum of all state-to-state reaction probabilities at a given total energy [78–80], and it is directly related to the rate constant $k^0(E)$ for a microcanonical ensemble with zero total angular momentum at the total energy E. In particular

$$k^0(E) = [h\rho^R(E)]^{-1}N^0(E) \tag{1}$$

where h is Planck's constant, and $\rho^R(E)$ is the reactant density of states, which carries no dynamical information. One advantage of modern quantum scattering theory as compared to experiment is that we can look at $k^0(E)$ and $N^0(E)$ which are impractical to *measure* for bimolecular reactions. The advantage of being able to do this is that $N^0(E)$ is more directly related to the structure and dynamical properties of the transition state than is the usual canonical-ensemble rate constant $k(T)$. In fact the step structure faintly observable in Fig. 1 is a direct consequence of the quantized nature of the transition state [80,98,99]. This structure is brought out more clearly in the derivative curve $\rho^0(E) = dN^0/dE$; $\rho^0(E)$ is called the density of reactive states [80,98,99] and may be considered to represent the Holy Grail of chemistry—the spectrum of the transition state. The density of reactive states for Cl + para H_2 is shown in Fig. 2, and it shows several quantized states clearly. The assignment of these states is most easily made by studying the vibrationally adiabatic potential energy curves defined by [91]:

$$V_a(v_1,v_2,K,s) = V_{MEP}(s) + \varepsilon(v_1,v_2,K,s) \tag{2}$$

where $V_{MEP}(s)$ is the Born-Oppenheimer potential energy along the minimum energy path (MEP) as a function of the reaction coordinate s, $\varepsilon(v_1,v_2,s)$ are the vibrational

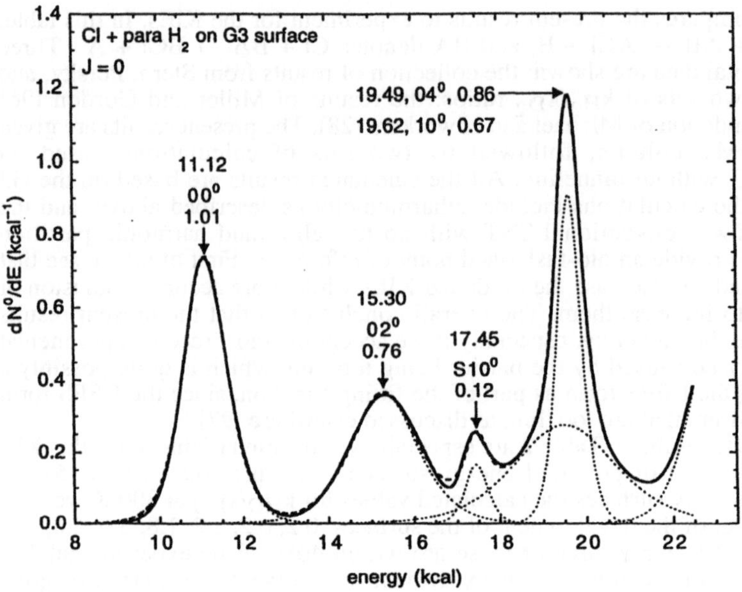

Figure 2: Density of reactive states, $\rho^0(E)$, for Cl + para-H_2 with $J = 0$ on the G3 surface. Heavy solid curve: accurate quantal; heavy dashed curve: fit; light dashed curve: components of the fit. The energies, state assignments, and transmission coefficients of the quantized transition state energy levels are indicated next to arrows at the energies of the peaks in the components of the fit.

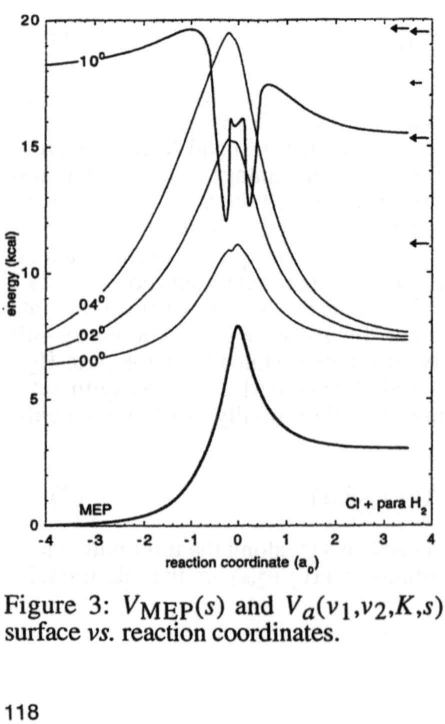

Figure 3: $V_{MEP}(s)$ and $V_a(v_1,v_2,K,s)$ for Cl + para-H_2 → HCl + H on the G3 surface *vs.* reaction coordinates.

energies of the generalized transition states, ν_1 is the stretching quantum number for the mode that adiabatically transforms from an H-H stretch to a quasisymmetric stretch of the Cl-H-H transition state to an H-Cl stretch, and ν_2 is the bending quantum number of the transition state. As usual [87], the quantum numbers are displayed as $\nu_1\nu_2^K$. The relation of the adiabatic energies to transition state theory has a long history, dating back to Hirschfelder and Wigner [100], and it was eventually clarified in 1979 by the proof that the adiabatic theory of reactions [101] is identical to microcanonical variational transition state theory [102]. Thus the maxima of the vibrationally adiabatic curves are identified with dynamical bottlenecks.

The density of reactive states was fit to a sum of contributions with line shapes [80,98] corresponding to parabolic effective barriers. The individual terms in the fit are shown in Fig. 2, and the area under each term corresponds to the transmission coefficient for an individual quantized level of the transition state [80,98]. To interpret the fit, we calculated the vibrationally adiabatic curves and scaled the vibrational energies so that the energies of the maxima agree with the energy levels obtained from the fits; these curves are shown in Fig. 3. The interpretation of the CRP is now very clear: The reaction threshold at 11 kcal corresponds to the ground state of the transition state (TS), which is a nearly ideal bottleneck since $\kappa \approx 1$. (The tail extending to lower energy is due to tunneling as are the low-energy sides of *all* the peaks.) At 15.3 kcal, the first excited state of the TS is accessed, which allows more flux to pass. However, although motion is locally vibrationally adiabatic near the TS, the curvature of the reaction path causes some transitions to $\nu_1 = 1$, and reflection occurs when the systems hits the local maximum

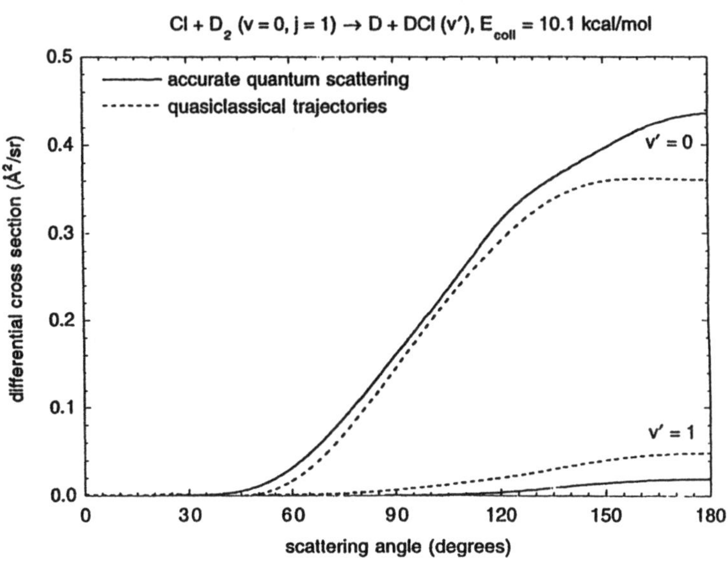

Figure 4: Differential cross section vs. scattering angle for $Cl + D_2(\nu = 0, j = 1) \rightarrow DCl(\nu') + D$ as a function of scattering angle. The results from quantum scattering theory are compared to those from quasiclassical trajectories for a collision energy of 10.1 kcal.

in $V_a(1,0,0,s)$ at $s \approx 0.6\ a_0$. At 17.4 kcal, this local bottleneck is surmounted, and the reflection mechanism closes down, but the CRP can only rise to &2 because the state count of 2 at the variational transition state near $s = 0$ is still globally limiting the flux. At about 19.6 kcal one encounters both the 04^0 state and the global maximum of the 10^0 curve. The latter is the 10^0 variational transition state, and the lower-energy 10^0 feature is a supernumerary transition state of the first kind, in the classification presented elsewhere [103]. The lifetimes of the dynamical bottlenecks in a quantum mechanical world (calculated from their widths in the energy domain) are consistent with the shapes of the vibrationally adiabatic curves: 14, 9, and 7 fs for peaks 1, 2, and 4, respectively, as the bending excitation makes the effective barriers thinner, and 26–29 fs for peaks 3 and 5 associated with broader barriers.

4 Cross Sections

Another valuable use of accurate quantum dynamics calculations is testing the validity of classical simulations for predicting product-state distributions, and reduced-dimensionality studies of this issue are available for both Cl + H_2 [67] and H + Cl_2 [104]. In the present case extensive quasiclassical trajectory (QCT) calculations have been carried out for the full-dimensional Cl + D_2 reaction by Aoiz and Bañares [105]. An example of how the QCT results compare to the accurate quantum ones is given in Fig. 4, which shows differential cross sections for Cl + $D_2(v{=}0,j{=}1) \rightarrow DCl(v') + D$, where v and v' are initial and final vibrational quantum number, respectively, j is initial rotational quantum number, and the results are summed over final rotational quantum number j'. The comparison in Fig. 4 is for an initial relative translational energy of 10.1 kcal. The agreement is quite good. Notice, however, that the QCT method overestimates the amount of vibrationally excited product.

Finally, we also note that *preliminary* comparison to Cl + H_2 molecular beam results [48] at low energy (6 kcal) shows good agreement for both angular and time-of-flight distributions of the products [106].

5 Acknowledgements

The assistance of Joan Gordon is gratefully acknowledged. The quantum scattering calculations were supported in part by the National Science Foundation under grant no. CHE94-23927, the VTST work was supported in part by the U.S. Department of Energy, Office of Basic Energy Sciences under grant no. FG02-86ER13579 and the electronic structure calculations were supported in part by a grant from the Air Force Office of Scientific Research (F49620-95-1-0077).

6 References

[1] M. Bodenstein and S. C. Lind: Z. phys. Chem. **57**, 168 (1906).
[2] J. A. Christiansen: K. Dan. Vidensk. Selsk. Mat. Phys. Medd. **1**, 14 (1919).
[3] K. F. Herzfeld: Ann. Phys. (Leipzig) **59**, 635 (1919).
[4] M. Polanyi: Z. für Elektrochem. **26**, 50 (1920).

[5] L. S. Kassel: ***The Kinetics of Homogeneous Gas Reactions*** (Chemical Catalog Co., New York, 1932), pp. 237–248, 267–276.
[6] N. Semenoff: ***Chemical Kinetics and Chain Reactions*** (Oxford University Press, London, 1935), pp. 89–122, 140–153.
[7] K. J. Laidler: ***Chemical Kinetics***, 3rd ed. (Harper & Row, New York, 1987), pp. 14, 288–298.
[8] J. W. Moore and R. G. Pearson: ***Kinetics and Mechanism***, 3rd ed. (John Wiley & Sons, New York, 1981), pp. 3, 390–394.
[9] G. G. Hammes: ***Principles of Chemical Kinetics*** (Academic, New York, 1978), pp. 76–79.
[10] G. C. Fettis and J. H. Knox: Prog. Reaction Kinetic. **2**, 1 (1964).
[11] M. Bodenstein and W. Dux: Z. phys. Chem. **85**, 297 (1913).
[12] M. Bodenstein: Z. für Elektrochem. **85**, 329 (1913).
[13] M. Bodenstein: Z. für Elektrochem. **22**, 52 (1916).
[14] W. Nernst: Z. für Elektrochem. **24**, 335 (1918).
[15] H. Eyring and M. Polanyi: Z. phys. Chem. **B12**, 279 (1931).
[16] A. Wheeler, B. Topley, and H. Eyring: J. Chem. Phys. **4**, 178 (1936).
[17] S. Sato: J. Chem. Phys. **23**, 2465 (1955).
[18] C. A. Parr and D. G. Truhlar: J. Phys. Chem. **75**, 1844 (1971).
[19] R. D. Levine and R. B. Bernstein: ***Molecular Reaction Dynamics and Chemical Reactivity*** (Oxford University Press, New York, 1987), pp. 396–411.
[20] J. H. Sullivan: J. Chem. Phys. **30**, 1292 (1959).
[21] J. G. A. Griffiths and R. G. W. Norrish: Proc. Roy. Soc. Lond., Ser. A **147**, 140 (1934).
[22] D. L. Chapman and L. K. Underhill: J. Chem. Soc. **103**, 496 (1913).
[23] R. G. W. Norrish and M. Ritchie: Proc. Roy. Soc. Lond. Ser. A **140**, 713 (1933).
[24] M. Bodenstein: Trans. Faraday Soc. **27**, 413, 427 (1931).
[25] M. Ritchie and R. G. W. Norrish: Proc. Roy. Soc. Lond. Ser. A **140**, 112 (1933).
[26] J. C. Miller and R. J. Gordon: J. Chem. Phys. **75**, 5305 (1981).
[27] W. H. Rodebush and W. C. Klingenhoefer: J. Amer. Chem. Soc. **55**, 130 (1933).
[28] S. S. Kumaran, K. P. Lim, and J. V. Michael: J. Chem. Phys. **101**, 9487 (1994).
[29] N. Semenoff: Phys. Z. d. Sowjetunion **1**, 718 (1932).
[30] Padoa and Buttironi: Gazz. Chim. Ital. (II) **47**, 6 (1917).
[31] F. Porter, D. C. Bardwell, and S. C. Lind: J. Amer. Chem. Soc. **48**, 2603 (1926).
[32] S. C. Lind and R. Livingston: J. Amer. Chem. Soc. **52**, 593 (1930).
[33] E. Hertel: Z. phys. Chem. B **15**, 325 (1932).
[34] G. K. Rollefson: J. Chem. Phys. **2**, 144 (1934).
[35] J. Bigeleisen, F. S. Klein, R. E. Weston, Jr., and M. Wolfsberg: J. Chem. Phys. **30**, 1340 (1959).
[36] G. Chiltz, R. Eckling, P. Goldfinger, G. Huybrechts, H. S. Johnson, L. Meyers, and G. Verbeke: J. Chem. Phys. **38**, 1053 (1963).
[37] A. Persky and F. S. Klein: J. Chem. Phys. **44**, 3617 (1966).

[38] Y. Bar Yaakov, A. Persky, and F. S. Klein: J. Chem. Phys. **59**, 2415 (1973).
[39] F. S. Klein and I. Veltman: J. Chem. Soc. Faraday Trans. II **74**, 17 (1978).
[40] R. E. Weston, Jr.: J. Phys. Chem. **83**, 61 (1979).
[41] J. C. Miller and R. J. Gordon: J. Chem. Phys. **78**, 3713 (1983).
[42] D. G. Truhlar, F. B. Brown, D. W. Schwenke, R. Steckler, and B. C. Garrett: in ***Comparison of Ab Initio Quantum Chemistry with Experiment for Small Molecules***, edited by R. J. Bartlett (Reidel, Dordrecht, 1985), p. 95.
[43] R. K. Boyd: Chem. Rev. **77**, 93 (1977).
[44] C. Lim and D. G. Truhlar: J. Chem. Phys. **79**, 3296 (1983).
[45] J. D. McDonald and D. R. Herschbach: J. Chem. Phys. **62**, 4740 (1975).
[45] C. A. Wight, F. Magnotta, and S. R. Leone: J. Chem. Phys. **81**, 3951 (1984).
[47] V. J. Barclay, B. A. Collings, J. C. Polanyi, and J. H. Wang: J. Phys. Chem. **95**, 2921 (1991).
[48] M. Alagia, N. Bulacani, P. Casavecchia, D. Stranges, and G. G. Volpi: J. Chem. Soc. Faraday Trans. **91**, 575 (1995).
[49] M. J. Stern, A. Persky, and F. S. Klein: J. Chem. Phys. **58**, 5697 (1973).
[50] M. Baer and I. Last: in ***Potential Energy Surfaces and Dynamics Calculations***, edited by D. G. Truhlar (Plenum, New York, 1981), p. 519.
[51] S. C. Tucker, D. G. Truhlar, B. C. Garrett, and A. D. Isaacson: J. Chem. Phys. **82**, 4102 (1985).
[52] F. B. Brown and D. G. Truhlar: Chem. Phys. Lett. **117**, 307 (1985).
[53] D. W. Schwenke, S. C. Tucker, R. Steckler, F. B. Brown, G. C. Lynch, and D. G. Truhlar, J. Chem. Phys. **90**, 3110 (1989).
[54] T. C. Allison, S. L. Mielke, G. C. Lynch, D. G. Truhlar, and M. S. Gordon, to be published.
[55] S. L. Mielke, G. C. Lynch, D. G. Truhlar, and D. W. Schwenke: Chem. Phys. Lett. **213**, 10 (1993), **217**, 173(E) (1994).
[56] G. C. Lynch, D. G. Truhlar, F. B. Brown, and J.-g. Zhao: J. Phys. Chem. **99**, 207 (1995).
[57] M. Salomon: Int. J. Chem. Kinet. **2**, 175 (1984).
[58] D. L. Thompson, H. H. Suzukawa, Jr., and L. M. Raff: J. Chem. Phys. **62**, 4727 (1975).
[59] A. Persky: J. Chem. Phys. **66**, 2932 (1977)
[60] A. Persky: J. Chem. Phys. **68**, 2411 (1978).
[61] A. Persky: J. Chem. Phys. **70**, 3910 (1979).
[62] T. Valencich, J. Hsieh, J. Kwan, T. Stewart, and T. Lenhardt: Ber. Bunsenges. Phys. Chem. **81**, 131 (1977).
[63] A. Persky and M. Broida: J. Chem. Phys. **84**, 2653 (1986).
[64] P. M. Aker and J. J. Valentini: Israel J. Chem. **30**, 157 (1990).
[65] M. Baer: Mol. Phys. **27**, 1429 (1974).
[66] A. Persky and M. Baer: J. Chem. Phys. **60**, 133 (1974).
[67] M. Baer, U. Halavee, and A. Persky: J. Chem. Phys. **61**, 5122 (1974).
[68] B. C. Garrett, D. G. Truhlar, and A. W. Magnuson: J. Chem. Phys. **74**, 1029 (1981).
[69] B. C. Garrett, D. G. Truhlar, and A. W. Magnuson: J. Chem. Phys. **76**, 2321 (1982).
[70] D. C. Clary: Chem. Phys. Lett. **80**, 271 (1981).

[71] J. M. Launay and S. B. Padkjær: Chem. Phys. Lett. **181**, 95 (1991).
[72] S. E. Branchett, S. B. Padkjær, and J. M. Launay: Chem. Phys. Lett. **208**, 523 (1993).
[73] S. Takada, K. Tsuda, A. Ohsaki, and H. Nakamura: Adv. Mol. Vib. Coll. Dyn. **2A**, 245 (1994).
[74] W. J. Hehre, L. Radom, P.v.R. Schleyer, and J. A. Pople: ***Ab Initio Molecular Orbital Theory*** (John Wiley & Sons, New York, 1986), pp. 38–40.
[75] A. D. McLean and G. S. Chandler: J. Chem. Phys. **72**, 5639 (1980).
[76] M. J. Frisch, J. A. Pople, and J. S. Binkley: J. Chem. Phys. **80**, 3265 (1984).
[77] M. S. Gordon and D. G. Truhlar: J. Amer. Chem. Soc. **108**, 5412 (1986).
[78] D. C. Chatfield, D. G. Truhlar, and D. W. Schwenke: J. Chem. Phys. **94**, 2040 (1991).
[79] W. H. Miller: J. Chem. Phys. **62**, 1899 (1975).
[80] D. C. Chatfield, R. S. Friedman, D. G. Truhlar, B. C. Garrett, and D. W. Schwenke: J. Amer. Chem. Soc. **113**, 486 (1991).
[81] Y. Sun, D. J. Kouri, D. G. Truhlar, and D. W. Schwenke: Phys. Rev. A **41**, 4857 (1990).
[82] Y. Sun, D. J. Kouri, and D. G. Truhlar: Nucl. Phys. **A508**, 41c (1990).
[83] D. W. Schwenke, S. L. Mielke, and D. G. Truhlar: Theor. Chim. Acta **79**, 241 (1991).
[84] G. J. Tawa, S. L. Mielke, D. G. Truhlar, and D. W. Schwenke: Adv. Mol. Vib. Coll. Dyn. **2B**, 45 (1994).
[85] S. L. Mielke, G. C. Lynch, D. G. Truhlar, and D. W. Schwenke: Chem. Phys. Lett. **216**, 441 (1993).
[86] S. L. Mielke, G. C. Lynch, D. G. Truhlar, and D. W. Schwenke: J. Phys. Chem. **98**, 8000 (1994).
[87] G. Herzberg: ***Infrared and Raman Spectra of Polyatomic Molecules*** (Van Nostrand, Princeton, 1945), pp. 210–211, 273–276.
[88] J. V. Michael and J. R. Fisher: J. Phys. Chem. **94**, 3318 (1990).
[89] D. G. Truhlar and C. J. Horowitz: J. Chem. Phys. **68**, 2466 (1978), **71**, 1514(E) (1979).
[90] A. J. C. Varandas, F. B. Brown, C. A. Mead, D. G. Truhlar, and N. C. Blais: J. Chem. Phys. **86**, 6258 (1987).
[91] B. C. Garrett, D. G. Truhlar, R. S. Grev, and A. W. Magnuson: J. Phys. Chem. **84**, 1730 (1980).
[92] B. C. Garrett and D. G. Truhlar: J. Chem. Phys. **79**, 4931 (1983).
[93] B. C. Garrett and D. G. Truhlar: J. Chem. Phys. **81**, 309 (1984).
[94] G. A. Natanson: J. Chem. Phys. **93**, 6589 (1990).
[95] B. C. Garrett and D. G. Truhlar: J. Phys. Chem. **95**, 10374 (1991).
[96] J. C. Miller and R. J. Gordon: J. Chem. Phys. **79**, 1252 (1983).
[97] D. G. Truhlar and R. E. Wyatt: Adv. Chem. Phys. **36**, 141 (1977).
[98] D. C. Chatfield, R. S. Friedman, D. W. Schwenke, and D. G. Truhlar: J. Phys. Chem. **96**, 2414 (1992).
[99] D. C. Chatfield, R. S. Friedman, S. L. Mielke, D. W. Schwenke, G. C. Lynch, T. C. Allison, and D. G. Truhlar: in ***Dynamics of Molecules and Chemical Reactions,*** edited by R. E. Wyatt and J. Z. H. Zhang (Dekker, New York), in press.
[100] J. O. Hirschfelder and E. Wigner: J. Chem. Phys. **7**, 616 (1939).

[101] D. G. Truhlar: J. Chem. Phys. **53**, 2041 (1970).
[102] B. C. Garrett and D. G. Truhlar: J. Phys. Chem. **83**, 1079 (1979), **87** 4553(E) (1983).
[103] D. C. Chatfield, R. S. Friedman, G. C. Lynch, D. G. Truhlar, and D. W. Schwenke: J. Chem. Phys. **98**, 342 (1993).
[104] D. G. Truhlar, J. A. Merrick, and J. W. Duff: J. Chem. Phys. **98**, 6771 (1976).
[105] F. J. Aoiz and L. Bañares: unpublished.
[106] P. Casavecchia: private communication, and in this volume.

Cross Sections and Rate Constants for Triatomic and Tetraatomic Reactions: *Three-Dimensional Quantum Mechanical Calculations*

M. Baer, and H. Szichman
Department of Physics and Applied Mathematics, Soreq NRC,
Yavne 81800, Israel

E. Rosenman, S. Hochman-Kowal, and A. Persky
Department of Chemistry, Bar-Ilan University,
Ramat-Gan 52900, Israel

Abstract

We are presenting a short review on our recent quantum mechanical calculations of reactive cross sections and rate constants. Seven different cases are considered: The two triatom systems, namely $F + H_2$ and $F + D_2$, were treated twice employing two different potential energy surfaces. The calculations were carried out within the coupled states approximation. Two atom-triatom reactions were considered: the $O + O_3$ reaction, for which the calculations were done employing the quasi-breathing-sphere approximation, and the $H + H_2O$ reaction for which the calculations were done employing the (cruder) infinite-order-sudden-approximation. One diatom-diatom system, namely $H_2 + OH$, was considered. For this system the calculations were carried out twice: once employing the infinite-order-sudden-approximation and once employing a mixed version of the coupled states approximation (with regard to the "reactive angle") and the infinite order sudden approximation (with regard to the other two angles). All calculations were carried out in the reagents' arrangement channel employing negative imaginary arrangement-decoupling potentials. In general very encouraging agreement with experiment was obtained.

1 Introduction

Temperature-dependent rate constants, (RCs), are fundamental parameters to be used in numerical simulations of macroscopic chemical processes such as combustion, distillation, atmospheric chemistry etc. [1-5]. Consequently, measuring RCs, although being very difficult and not always rewarding, is one of the most important tasks many experimental chemists have been carrying out with a lot of devotion for several decades [6-8].

Over the years, in parallel, a different way to obtain RCs has been making progress, namely deriving them numerically, from first principles, employing accurate quantum mechanical (QM) methods. In general there are several approa-

Springer Series in Chemical Physics, Volume 61
Gas Phase Chemical Reaction Systems
Eds.: J. Wolfrum, H.-R. Volpp, R. Rannacher, and J. Warnatz

ches to calculate RCs; these range from the approximate transition state [9,10] and statistical [11,12] theories, through the quasi-classical trajectory (QCT) [13] and the different semi-classical approaches [14,15] ending with quantum mechanical (QM) treatments [16-18]. Until recently the QM approach, unless applied to the simplest systems such as H + H_2 or its isotopic analogues, was employed following crude approximations. However, in recent years significant progress has been made in developing new methods [17-20] to treat such problems and these, together with the newly advanced means of computation, made the numerical calculation of RCs a reliable way to obtain the much required temperature-dependent RCs. In this short review we will describe our QM approach to perform such calculations and present results, cross sections (CSs) and RCs for triatom and tetra-atom exchange processes.

Two triatom systems will be treated here:

$$\left.\begin{array}{l} F + H_2 \rightarrow HF + H \\ F + D_2 \rightarrow DF + D \end{array}\right. \qquad \Delta H_0 = -1.23\ \text{eV}, \qquad \text{(I)}$$

and results will be presented for two different potential energy surfaces (PES); the one is the 6SEC PES [21] and the other is the Stark-Werner (S-W) PES [22-24]. It is important to mention that the two PESs are based on *ab initio* points.

We will consider three tetra-atom systems: two atom-triatom systems namely the

$$O + O_3 \rightarrow 2\,O_2 \qquad \Delta H_0 = -4.1\ \text{eV} \qquad \text{(II)}$$

and the

$$H + H_2O \rightarrow H_2 + OH \qquad \Delta H_0 = 0.81\ \text{eV} \qquad \text{(III)}$$

reactions and one diatom-diatom system namely the

$$H_2 + OH \rightarrow H_2O + H \qquad \Delta H_0 = -0.81\ \text{eV} \qquad \text{(IV)}$$

reaction. The calculations for the O_4 system were carried out on a PES obtained from the double-many-body-expansion method (DMBE) [25], whereas the calculations for the HHOH system were performed on a "semi-*ab initio*" PES [26].

In what follows we first present the general theory, next describe, briefly, the approach we applied in each case, then discuss the energy-dependent CSs and finally present the temperature-dependent RCs.

2 Theory

The approach to be applied has three basic components [27]:

(a) Application of the perturbative-type Schrödinger equation (SE) to treat nonreactive collisions taking place in the reagents arrangement channel (AC) [27a]:

$$(E - H)\,\chi_{t_o} = V_{t_o}\,\psi_{t_o} \qquad (1)$$

Here E is the total energy, H is the hamiltonian, χ_{t_o} is the short range part of the total wave function Ψ_{t_o}, V_{t_o} is a t_o dependent perturbation where t_o (and t) stands for the set of relevant quantum numbers and ψ_{t_o}, the unperturbed part of Ψ_{t_o} (thus, $\Psi_{t_o} = \chi_{t_o} + \psi_{t_o}$) representing essentially the asymptotic incoming wave, is a solution of the unperturbed SE:

$$(E - H_0)\,\psi_{t_0} = 0 \tag{2}$$

(b) Conversion of Eq. (1), which treats nonreactive processes, into an equation that treats reactive processes. This is achieved by adding to the Hamiltonian H in Eq. (1), in an *ad hoc* way, a negative imaginary potential (NIP) [28] in such a manner as to decouple all the reactive ACs.
(c) Conversion of the scattering problem into a bound system problem by introducing one additional NIP in the (reagents') asymptotic region, thus permitting the expansion of χ_{t_0} in terms of L^2 basis sets [27].

Throughout the whole present study we used the Neuhauser-Baer linear ramp NIP given in the form [28]:

$$U_I(r) = \begin{cases} -iU_{0I}\dfrac{r - r_0}{\Delta r} & ;\ r_0 \le r \le r_0 + \Delta r \\ 0 & ;\ \text{otherwise} \end{cases} \tag{3}$$

where U_{0I}, the height of the NIP, and Δr, its width, are chosen in such a way as to ensure that reflection will be negligibly small; the value of r_0 is chosen large enough to ensure that the NIP does not interfere in the processes taking place in the strong interaction region.

The aim of the calculation is to obtain the following (nonreactive) S-matrix elements:

$$S(t \leftarrow t_0) = \left(\delta_{tt_0} + \frac{m}{i\hbar^2}\left\langle \psi_t \middle| V_t \middle| \Psi_{t_0} \right\rangle\right)\exp(i\varphi_t) \tag{4}$$

where δ_{tt_0} is the Kronecker delta function and φ_t is the t-th (elastic) phase shift related to ψ_t. These matrix elements, once derived as a function of various initial conditions, are used to calculate the reactive, state selected, total cross sections (CS) according to the following formula:

$$\sigma(n_0 j_0) = \frac{\pi}{k^2(n_0 j_0)(2j_0 + 1)} \sum_{|\Omega| \le j^0} \sum_{J \ge |\Omega|} (2J + 1)\left\{1 - \sum_n \sum_{j \ge |\Omega|} \left|S(nj \leftarrow n_0 j_0 | J\Omega)\right|^2\right\} \tag{5}$$

where n (n_0) and j (j_0) are vibrational and rotational quantum numbers, J and Ω are the total angular momentum and its (body-fixed) z-component, $k(n_0 j_0)$ is the intial wave vector and $S(nj \leftarrow n_0 j_0 | J\Omega)$ stands for $S(t \leftarrow t_0)$.

The advantage of performing the calculations only in the reagents AC becomes even more apparent when approximate treatments are applied. The various approximations may yield reliable results for nonreactive systems [29], namely single AC-systems, but they may become less reliable when applied to cases that require considering more than one AC [30,31]. In the present study, we applied several approximations that were found in many nonreactive studies to be reliable

[29] (depending on the type of the approximation and the particular system) even for some of the more detailed magnitudes.

3 The Triatom System

The numerical treatment of the triatom reactions (I) was carried out within the coupled states approximation. Consequently the Hamiltonian takes the form:

$$H = -\frac{\hbar^2}{2MR}\frac{\partial^2}{\partial R^2}R - \frac{\hbar^2}{2\mu r}\frac{\partial^2}{\partial r^2}r + \mathbf{j}^2\left(\frac{1}{2\mu r^2} + \frac{1}{2MR^2}\right) +$$
$$\frac{\hbar^2}{2MR^2}\left[J(J+1) - 2\Omega^2\right] + U(Rr\gamma) \tag{6}$$

Here R and r are the translational and the vibrational coordinates, respectively, γ is the Jacobi angle defined as ($\hat{\mathbf{R}}\cdot\hat{\mathbf{r}}$), μ and M are the reduced masses of the reagent diatomic and the triatom systems, respectively, and $U(Rr\gamma)$ is the potential that governs the motion of the three interacting atoms. More details can be found in Ref. 27c, but here we would like to add a few words regarding $W_t(Rr\gamma)$, the unperturbed potential which is used to calculate the unperturbed function ψ_t. It will be defined as an isotropic separable potential in R and r, namely:

$$W_t(Rr\,\gamma) = w_t\,(R) + v(r) \tag{7}$$

where v(r) is the diatomic potential and $w_t(R)$ is a translational potential defined as:

$$w_t(R) = \langle\phi_t|[U(Rr\gamma) - v(r)]|\phi_t\rangle \tag{8}$$

Here ϕ_t is the t-th vibrational-rotational eigenfunction of the diatomic. Once $W_t(Rr\gamma)$ is introduced we can now also present, more explicitly, the perturbation potential, V_t, mentioned in Eq. (1) and (4).

$$V_t(Rr\gamma) = U(Rr\gamma) - W_t(Rr\gamma) \tag{9}$$

In Figs. 1 and 2 the state-selected total CSs for reactions (I) as a function of translational energy are presented, calculated for different initial rotational states j, namely for j = 0,...,3 in case of H_2 and for j = 0,...,4 in case of D_2. In addition to the coupled states results we also present the QCT CSs [32a] and a few formally accurate (exact) QM results [32b]. Regarding the comparison with the accurate results (which are only available for F + H_2), it is seen that the two QM treatments yield identical results. As for the comparison with the QCT results, the following can be said (for more details see Ref. 32c):

(a) Along most of the energy range the QM (j = 0) and to a certain extent the (j = 1) CSs are larger than the respective QCT ones. This applies to both reactions but the differences in case of F + H_2 are more significant.

(b) Along most of the energy range the QM (j > 1) CSs overlap very nicely with the corresponding QCT ones. The only difference between the two types of the results is seen at the threshold region.

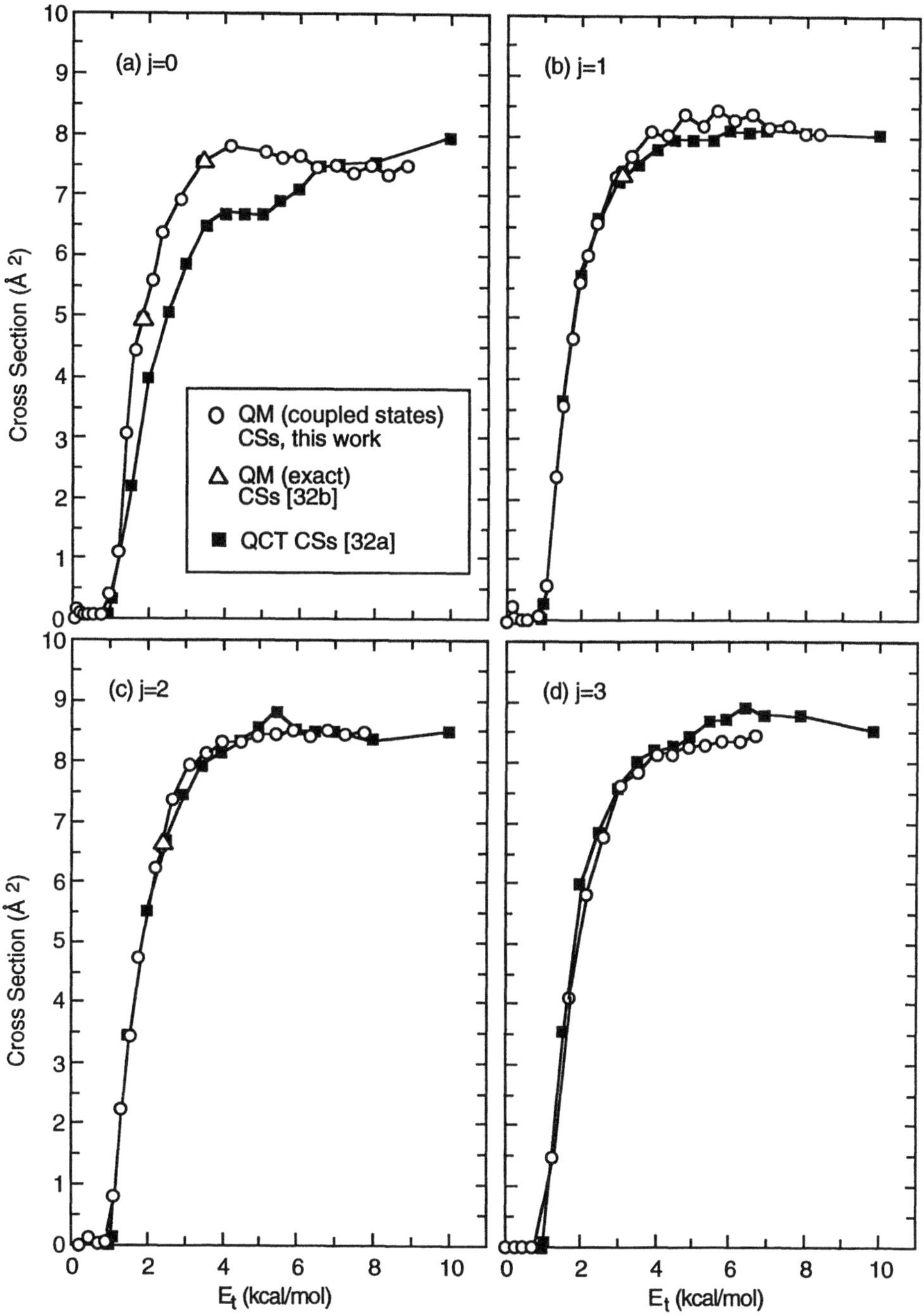

Figure 1: State-selected cross sections (CSs) as a function of translational energy for the F + H_2(v = 0,j) → HF + H reaction calculated on the 6SEC PES.

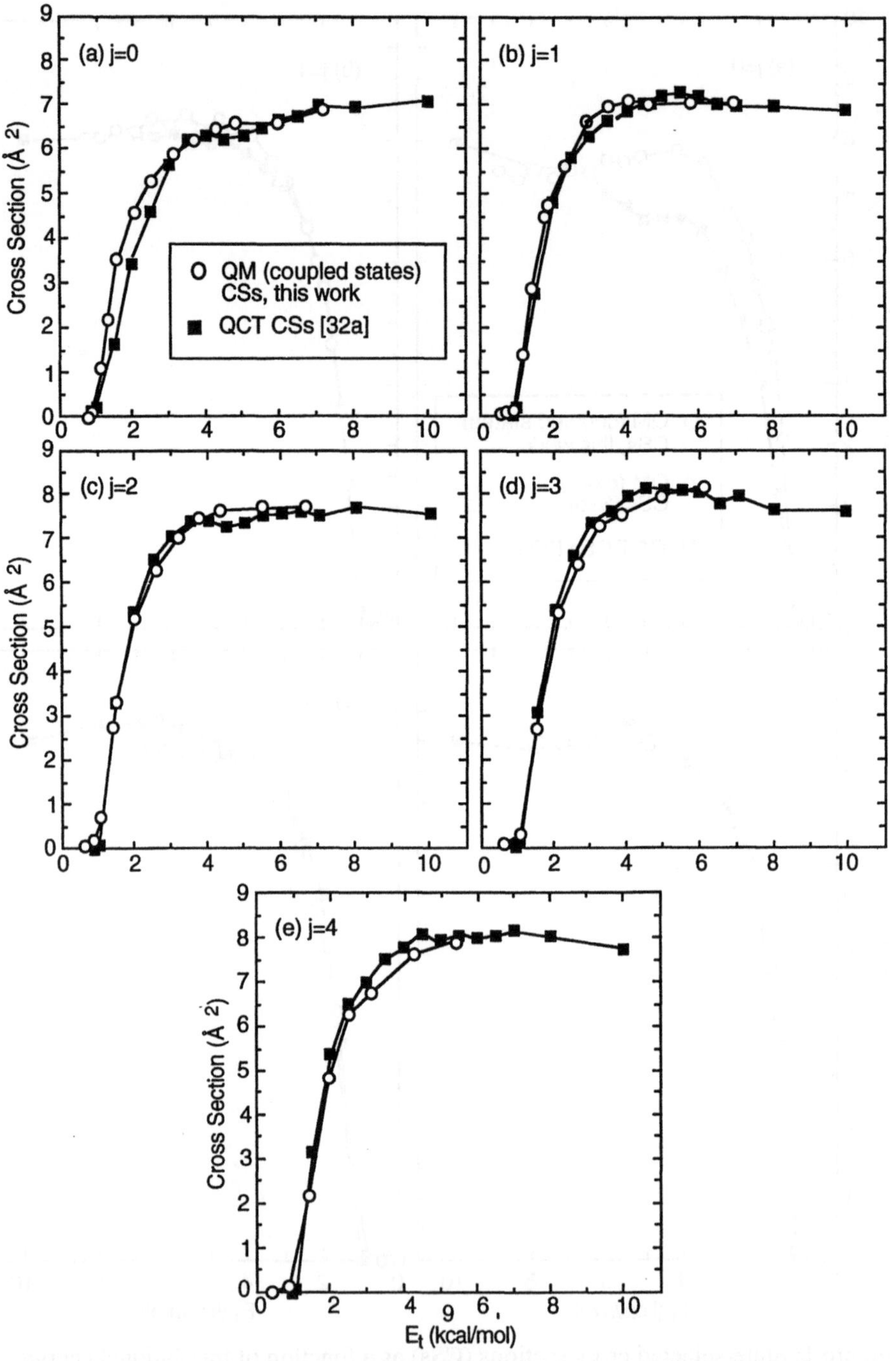

Figure 2: State-selected cross sections (CSs) as a function of translational energy for the $F + D_2(v = 0,j) \rightarrow DF + D$ reaction calculated on the 6SEC PES.

RCs, as a function of (1000/T), are presented in Fig. 3. Here, in addition to the QCT and QM RCs, we show also the experimental ones [7b-d]. Regarding the $F + H_2$ RCs given in Fig. 3a the following is to be noted: for all practical purposes the QM and the QCT RCs are identical, but they are somewhat too low as compared to the experiments – in particular in the low temperature region. As for the $F + D_2$ RCs shown in Fig. 3b, again, the two calculations yield similar results but this time they fit the experimental results quite well.

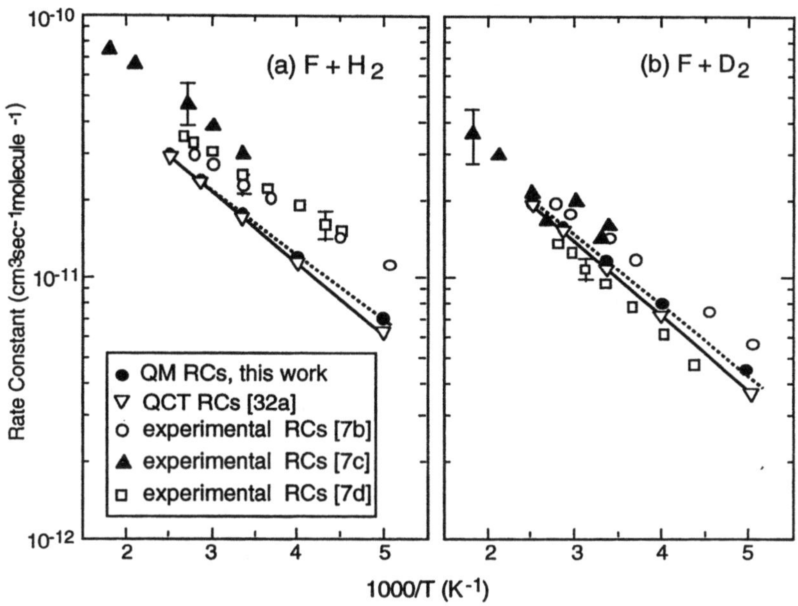

Figure 3: Rate constants (RCs) as a function of 1000/T for the $F + H_2 \rightarrow HF + H$ (a), and for the $F + D_2 \rightarrow DF + D$ (b) reactions calculated on the 6SEC PES.

Calculations for the same two reactions were repeated employing the S-W PES. The energy-dependent CSs are presented in Fig. 4 (for $F + H_2$) and Fig. 5 (for $F + D_2$). Also shown are a few (available) QCT CSs [32d]. The main difference between the S-W and the 6SEC results is that the S-W ones do not exhibit any potential barrier effects; thus the reactive CS curve starts to rise from zero (at a fast rate) at the point at which the kinetic energy becomes non-zero. This behaviour is not only different from anything that is known about this system, but came as a surprise because the PES itself has a barrier. However, it turns out that this barrier, is relatively thin, thus permitting transitions due to tunneling. In Fig. 6 are compared the temperature-dependent S-W RCs with experiment. As can be seen the theoretical results fit the experimental ones for both systems reasonably well although they are somewhat too high, especially at low temperatures.

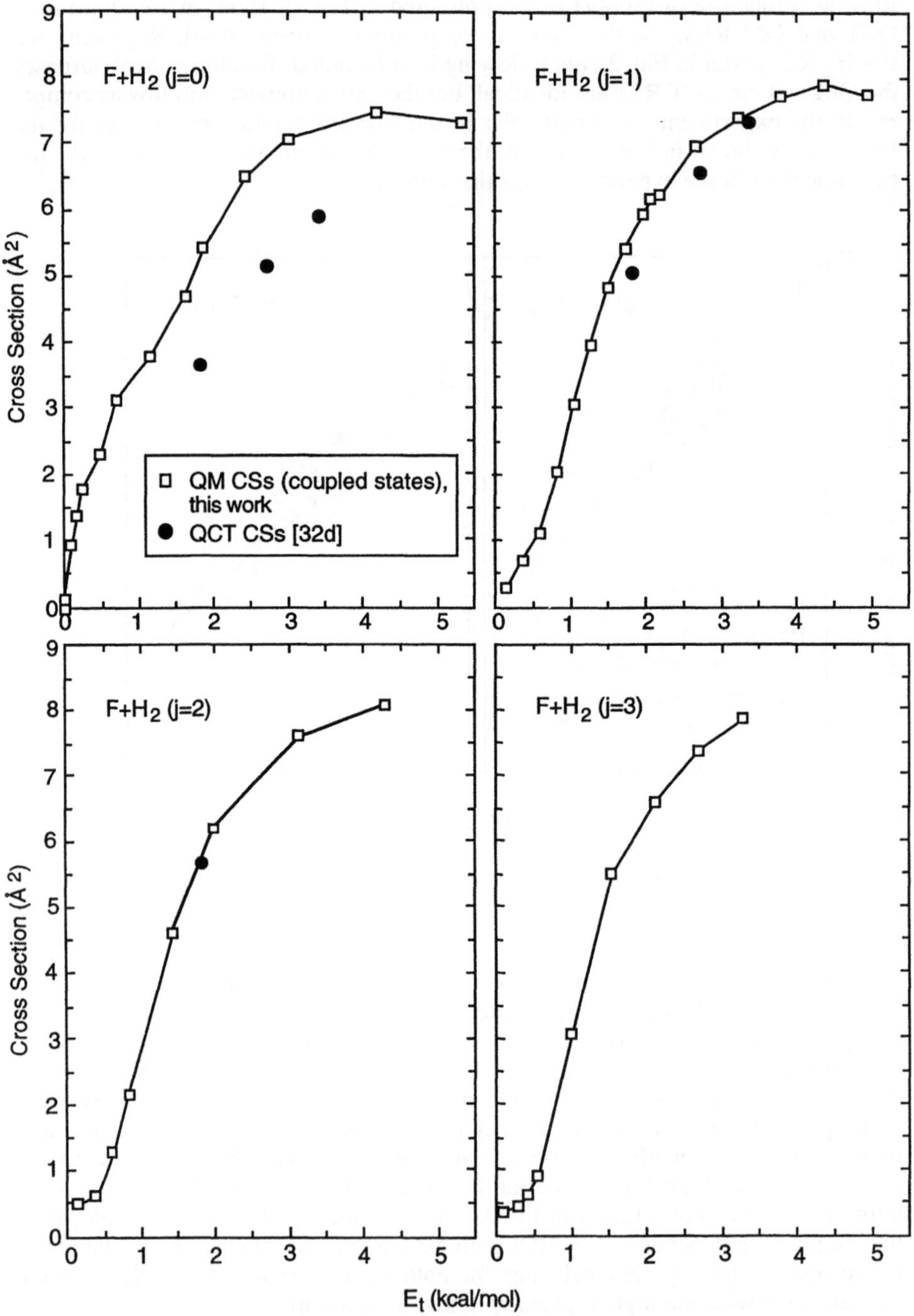

Figure 4: State-selected cross sections (CSs) as a function of translational energy for the $F + H_2(v = 0,j) \rightarrow HF + H$ reaction calculated on the S-W PES.

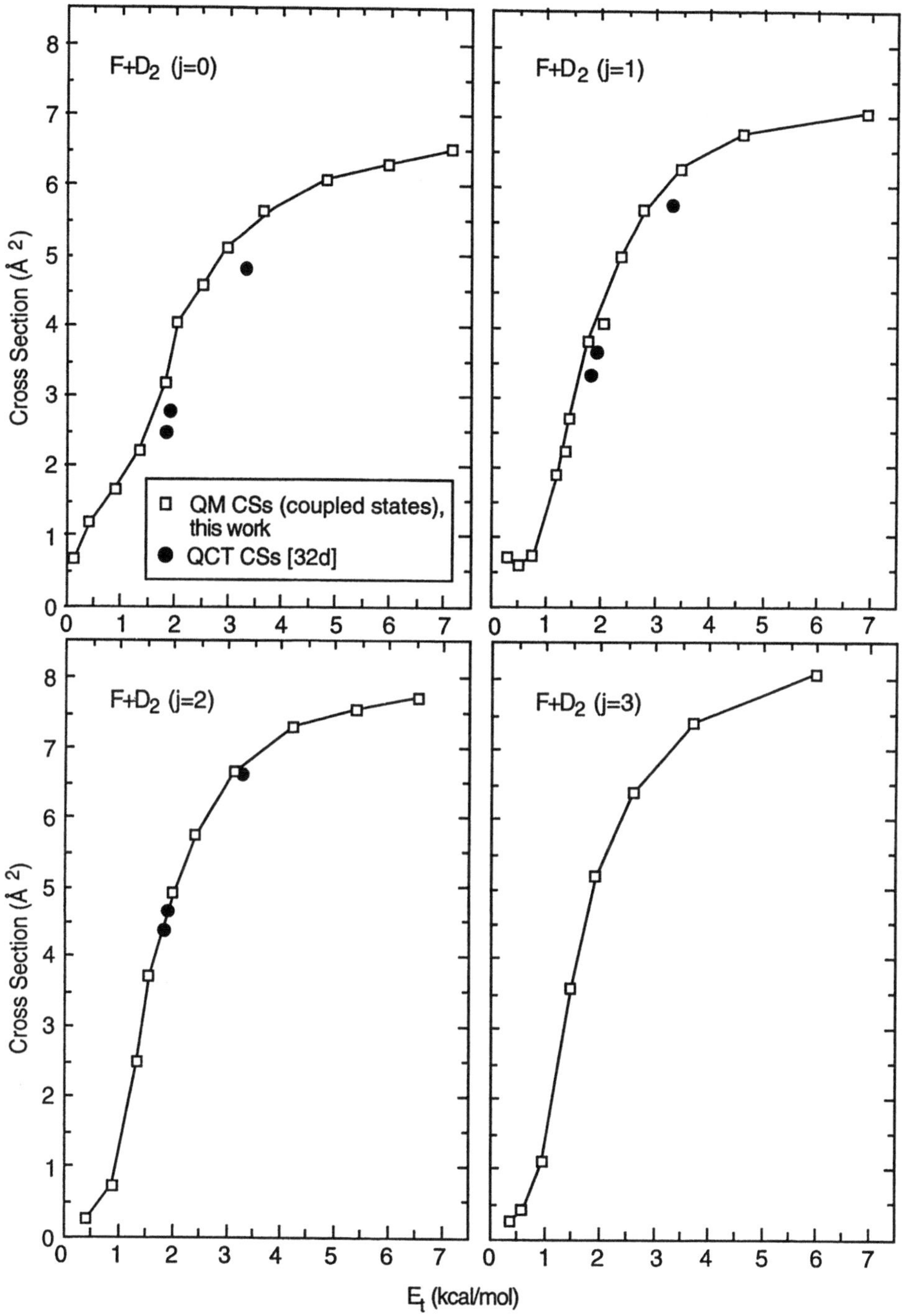

Figure 5: State-selected cross sections (CSs) as a function of translational energy for the $F + D_2(v = 0,j) \rightarrow DF + D$ reaction calculated on the S-W PES.

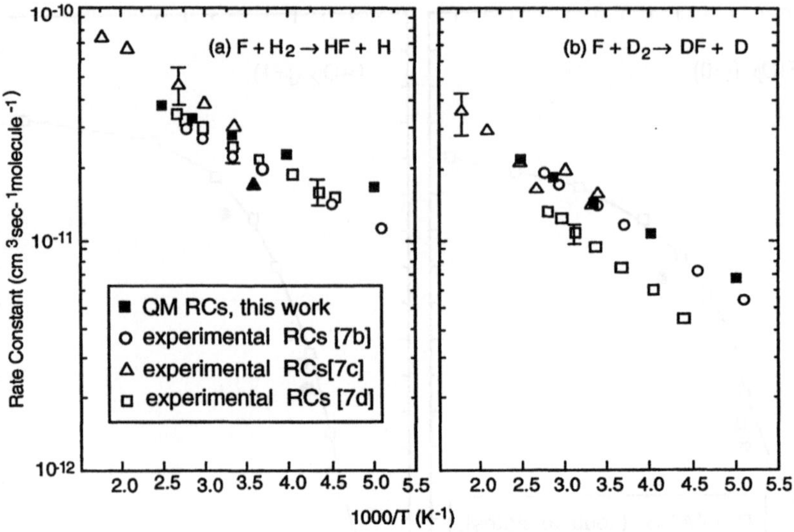

Figure 6: Rate constants (RCs) as a function of 1000/T for the $F + H_2 \rightarrow HF + H$ (a), and for the $F + D_2 \rightarrow DF + D$ (b) reactions calculated on the S-W PES.

4 The Atom-Triatom System

4.1 The $O + O_3$ System

The numerical treatment of reaction (II) is carried out employing what we termed the "quasi-breathing sphere approximation" (QBSA). Within the QBSA we distinguish between two regions in configuration space: (a) the asymptotic region where the numerical treatment is performed employing the coupled states approximation and (b) the close interaction region which is assumed to be short enough so that the interaction takes place without the angles varying significantly; in other words, in this region the numerical treatment is carried out employing the infinite order sudden approximation (IOSA).

The starting point of our theoretical treatment is the following atom-triatom Hamiltonian:

$$H = -\frac{\hbar^2}{2MR}\frac{\partial^2}{\partial R^2}R - \frac{\hbar^2}{2mr}\frac{\partial^2}{\partial r^2}r - \frac{\hbar^2}{2\mu\rho}\frac{\partial^2}{\partial \rho^2}\rho + \frac{\mathbf{j}^2}{2mr^2} + \frac{(\mathbf{K}-\mathbf{j})^2}{2\mu\rho^2} + \frac{\mathbf{l}^2}{2MR^2} + U(Rr\rho\theta\gamma\beta) \qquad (10)$$

where R and M are the translational coordinate and the reduced mass of the atom-triatom system, r and m are the vibrational coordinate and the reduced mass of the non-dissociating diatom, ρ and μ are the distance and the reduced mass of the third atom with respect to the diatom, θ and γ are defined as ($\hat{\boldsymbol{\rho}}\cdot\hat{\mathbf{r}}$) and ($\hat{\mathbf{R}}\cdot\hat{\boldsymbol{\rho}}$)

respectively, β is the out-of-plane angle relating **R** to **ρ**, **K** is the total angular momentum of the triatom molecule, **l** is the orbital angular momentum and $U(Rr\rho\theta\gamma\beta)$ is the potential that governs the motion of the four interacting atoms. The asymptotic region is treated by considering the SE given in Eq. (2). The asymptotic (unperturbed) elastic wave function ψ_t will be calculated employing the following (coupled-states) Hamiltonian:

$$H_0 = -\frac{\hbar^2}{2MR}\frac{\partial^2}{\partial R^2}R - \frac{\hbar^2}{2mr}\frac{\partial^2}{\partial r^2}r - \frac{\hbar^2}{2\mu\rho}\frac{\partial^2}{\partial \rho^2}\rho + \frac{\mathbf{j}^2}{2}\left(\frac{1}{mr^2}+\frac{1}{\mu\rho^2}\right) + \frac{\hbar^2\left[K(K+1)-2\Omega_K^2\right]}{2\mu\rho^2} + \frac{\hbar^2\left[J(J+1)+K(K+1)-2\Omega^2\right]}{2MR^2} + W(Rr\rho\theta) \quad (11)$$

where J is the total angular momentum quantum number, Ω is the its z-component in the body fixed system, Ω_K is the z-component of both J and K in the triatom body fixed system and $W(Rr\rho\theta)$, the (unperturbed) potential, is given in the form

$$W(Rr\rho\theta) = v(r\rho\theta) + w(R). \quad (12)$$

Here $v(r\rho\theta)$, which is defined as:

$$v(r\rho\theta) = \lim_{R\to\infty} U(Rr\rho\theta\gamma\beta) \quad (13)$$

is the potential of the O_3 molecule and w(R), a translational distortion potential, is given in the form

$$w(R) = U(R{:}r_e\rho_e|\theta\gamma\beta) \quad (14)$$

where r_e and ρ_e are equilibrium distances.
To solve for χ_{t_0} – see Eq. (1) – we employ an extended IOSA where the three Jacobi angles θ, γ, and β are considered to be parameters as in the ordinary IOSA, but the Hamiltonian is more general.

Thus:

$$H(Rr\rho|\theta\gamma\beta) = -\frac{\hbar^2}{2MR}\frac{\partial^2}{\partial R^2}R - \frac{\hbar^2}{2mr}\frac{\partial^2}{\partial r^2}r - \frac{\hbar^2}{2\mu\rho}\frac{\partial^2}{\partial \rho^2}\rho + \frac{\hbar^2\left[J(J+1)+K(K+1)-2\Omega^2\right]}{2MR^2} + \frac{\hbar^2\left[K(K+1)-\Omega_K^2\right]}{2\mu\rho^2} + \frac{\hbar^2}{2}j^*\left(j^*+1\right)\left[\frac{1}{mr^2}+\frac{1}{\mu\rho^2}\right] + \hat{U}(Rr\rho|\theta\gamma\beta) \quad (15)$$

where j^* is a parameter (assumed to be zero in the present calculation) and $\hat{U}(Rr\rho|\theta\gamma\beta)$ is the potential defined as a function of the three distances R, r, and ρ, only. To solve for χ_{t_0} we also had to introduce V, the perturbation potential, which is defined as:

$$V(Rr\rho) = \hat{U}(Rr\rho|\theta\gamma\beta) - W(Rr\rho|\theta\gamma) \quad (16)$$

The numerical treatment was detailed elsewhere [33]. Here we shall only mention that the calculated probabilities for each J, K and Ω_K were based on 20 sets of (γ, β) angles selected randomly by the Monte Carlo method. As for the angle θ, it was held fixed and its value was assumed to be equal to the equilibrium angle θ_e throughout the whole numerical treatment.

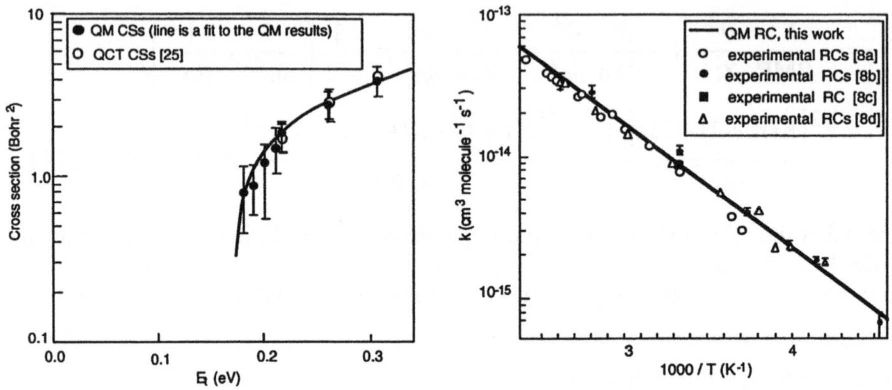

Figure 7: Results for the reaction $O + O_3(v = 0, j = 0, v' = 0, K = 0) \rightarrow 2\,O_2$. Left (a), cross section as a function of translational energy. Right (b), rate constant as a function of 1000/T.

The results are given in Fig. 7. In Fig. 7(a) the QM CSs are presented as a function of translational energy and compared with QCT CSs [25]. An encouraging fit is obtained. It is important to emphasize that since the four atoms taking part in the exchange process are heavy, strong quantum effects are not anticipated. Thus, the nice agreement between the two types of results serves as a probe for the quality of the QM treatment. This significant achievement must be appreciated not only because four heavy atoms are treated in the numerical study, but also because reaction (II) is highly exothermic.

To calculate the QM RCs we first fitted the (translational) energy-dependent QM CSs to a known (relevant) analytical form of the following kind [34]:

$$\sigma(E_t) = \begin{cases} C\left(E_t - E_{t_0}\right)^n \exp\left(-ß\left(E_t - E_{t_0}\right)\right) & ;\ \text{for } E_t \geq E_{t_0} \\ 0 & ;\ \text{for } E_t \leq E_t \end{cases} \qquad (17)$$

C, n, ß and E_{t_0} were determined such that $\sigma(E_t)$ passes, as closely as possible, through the calculated points. More details can be found in ref. [34]. The calculated curve is also shown in Fig. 7(a). Eq. (17) was used to calculate the temperature-dependent RCs which are presented in Fig. 7(b). Also presented are experimental results due to several groups [8]. As can be seen a nice fit was obtained between the QM and the experimental RCs.

4.2 The H + H_2O System

In contrast to the previous case, the reaction H + H_2O was treated within the IOSA. The main reason is that we could not get well-converged results within the QBSA. The IOSA Hamiltonian assumed for this case is [35]:

$$H = -\frac{\hbar^2}{2MR}\frac{\partial^2}{\partial R^2}R - \frac{\hbar^2}{2mr}\frac{\partial^2}{\partial r^2}r - \frac{\hbar^2}{2\mu\rho}\frac{\partial^2}{\partial \rho^2}\rho + \hbar^2\frac{j^*\left(j^*+1\right)}{2mr^2} +$$

$$\hbar^2\frac{K^*\left(K^*+1\right)}{2\mu\rho^2} + \hbar^2\frac{J(J+1)}{2MR} + U^*\left(Rr\rho|\theta\gamma\beta\right) \qquad (18)$$

where j* and K* are the angular quantum numbers of the diatom and the triatom, respectively (both are parameters and in the present calculations they were assumed to be zero) and J is the total angular momentum quantum number. This Hamiltonian was substituted in Eq. (1) to calculate χ_{t_0}. The same Hamiltonian, but where $W(Rr\rho|\theta\,\gamma\beta)$ – see Eq. (12) – replaces $U^*(r\rho R|\theta\,\gamma\beta)$, is employed to calculate the relevant unperturbed (elastic) functions ψ_{t_0} and ψ_t.

The calculations were carried out employing the Walch-Dunning-Schatz-Elgersma PES [26]. As in the O + O_3 case, each calculation was carried out for a given set of angles. These, except for θ (that was assumed to be equal to θ_e, the bending angle of H_2O), were selected quasi-randomly employing the Monte Carlo method. About 50 sets of angles were used to calculate each CS.

The results are shown in Fig. 8. Energy-dependent CSs are presented in Fig. 8(a) where they are compared with QCT [36a], with approximate QM [36b], and with experimental results [37]. It is noticed that, in general, the IOSA CSs are smaller then any of the other ones presented. The closest are Clary's CSs [36b] which, at the low energy region, overlap reasonably well with ours. As for the

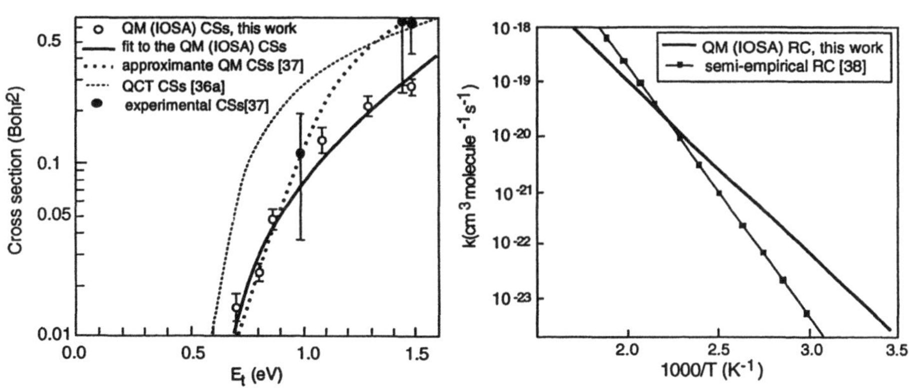

Figure 8: Results for the reaction H + $H_2O(v = 0, j^* = 0, v' = 0) \rightarrow H_2$ + OH. Left (a), cross section as a function of translational energy. Right (b), rate constant as a function of 1000/T.

QCT CSs, they are much larger than the IOSA ones along the whole energy range studied. Considering the experimental CSs, the lower energy value fits our results reasonably well, whereas the higher energy one is about two times larger.

To calculate the IOSA RCs the energy-dependent CSs were fitted to a known analytical form, see Eq. (17). The fit is also shown in Fig. 8(a) and the temperature-dependent RCs are presented in Fig. 8(b). Also given are the extracted rate constants based on experimental measurements of the reverse reaction [38]. As is noticed, the fit between the experimental rate constants and the QM IOSA ones is reasonable.

5 The Diatom-Diatom System

The previous two tetra-atom systems were treated assuming the three Jacobi angles θ, γ and β to be "frozen" either throughout the whole interaction region (within the IOSA) or in the close interaction region (within the QBSA). These approximations were found to be inadequate for calculating low energy CSs (and consequently also low temperature RCs) in case of reaction (IV). The IOSA CSs for this reaction were found to be reasonably close to the coupled states CSs (considered, so far, to be the most accurate ones) as long as the kinetic energy is not too low – see Fig. 9(a) – but then its rate of decrease as the kinetic energy decreases to zero is too slow, thus yielding CSs and RCs that are much too large. Therefore the IOSA, as discussed above, has to be changed and this is done as follows: the diatom-diatom interaction in the reagents AC is described in terms of the three radial distances R, r_1 (the H-H distance) and r_2 (the O-H distance) and the corresponding three Jacobi angles γ_1, γ_2 and φ. In what follows, each of the three angles will be treated differently. The angle $\gamma_1 = (\hat{\mathbf{R}} \cdot \hat{\mathbf{r}}_1)$ will be considered as an ordinary free coordinate, the angle $\gamma_2 = (\hat{\mathbf{R}} \cdot \hat{\mathbf{r}}_2)$ will be treated, as before, as a frozen IOSA angle and the angle φ will be eliminated altogether by employing a φ ave-

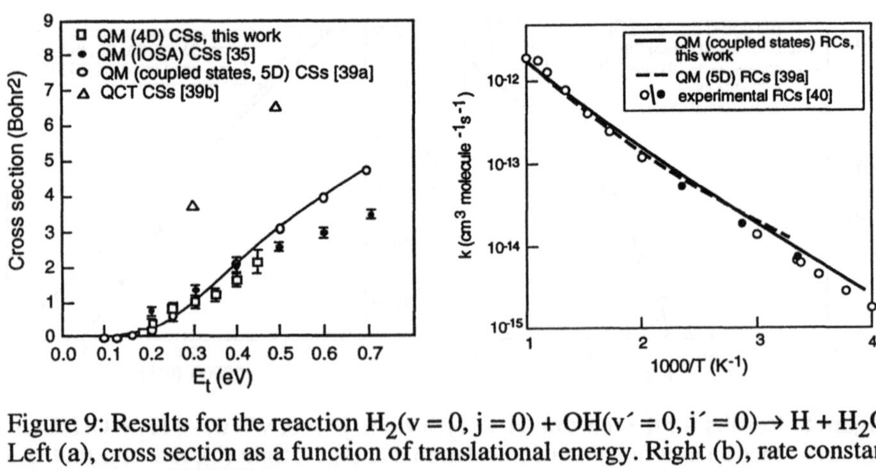

Figure 9: Results for the reaction $H_2(v=0, j=0) + OH(v'=0, j'=0) \rightarrow H + H_2O$. Left (a), cross section as a function of translational energy. Right (b), rate constant as a function of 1000/T.

raged potential. Consequently this approach is termed a 4-dimensional (4D) treatment.

With the above procedure the Hamiltonian that was used is:

$$H = -\frac{\hbar^2}{2\mu_1 r_1}\frac{\partial^2}{\partial r_1^2} r_1 - \frac{\hbar^2}{2\mu_2 r_2}\frac{\partial^2}{\partial r_2^2} r_2 - \frac{\hbar^2}{2\mu R}\frac{\partial^2}{\partial R^2} R + \frac{\hbar^2}{2\mu_1 r_1^2} \mathbf{j}_1^2 +$$

$$\frac{\hbar^2}{2\mu_2 r_2^2} j_2^*\left(j_2^* + 1\right) + \frac{\hbar^2}{2\mu R^2} \mathbf{l}^2 + U^*\left(R r_1 r_2 | \gamma_1 \gamma_2\right) \qquad (19)$$

where μ_i, i = 1, 2, are the reduced masses of the two diatomics, $\mathbf{j}_1$ is the H_2 angular momentum operator, $j_2{}^*$ is the OH angular momentum quantum number, $\mathbf{l}$ is the orbital angular momentum operator, R is the translational distance, μ is the corresponding reduced mass of the whole system and $U^*(Rr_1r_2|\gamma_1\gamma_2)$ is the above mentioned φ-averaged potential.

Employing now the coupled states approximation, Eq. (19) becomes:

$$H = -\frac{\hbar^2}{2\mu_1 r_1}\frac{\partial^2}{\partial r_1^2} r_1 - \frac{\hbar^2}{2\mu_2 r_2}\frac{\partial^2}{\partial r_2^2} r_2 - \frac{\hbar^2}{2\mu R}\frac{\partial^2}{\partial R^2} R + \frac{\hbar^2}{2} \mathbf{j}_1^2 \left(\frac{1}{\mu_1 r_1^2} + \frac{1}{\mu R^2}\right) +$$

$$\frac{\hbar^2}{2\mu_2 r_2^2} j_2^*\left(j_2^* + 1\right) + \frac{\hbar^2}{2\mu R^2}\left[J(J+1) - 2\Omega^2\right] + U^*\left(R r_1 r_2 | \gamma_1 \gamma_2\right) \qquad (20)$$

The Hamiltonian given in Eq. (20) is the one to be substituted in Eq. (1) and in Eq. (2) – with the obvious changes.

Preliminary results using this approach are shown in Fig. 9. In Fig. 9(a) the energy-dependent CSs are shown and compared with IOSA [35], full coupled states [39a], and QCT CSs [39b]. It is noticed that whereas all QM CS curves are reasonably close to each other, the QCT curve lies much above them. The somewhat unexpected result here is that at the higher energy region the IOSA and the present 4D cross sections overlap reasonably well, whereas the full coupled states ones are somewhat larger.

Temperature-dependent RCs are presented in Fig. 9(b) where they are compared to the full coupled states RCs and experiment [40]. All three types of results are relatively close to each other.

6 Conclusions

In this review seven chemical reactive systems are discussed for which quantum mechanical energy-dependent initial state-selected (total) cross sections and temperature-dependent rate constants were calculated. In all cases PESs based on *ab initio*values were used. The calculations are characterized by two features namely: (a) in each case they were done in the reagents AC only, applying reagents Jacobi coordinates and (b) the reactive ACs were eliminated using the appropriate NIPs. Two triatom systems, namely F + H_2 and F +

D_2 were treated twice using the two most up-to-date PESs. The two PESs yield rate constants which are in reasonable agreement with experiments. The calculations were done within the (nonreactive) coupled-states approximation. Two atom-triatom systems were considered. For the $O + O_3$ system the calculated rate constants fitted the experiment very nicely although the calculations were carried out within the crude QBSA. As for the $H + H_2O$ system, the calculations were done within IOSA (a somewhat cruder approximation than the QBSA) and still the cross sections were found to be reasonable: they fit quite well a low-energy measured value but are about two times too small as compared to a higher-energy measurement. The calculated rate constants were in an approximate agreement with the (estimated) measured ones. The most troublesome system was the $H_2 + OH$ system. Here, again, the IOSA was employed and although, in some cases, the cross sections were in a reasonable agreement with the Zhang & Zhang coupled states results (the best so far available) the rate constants were entirely off. We reported also on results from an improved version where the "reactive" Jacobi angle was treated within the coupled states approximation and the others, employing the IOSA. And in this case, indeed the results improved significantly. The calculated rate constants are in good agreement with both the more accurate 5D results and with experiment. However, these are only preliminary results.

7 Acknowledgement

M. Baer would like to thank the Israeli Ministry of Science and Arts and the German Ministry of Science and Technology for partly supporting the research reported here under grant No. E1447.

8 References

[1] Warnatz, J. and Jäger W., (Eds.) ***Complex Chemical Reaction Systems: Mathematical Modelling and Simulation***, Springer Series in Chem. Phys. (Berlin, Heidelberg, Springer-Verlag, 1987), Vol. **47**.

[2] Miller, J. and Fisk, G.A., Chem. Eng. News 1987, **65**, 22.

[3] Basevich, V. Ya., Prog. Energy Combust. Sci. 1987, **13**, 199.

[4] Miller, J., Kee, R.J. and Westbrook C.K., Annu. Rev. Phys. Chem. 1990, **41**, 345.

[5] Schuck, E.A. and Stephens, E.R., Environ. Sci. 1969, **1**, 73.

[6] a) Levine, R.D. and Bernstein, R.B., ***Molecular Reaction Dynamics and Chemical Reactivity*** (Oxford Univ. Press, Oxford, 1987); b) Gerlach-Mayer, U., Kleinermanns, K., Linnebach, E. and Wolfrum, J., J. Chem. Phys. 1987, **86**, 3047; c) Bigeleisen, J., Klein, F.S., Weston, R.E. and Wolfsberg, M., J. Chem. Phys. 1959, **30**, 1340; d) Lillich, H., Schuck, A., Volpp, H.–R., Wolfrum, J. and Naik, P.D., *25th Symposium (Int.) on Combustion,* The Combustion Institute, Pittsburgh, 1994, p. 993; e) Miller, J.A. and Bowman, C. T., Prog. Energy Combust. Sci. 1989, **15**, 287; f) Hack, W.,

Wagner, H.Gg. and Zasypkin, A., Ber. Bunsenges. Phys. Chem. 1994, **98**, 156; g) Wagner, H.Gg., and Zetzsch, C., Ber. Bunsenges. Phys. Chem. 1972, **76**, 526; h) Atkinson, R., Baulch, D.L., Cox, R.A., Hampson, R.F., Jr., Kerr, J.A. and Troe, J., J. Phys. Chem. Ref. Data 1989, **18**, 881.

[7] a) Wagner, H.Gg., Warnatz, J. and Zetzsch, C., Ber. Bunsenges. Phys. Chem. 1971, **75**, 119; b)Wurzberg, E. and Houston, P.L., J. Chem. Phys. 1980, **72**, 4811; c) Heidner, R.F., Bott, J.F., Gardner, C.E. and Meizer, J.E., J. Chem. Phys. 1980, **72**, 4815; d) Stevens, P.S., Brune, W.H. and Anderson, J.G., J. Phys. Chem. 1989, **93**, 4068; e) Persky, A., J. Chem. Phys. 1973, **59**, 3612, 5578.

[8] a) McCrumb, J.L. and Kaufman, F., J. Chem. Phys. 1972, **57**, 1270; b) Davis, D.D., Wong, W. and Lephardt, J., Chem. Phys. Lett. 1973, **22**, 273; c) West, G.A., Weston, R.E. Jr. and Flynn, G.W., Chem. Phys. Lett. 1978, **56**, 429; d) Wine, P.H., Nicovich, J.M., Thompson, R.L. and Ravishankara, A.R., J. Phys. Chem. 1983, **87**, 3948; e) Chekin, S.K., Gershenzon, Y.M., Konoplyov, A.V. and Rozenshtein, V.B., Chem. Phys. Lett. 1979, **68**, 386.

[9] Truhlar, D.G., Isaacson, A.D., Garret, B.C., in ***Theory of Chemical Reaction Dynamics***, Ed. Baer, M. (CRC press, Inc. Boca-Raton, FL, 1985) Vol. IV, Chap. 2.

[10] Child, M.S., in ***Theory of Chemical Reaction Dynamics***, Ed. Baer, M. (CRC press, Inc. Boca-Raton, FL, 1985) Vol. III, Chap. 3.

[11] Levine, R.D., in ***Theory of Chemical Reaction Dynamics***, Ed. Baer, M. (CRC press, Inc. Boca-Raton, FL, 1985) Vol. IV, Chap. 1.

[12] Troe, J., in ***State-Selected and State-To-State Ion Molecule Reaction Dynamics***, Eds. Baer, M. and Ng. C-Y. (John Wiley & Sons, Inc. N.Y., 1992), Vol. II, Chap. 8.

[13] Raff, L.M. and Thompson, D.L. in ***Theory of Chemical Reaction Dynamics***, Ed. Baer, M. (CRC press, Inc. Boca-Raton, FL, 1985) Vol. III, Chap. 1.

[14] a) Billing, G.D., Comput. Phys. Rept. 1984, **1**, 237; b) Billing, G.D., Chem. Phys. 1990, **146**, 63.

[15] Pollak, E., in ***Theory of Chemical Reaction Dynamics***, Ed. Baer, M. (CRC press, Inc. Boca-Raton, FL, 1985) Vol. III, Chap. 2.

[16] a) Kuppermann, A., Schatz, G.C. and Baer, M., J. Chem. Phys. 1976, **65**, 4596; b) Webster, F. and Light, J.C., J. Chem. Phys. 1989, **90**, 265, 300; c) Pack, R.T. and Parker, G.A., J. Chem. Phys. 1987, **87**, 3888; d) Zhang, J.H.Z., Kouri, D.J., Haug, K., Schwenke, D. W., Shima, Y. and Truhlar, D.G., J. Chem. Phys. 1988, **88**, 2492; e) Zhang, J.H.Z. and Miller, W.H., J. Chem. Phys. 1988, **88**, 3888; f) Baer, M., J. Chem. Phys. 1989, **90**, 3048; g) Schatz, G.C., Amaee, B. and Connor, J.N.L., J. Chem. Phys. 1990, **92**, 4893; h) Mohan, V. and Sathyamourthy, N., Comput. Phys. Rep. 1988, **7**, 7317.

[17] Bowman, J.M., (Ed.), ***Advances in Molecular Vibrations and Collision Dynamics*** (JAI Press Inc. Greenwich Conn., 1994) Vol. I, II.

[18] Kulander, K.C. (Ed.), ***Computer Physics Communications***, (North-Holland Physics Publishing, Netherland, 1991), Vol. 63.

[19] a) Neuhauser, D. and Baer, M., J. Phys. Chem. 1990, **94**, 185; b) Neuhauser, D. and Baer, M., J. Chem. Phys. 1990, **92**, 3419; c) Seideman, T.

and Miller, W.H., J. Chem. Phys. 1992, **96**, 4412; d) Kouri, D.J., Arnold, M. and Hoffman, D.K., Chem. Phys. Lett. 1993, **203**, 166.
[20] a) Neuhauser, D. and Baer, M., J. Chem. Phys. 1989, **91**, 4651; b) Kosloff, R. and Kosloff, D., J. Comp. Phys. 1986, **52**, 35; c) Balint-Kurti, G.B., Gotgas, F., Mort, S.P., Offer, A.R., Lagana, A. and Gervasi, O., J. Chem. Phys. 1993, **99**, 9567; d) Neuhauser, D., J. Chem. Phys. 1994, **100**, 9272; e) Zhang, D.H. and Zhang, J.H., J. Chem. Phys. 1994, **100**, 2697.
[21] a) Steckler, R., Truhlar, D.G., and Garret, B.C., J. Chem. Phys. 1985, **82**, 5499; b) Lynch, G.C., Steckler, R., Schwenke, D.W., Varandas, A.J.C., and Truhlar, D.G., J. Chem. Phys. 1991, **94**, 7136; c) Mielke, S.L., Lynch, G.C., Truhlar, D.G., and Schwenke, D.W., Chem. Phys. Lett. 1993, **213**, 10.
[22] Stark, K. and Werner, H.-J., J. Chem. Phys. (in press).
[23] Baer, M., Faubel, M., Martinez-Haya, B., Rusin, L.Y., Tappe, U., Toennies, J.P., Stark, K. and Werner, H.-J. (submitted for publication).
[24] Manolopoulos, D., Stark, K., Werner, H-J., Arnold, D.W., Bradforth, S.E., Neumark, D.M., Science 1993, **262**, 1852.
[25] Varandas, A.J.C. and Pais, A.C., in ***Theoretical and Computational Models for Organic Chemistry***, Eds. Formosinho, S.J., Czimadia, I.G. and Arnaut, L.G. (Kluver, Dordrecht, 1991), p. 55.
[26] a) Walch, S.P. and Dunning, T.H., J. Chem. Phys. 1980, **72**, 1303; b) Schatz, G.C. and Elgersma, H., Chem. Phys. Lett. 1980, **73**, 21.
[27] a) Baer, M., Neuhauser, D. and Oreg, Y., J. Chem. Soc. Faraday Trans. 1990, **86**, 1721; b) Baer, M. and Nakamura, H., J. Chem. Phys. 1992, **96**, 6565; c) Baer, M., Last, I. and Loesch, H.-J., J. Chem. Phys. 1994, **101**, 9648.
[28] a) Neuhauser, D. and Baer, M., J. Chem. Phys. 1989, **90**, 4351; b) Child, M.S., Mol. Phys. 1991, **72**, 89.
[29] a) McGuire, P. and Kouri, D.J., J. Chem. Phys. 1974, **60**, 2488; b) Pack, R.T., J. Chem. Phys. 1974, **60**, 633; c) Secrest, D., J. Chem. Phys. 1975, **62**, 710; d) Chu, S.I. and Dalgarno, A., Proc. Roy. Soc. A 1975, **342**, 191; e) Goldflam, R., Green, S. and Kouri, D.J., J. Chem. Phys. 1977, **67**, 4149; f) Schinke, R. and McGuire, P., J. Chem. Phys. 1979, **71**, 4201; g) Tsien, T.P., Parker, G.A. and Pack, R.T., J. Chem. Phys. 1973, **59**, 5373; h) Kouri, D.J., in ***Atom Molecule Collision Theory: A Guide for the Experimentalist,*** Ed. Bernstein, R.B. (Plenum press, N.Y., 1979), p. 301; i) Pack, R.T., published in Ref. 17, Vol IIA, p. 111;
[30] a) Khare, V., Kouri, D.J. and Baer, M., J. Chem. Phys. 1979, **71**, 1188; b) Barg, G.D. and Drolshagen, G., Chem. Phys. 1980, **17**, 209; c) Bowman, J.M. and Lee, K.T., J. Chem. Phys. 1980, **72**, 5071; d) Grossi, G., J. Chem. Phys. 1984, **81**, 3355; e) Nakamura, H., Ohsaki, A. and Baer, M., J. Phys. Chem. 1986, **90**, 676; f) Clary, D.C., Chem. Phys. Lett. 1981, **80**, 271; g) Nakamura, H., Reps. 1990, **187**, 1; h) Lagana, A., Aguilar, A., Gimenez, X. and Lucas, J.M., published in Ref. 17, Vol IIA, p. 183; i) Takayanagi, T., Tsunashima, S. and Sato, S., J. Chem. Phys. 1990, **93**, 2487; j) Aguilar, A., Gilibert, M., Gimenez, X., Gonzales, M. and Sayos, R., J. Chem. Phys. 1995, **103**, 4496.
[31] a) Hays, E.F., and Walker, R.B., J. Phys. Chem. 1983, **87**, 1255; b) Hays,

E.F., and Walker, R.B., J. Phys. Chem. 1984, **88**, 3318; c) Schatz, G.C., Hubbard, L.M., Dardi, P.S., and Miller, W.H., J. Chem. Phys. 1984, **81**, 231; d) Jansen Op der Haar, B.M.D.D. and Balint-Kurti, G.C., J. Chem. Phys. 1986, **85**, 2614; e) Schatz, G.C., Chem. Phys. Lett. 1984, **108**, 532; f) Tang, K.T. in ***Theory of Chemical Reaction Dynamics***, Ed. Baer, M. (CRC press, Inc. Boca-Raton, FL, 1985) Vol. II, Chap. 2; g) Clary, D.C. and Connor, J.N.L., Chem. Phys. Lett. 1979, **66**, 493.

[32] a) Rosenman, E. and Persky, A., Chem. Phys. 1995, **195**, 291; b) Truhlar, D.G., private communication (reported in: Aoiz, F.Z., Bañares, L., Herrero, V.J. and Sáez Rábanos, V., Chem. Phys. Lett 1994, **218**, 422; c) Rosenman, E., Hochman-Kowal, S., Persky, A. and Baer, M., J. Phys. Chem. 1995, **95**, 16523; d) Aoiz, F.J., Bañares, L., Herrero, V.J., Sáez Rábanos, V., Stark, K. and Werner, H.-J., Chem. Phys. Lett. 1994, **223**, 215.

[33] a) Szichman, H. and Baer, M., J. Chem. Phys. 1994, **101**, 2081; b) Szichman, H., Varandas, A.J.C. and Baer, M., Chem. Phys. Lett. 1994, **231**, 253; c) Szichman, H., Varandas, A.J.C. and Baer, M., J. Chem. Phys. 1995, **102**, 3474.

[34] LeRoy, R.L., J. Chem. Phys. 1969, **49**, 4338.

[35] Szichman, H. and Baer, M., Chem. Phys. Lett. 1995, **242**, 285.

[36] a) Kudla, K. and Schatz, G.C., J. Chem. Phys. 1993, **98**, 4645; b) Clary, D.C., J. Chem. Phys. 1991, **95**, 7248.

[37] Jacobs, A., Volpp, H.-R. and Wolfrum, J., J. Chem. Phys. 1994, **100**, 1936.

[38] Dixon-Lewis, G. and Williams, D.J., Comp. Chem. Kinet. 1977, **17**, 1.

[39] a) Zhang, D.H. and Zhang, J.H., J. Chem. Phys. 1994, **100**, 2697; b) Schatz, G.C., J. Chem. Phys. 1981, **74**, 1133.

[40] a) Tully, F.P. and Ravishankara, A.R., J. Phys. Chem. 1980, **84**, 3126; b) Ravishankara, A.R., Nicovich, J.N., Thompson, R.L. and Tully, F.P., *ibid*, 1981, **85**, 2498.

Mode-Specific Chemistry in the H + HCN and H + N_2O Reactions

M. ter Horst, K.S. Bradley, and G.C. Schatz
Department of Chemistry, Northwestern University,
Evanston, IL 60208-3113, U.S.A.

Abstract

We present a theoretical study of the effect of reagent vibrational excitation on cross sections and rate constants for the reactions H + HCN ($\nu_1\nu_2\nu_3$) $\rightarrow$ H_2 + CN and H + $N_2O(\nu_1\nu_2\nu_3)$ $\rightarrow$ N_2 + OH, NH + NO. For H + HCN, we study the states (000), (004) and (302), and we find that C–H stretch excitation is much more effective than C–N stretch in lowering the reactive threshold energy and thus enhancing the rate constant. For H + N_2O we study the states (000), (100), (01^10) and (001). Here N–N stretch excitation is more effective in enhancing the cross section at low energy where the reaction mechanism involves HNNO complex formation. At higher energy (3.3 kcal/mol above threshold and higher), N–O stretch excitation is more effective in enhancing the reaction cross section, due to the dominance of a direct mechanism involving a N-N-O-H transition state.

1 Introduction

The influence of reagent vibrational excitation on the cross sections and rate constants for bimolecular reactions involving triatomic and larger molecules has been a topic of great interest in the field of reaction kinetics in recent years. The first successful experimental demonstration of vibrational mode-specific reaction rate enhancements was for the H + H_2O and H + HOD reactions based on work by the Crim [1] and Zare [2] groups, and there have now been detailed theoretical studies of these experiments [4-7]. H + HOD is probably the simplest case of mode-specific chemistry in that reaction is accelerated by exciting the bond (i.e., the O–H stretch or O–D stretch) that breaks during reaction (i.e., exciting the reaction coordinate), while at the same time there is little or no change in the rate of reaction with the unexcited bond. The influence of excitation on the cross sections for reaction with the excited bond is both to enhance the magnitude of the cross section at fixed energy and to lower the effective reactive threshold. The Zare experiment observed cross section enhancement by factors of 20 or more, while the Crim group observed enhancement in the thermal rate constant, due mostly to lowering of the reactive threshold energy. Because of the exponential dependence of rate constant on threshold energy, the rate constant enhancement is very large, i.e., several orders of magnitude, but the Crim measurements did not determine a precise value. The H + HOD reaction is also simple because the only significant reaction mechanisms are H or D abstraction. The only possible competing mechanism at the energies of the experiments would be insertion to form an H_2OD intermediate followed by elimination of H_2 or HD, or by emission

Springer Series in Chemical Physics, Volume 61
Gas Phase Chemical Reaction Systems
Eds.: J. Wolfrum, H.-R. Volpp, R. Rannacher, and J. Warnatz

of H or D. However, this intermediate occurs at relatively high energies [8,9], and it is very weakly bound, so the cross section for its production is well below that for H or D abstraction.

In this paper we turn our attention to mode-specific chemistry in two other atom-triatom reactions, H + HCN and H + N_2O. Our motivation is that both of these reactions have more complex reaction mechanisms than does H + HOD, and thus the question of what kind of mode-specific propensities should be observed is less obvious.

For H + HCN, Bair and Dunning [10] have located two important reaction pathways, as illustrated in Fig. 1. These are:

(a) H-abstraction: $H + HCN \rightarrow H_2 + CN$ (R1)

(b) H-addition: $H + HCN \rightarrow H_2CN$ (R2)

After H-addition, several processes are possible, including (1) dissociation back to H + HCN (either with or without H-exchange), (2) isomerization to *cis* or *trans* HCNH (which then can emit H to form HNC), (3) H_2 elimination to form H_2 + CN. H_2 elimination would provide an alternative mechanism to abstraction for H_2 + CN formation. The barrier to H-addition (i.e., reaction R2) is much lower than that for H-abstraction (reaction R1), so addition should be facile, but what is not clear is whether H_2 + CN can be produced subsequent to addition. Bair and Dunning [10], and more recently Harding [11], have been unable to find reaction pathways leading between H_2CN and H_2 + CN, however reaction is still possible, as a ridge that separates these species is energetically below the reactive threshold for abstraction. Another important issue for the addition-elimination mechanism is the lifetime of the H_2CN complex. If this is sufficiently long to allow for scrambling of initial HCN vibrational excitation, then mode-specific behaviour could be quenched. This would provide a distinct signature of the addition-elimination mechanism, but it is not known at this point what is the H_2CN lifetime.

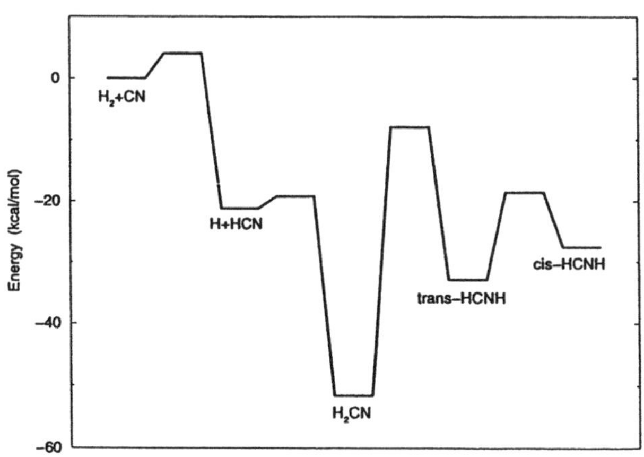

Figure 1: Schematic energy level diagram for CN + H_2, showing stationary points and the reaction paths that connect them.

Crim's group [12] has recently studied the reactions H + HCN and Cl + HCN, with HCN initially in either the states with four quanta of CH stretch, i.e. (004), or in the nearly degenerate state with 2 quanta of CH stretch and 3 quanta of CN stretch, i.e., (302). They find that H + HCN exhibits what appears to be simple abstraction behaviour, with the (004) state being much more reactive than (302), but Cl + HCN seems to involve complex formation, such as might occur via an HClCN intermediate. In this paper we will consider these two states for H + HCN.

The reaction H + N_2O has three reaction pathways [13,14], as shown in Fig. 2. These include:

(a) H addition to the N-end of N_2O to form HNNO and then N-N bond fission to produce NH + NO, i.e.,

$$H + N_2O \rightarrow HNNO \rightarrow NH + NO \quad \text{(R3)}$$

(b) H addition to form HNNO as in (a), then 1,3 H-migration to form $OH + N_2$

$$H + N_2O \rightarrow HNNO \rightarrow OH + N_2 \quad \text{(R4)}$$

(c) H addition to the O-end of N_2O, leading directly to $OH + N_2$

$$H + N_2O \rightarrow H\cdots ONN \rightarrow OH + N_2 \quad \text{(R5)}$$

Reactions (R4) and (R5) both lead to the same product, but (R4) involves formation of an intermediate complex while (R5) is direct. The 1,3 H-migration barrier is the highest bottleneck in (R4), being 3.4 kcal/mol above the H-NNO addition barrier, and 3.3 kcal/mol *below* the H-ONN addition barrier associated with (R5). This means that (R4) is the lowest energy path to $OH+N_2$ production,

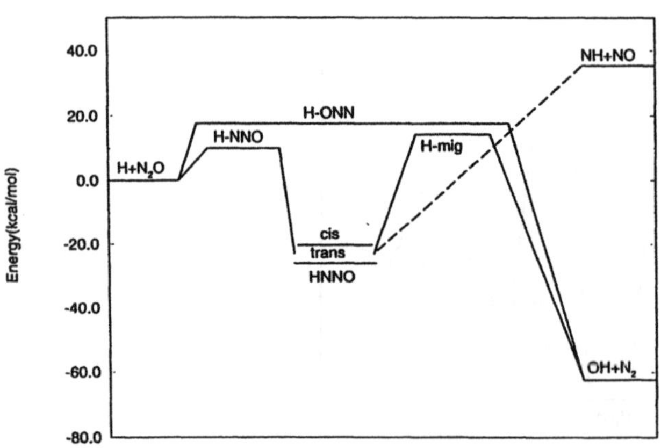

Figure 2: Schematic energy level diagram for H + N_2O, showing stationary points and reaction paths.

but (R4) and (R5) provide competitive reaction pathways at energies only 3.3 kcal/mol above the reactive threshold. The formation of NH + NO products takes place via (R3), and is substantially endothermic. Since (R4) provides a lower energy path for HNNO decay, NH + NO is a minor product of H + N_2O even at energies above threshold.

The reaction H + N_2O has been the subject of thermal kinetics studies [15] and hot-atom studies [16], but not mode-specific dynamics studies. Ravichandran and Fletcher [17] have recently reported mode-specific studies of SH + $N_2O(v_1v_2v_3)$, which shows reaction paths similar to (R4) and (R5) that lead to HSO + N_2 production. They have studied the nearly isoenergetic (100) and (020) states of $N_2O(100)$, and they find that (100) excitation, i.e., N–O stretch, enhances the rate constant by a factor of four, while (020) excitation (bend) has no effect on the reaction rate. In this paper we will consider the states (001), (01^10) and (100).

The approach used for the present study is quasiclassical trajectory (QCT) calculations based on global potential surfaces that have been derived from high quality *ab initio* calculations. QCT calculations have been used in the past to study mode-specific reactivity in H + HOD and related reactions [4,7]. Correspondence of the results with quantum dynamics calculations [6] has been good at the qualitative level, and sometimes even quantitative comparisons have been obtained. There have been quantum dynamics studies of H + HCN in the past [18-20], but these studies used reduced dimensionality approximations that did not allow for formation of H_2CN intermediate complexes. There have not been previous QCT studies of either H + HCN or H + N_2O, but Bradley *et al.* [14] have studied the NH + NO reaction using the same potential surface that we now use to study H + N_2O.

2 Details of Calculations

2.1 Quasiclassical trajectory calculations

The trajectory method used for these calculations has been described in the past [20,21]. The only non-standard aspect of it is the vibrational state determination for the reagent triatomic molecule. For this, we use a semiclassical eigenvalue calculation due to Eaker and Schatz [22] in which good actions are evaluated from Fourier representations of the coordinates and momenta of the triatomic motions. The present applications to the HCN and N_2O molecules are new, but they are based on our past work with other linear triatomic molecules [22,23]. In the present applications, we are able to obtain converged semiclassical eigenvalues without difficulty for all the states considered, including high overtones like HCN(004). For most of the states studied, we verified the "goodness" of the calculated actions by repeating the action calculation using a new time segment of the triatomic motions as input.

2.2 Potential Energy Surfaces

The HHCN potential surface is taken from Surface I of ter Horst *et al.* [25], and is based on extensive *ab initio* calculations [11]. The surface is represented using the following modified many-body expansion:

$$V(R_{H1H2},R_{H1C},R_{H1N},R_{H2C},R_{H2N},R_{CN}) =$$
$$V_{HCN}(R_{H1C},R_{H1N},R_{CN}) + V_{HCN}(R_{H2C},R_{H2N},R_{CN}) - V_{CN}(R_{CN}) +$$
$$V_{HH}(R_{H1H2}) + V^{(3)}_{CHH}(R_{H1C},R_{H2C},R_{H1H2}) +$$
$$V^{(3)}_{NHH}(R_{H1N},R_{H2N},R_{H1H2}) + V_4$$

In this expression V_{HCN} is an accurate global potential for HCN [26], one which determines spectroscopically accurate energy levels for HCN and HNC, and which dissociates correctly. V_{CN} and V_{HH} are experimentally determined Morse potentials for CN and H_2, respectively. The specific combination of these diatomic potentials with the two HCN potentials is such that all six pair potentials associated with HHCN are included. $V^{(3)}_{CHH}$ and $V^{(3)}_{NHH}$ are 3-body LEPS potentials for CHH and NHH, respectively. They are included so that the four possible 3-body potentials in HHCN are included. V_4 is the four-body potential for HHCN, and is generally represented by sums of Gaussian functions which are centered on the stationary points of the full potential.

Further details of the potential are described in [25]. Reaction path information is presented in Fig. 1, and Table 1 gives energies, structures and frequencies for the most important stationary points. Included in the Table are comparisons with the *ab initio* results of Bair and Dunning [10], with Harding [11], and with experimentally derived "Best Estimates." The agreement of the energetic and structural results is generally quite good, but some of the vibrational frequencies (those associated with the H-H-C-N barrier and with the H_2CN minimum) are too high. This error in frequencies may cause certain adiabatic barriers to be too high, but for the high energies that we are considering, this should not affect our qualitative conclusions.

The HNNO surface that we used for our H + N_2O calculations has been described in detail previously [14]. The relevant reaction pathways are illustrated in Fig. 2. This surface has been successfully used to describe the NH + NO reaction dynamics, particularly branching between the N_2 + OH and H + N_2O products [14,24].

3 Results

3.1 H + HCN

Figure 3 presents the reactive cross section σ versus translational energy E for H + HCN($v_1v_2v_3$) with ($v_1v_2v_3$) = (000), (004) and (302). In calculating these cross sections we have omitted trajectories for which the H_2-vibrational energy is less than zero point (following a procedure described in [20] in which CN is assumed to be a spectator bond and thus not subject to zero point constraints), thus insuring that the reactive threshold is not below its minimum possible value for each state. For each translational energy considered we integrated 10000 trajectories, using a maximum impact parameter of b_{max} = 4.0 a_0. The reaction probability is quite small for this reaction (<1% for the HCN ground state) so the error bars are substantial, but the trends in the energy dependence of the cross section are clearly visible.

Table 1. Energies, Geometries and Frequencies of Important HHCN Stationary Points. Energies are in kcal/mol, distances in bohr, angles in degrees, and frequencies in wavenumbers.

	Surface 1	BD	Harding
1) **H + HCN**			
Energy (reference)	0	0	0
R(CH)	2.02	2.04	2.02
R(CN)	2.19	2.21	2.19
frequencies	677	811	707
	2105	2264	2152
	3424	3533	3524
2) **H-H-C-N**			
Energy	25.3	31.3	26.6
R(CH)	3.01	2.93	3.10
R(CN)	2.19	2.23	2.21
R(HH)	1.52	1.57	1.49
frequencies	819i	1065i	664i
	333	165	112
	821	761	573
	2186	2172	2172
	4097	2808	3181
3) **H_2CN**			
Energy	-30.3	-24.6	
R(CH)	2.25,2.21	2.10	
R(CN)	2.29	2.40	
θ(HCN)	137	121	
frequencies	776	1007	
	887	1012	
	1131	1407	
	1814	1722	
	1979	3031	
	5743	3098	
4) **H_2 + CN**			
Energy	21.2	25.3	23.2
R(HH)	1.40	1.44	1.41
R(CN)	2.21	2.23	2.22
frequencies	2058	2256	2089
	4405	4443	4405

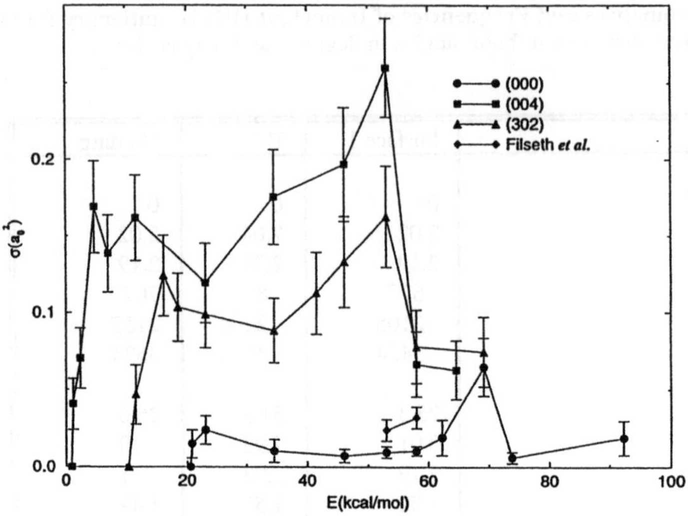

Figure 3: Cross section σ (in a_0^2) versus translational energy E (in kcal/mol) for H + HCN, with filled circles showing results for HCN(000), triangles for HCN(302) and squares for HCN(400). In all calculations, reactive trajectories were omitted if the product H_2 had less than zero point energy. Measured results by Filseth *et al.* [27] for (000) are plotted using diamonds.

Substantial enhancements in the reactive cross section, and lowering of the reactive threshold energy, upon vibrational excitation of HCN – with the (004) enhancement and lowering being significantly larger than the (302) – is shown in Fig. 3. The (000) state has a threshold at 20.5 kcal/mol, corresponding to the minimum possible energy needed to form CN + H_2 in their ground states. The corresponding thresholds for (302) and (004) are 10.2 and 0.2 kcal/mol, respectively, which means that the threshold is lowered by 10.3 kcal/mol in going from (000) to (302), and by 10.0 kcal/mol in going from (302) to (004). This is the expected behaviour if only C–H stretch is effective in lowering the reactive threshold. The excitation energy associated with two quanta of C–H stretch excitation is about 17 kcal/mol, so ≥50% of this energy is used for lowering the threshold energy.

Figure 3 shows that each cross section rises to a maximum a few tenths of a kcal/mol above threshold, then it drops to a local minimum before showing a second maximum about 40 kcal/mol above threshold. The origin of this unusual second peak is unknown. We have examined our trajectories to determine the reaction mechanism, and we find that direct abstraction dominates at all energies. H_2CN complex formation does become appreciable at 40 kcal/mol and above.

One comparison with experiment that can be made refers to absolute cross section measurements by Filseth *et al.* [27] at E = 53 and 58 kcal/mol. The results are included in Fig. 3, and they are in factor-of-two agreement with the calculated results.

The mode-specific measurements of Metz *et al.* [12] on H + HCN refer to the thermal rate constant at 300 K for reaction with the (004) state. A precise value of the thermal rate constant was not reported, but the value would have to be a reasonable fraction of the gas kinetic collision rate constant to be measurable. In addition, Crim reports [28] that the (302) state is not detectable under the same conditions. In our results, we find that the rate constant for (004) is 3 x 10^{-13} cm^3/s at 300 K with very small activation energy, while that for (302) is smaller by at least a factor of 10^8. This comparison with experiment is clearly rather crude, but at least the trends are consistent.

3.2 $H + N_2O$

Figure 4 presents the cross section versus energy for $H + N_2O(v_1v_2v_3)$ for $(v_1v_2v_3)$ = (000), (100), (01^10) and (001). Only the N_2 + OH product is shown, but we did calculate a small cross section for NH + NO formation at energies above 35 kcal/mol. Each cross section in Fig. 4 is based on 1500-5000 trajectories and using a maximum impact parameter $b_{max} = 4.0\ a_0$. Both OH and N_2 zero point energies were constrained in calculating the N_2 + OH cross sections, as it is not clear that either diatomic is a spectator when HNNO complex formation is involved.

Figure 4 shows that the effect of reagent vibrational excitation on the cross sections is small, but with several well-defined propensities. For (000) the reactive threshold is at 14.4 kcal/mol, which is close to the vibrationally adiabatic barrier for 1,3 H-migration. The thresholds for the other states are 12.9 kcal/mol for (100), 14.4 kcal/mol for (01^10) and 11.7 kcal/mol for (001). At energies well above threshold, the cross sections all rise to a value of about 3 a_0^2. The excited state cross sections rise somewhat more quickly than the ground state with the largest effect being associated with the (100) state.

The results in Fig. 4 can be understood based on a consideration of the reaction paths in Fig. 2. At energies near the N_2 + OH threshold, only reaction *via*

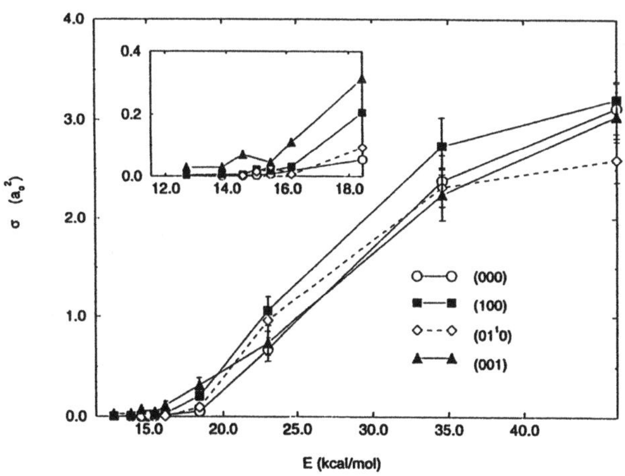

Figure 4: Cross section σ (in a_0^2) versus translational energy E (in kcal/mol) for $H + N_2O \rightarrow OH + N_2$, showing results for the states (000) (open circles), (100) (filled squares), (01^10) (open diamonds connected by dotted lines), (001) (filled triangles).

the HNNO intermediate complex is possible. The barrier to produce this species from H + N_2O involves elongation of the N-N bond, so the (001) state enhances the cross section more than the others. The effect is not very large, however, as the more significant bottleneck to reaction involves 1,3 H-migration, and this bottleneck occurs after the HNNO complex formation allows for scrambling of the initial N_2O excitation.

At energies above the threshold for the direct reaction pathway (via the H···ONN transition state), the contribution of this mechanism to the production of OH + N_2 quickly exceeds that from the HNNO complex mechanism. As a result, the dominant mode-specific propensities change so that the (100) state (N–O stretch) is the most important enhancer. Indeed in this limit, (001) excitation gives a cross section that is almost indistinguishable from the ground state. Bend excitation generally leads to little or no enhancement in the reactive cross section.

The calculated rate constants based on the cross sections in Fig. 4 are all very small, as might be expected given the high reactive threshold energies. For the ground state, the rate constant at 300K is 3×10^{-22} cm^3/s. The rate constant is not enhanced by bend excitation, but (100) excitation increases it by factor of 7 and (001) excitation increases it by a factor of 120. These results are qualitatively similar to those reported by Ravichandran and Fletcher for SH + N_2O.

The NH + NO results (not plotted) show little effect from reagent vibrational excitation, with the (001) state slightly favoured. This result is consistent with the reaction path involving HNNO formation.

One measurement that relates directly to our H + N_2O results is the Wittig group [16] study of the OH + N_2/NH + NO branching ratio for H + N_2O(000) at 46 kcal/mol. Although absolute cross sections were not measured, the cross section ratio OH + N_2/NH + NO was estimated to be about 10. Our ratio is also 10. We find that at this energy, the OH + N_2 product comes mostly from H attack on the O-end of N_2O while the NH + NO product comes from HNNO complex formation. Thus this ratio reflects the efficiency of two very different reaction mechanisms.

4 Conclusion

The present study has demonstrated several important points about mode-specific chemistry. For H + HCN we found what may be characterized as "strong" mode-specific effects in which the cross section is greatly enhanced and the reactive threshold greatly lowered by C–H stretch excitation. In agreement with earlier dynamics studies, we find that the nearly degenerate (004) and (302) states have very different enhancements, indicating that C–N stretch excitation is less effective than C–H excitation in enhancing reactivity. All of this reflects the dominance of direct abstraction in the reactive collisions. H_2CN intermediate complexes are easily formed in H + HCN collisions, but do not lead to CN + H_2 formation except at high energies. All of these conclusions are consistent with experimental results for H + HCN obtained by Metz *et al.* [12]. These studies also obtained information about CN product state distributions that will be reported elsewhere [25]. An important question for future work is whether the Cl + HCN reaction will show the dramatically different behaviour that has been observed in the experiments.

The H + N_2O results show weaker mode-specific effects, although we should remember that the states we considered have much smaller excitation energies than in H + HCN. The most interesting feature of H + N_2O is the change in reaction mechanism (and hence in mode-specific propensities) as energy is increased. At low energies the reaction to produce N_2 + OH is dominated by H addition to the N atom, and so N–N stretch excitation is most effective in enhancing reactivity. At high energies, addition switches to the O atom so N–O stretch excitation is more effective. We also saw that HNNO complex formation serves to quench mode-specific effects, though not completely. Further studies of the H + N_2O reaction will be reported elsewhere [29].

5 Acknowledgements

This research was supported by NSF Grant CHE-9016490 and by NASA Grant NCC2-478. We thank Larry Harding, and Stephen Walch for helpful discussions. GCS thanks Hansi Volpp and Jürgen Wolfrum for their kind hospitality during the Bodenstein Conference, and the Humboldt Foundation and Max Planck Society for support of the Bodenstein Conference through the Max Planck Research Award.

6 References

[1] A. Sinha, M. C. Hsiao and F. F. Crim. J. Chem. Phys. **92**, 6333 (1990); A. Sinha, J. Phys. Chem. **94**, 4391 (1990); A. Sinha, M. C. Hsiao and F. F. Crim, J. Phys. Chem. **94**, 4928 (1991); M. C. Hsiao, A. Sinha and F. F. Crim, J. Phys. Chem. **95**, 8263 (1991); R. B. Metz, J. D. Thoemke, J. M. Pfeiffer, F. F. Crim, J. Chem. Phys. **99**, 1744 (1993).

[2] M. J. Bronikowski, W. R. Simpson, B. Girard and R. N. Zare, J. Chem. Phys. **95**, 8647 (1991); M. J. Bronikowski, W. R. Simpson and R. N. Zare, J. Phys. Chem. **97**, 2194, 2204 (1993); D. E. Adelman, S. V. Filseth and R.N. Zare, J. Chem. Phys. **98**, 4636 (1993).

[3] G. C. Schatz, M. C. Colton and J. L. Grant, J. Phys. Chem. **88**, 2971 (1984).

[4] K. Kudla and G. C. Schatz, Chem. Phys. Lett. **193**, 507 (1992).

[5] K. Kudla and G. C. Schatz, Chem. Phys. **175**, 71 (1993).

[6] J. M. Bowman and D.-S. Wang, J. Chem. Phys. **96**, 7852 (1992); D.-S. Wang and J. M. Bowman, J. Chem. Phys., **98**, 6235 (1993); D.-S. Wang and J. M. Bowman, J. Chem. Phys. **96**, 8906 (1992); D. C. Clary, Chem. Phys. Lett. **192**, 34 (1992); J. Chem. Phys. **95**, 7298 (1991); **96**, 3656 (1992); H. Szichman, I. Last, M. Baer, J. Phys. Chem. **97**, 6436 (1993); **98**, 828 (1993); N. Balakrishnan, G. D. Billing, J. Chem. Phys. **101**, 2785 (1994).

[7] J. M. Bowman and G. C. Schatz, Annu. Rev. Phys. Chem., **46**, 169 (1995).

[8] D. Talbi and R. P. Saxon, J. Chem. Phys. **91**, 4396 (1992).

[9] K. Kudla and G. C. Schatz, J. Chem. Phys. **98**, 4644 (1993).

[10] R. A. Bair and T. H. Dunning, Jr., J. Chem. Phys. **82**, 2280 (1988).

[11] L. B. Harding, private communication.

[12] R. B. Metz, J. M. Pfeiffer, J. D. Thoemke, and F. F. Crim, Chem. Phys. Lett. **221**, 347 (1994).

[13] S. P. Walch, J. Chem. Phys. **98**, 1170 (1993).

[14] K. S. Bradley, P. McCabe, G. C. Schatz and S. P. Walch, J. Chem. Phys. **102**, 6696 (1995).

[15] P. Marshall, T. Ko, and A. Fontijn, J. Phys. Chem. **93**, 1922 (1989).

[16] E. Bohmer, S. K. Shin, Y. Chen and C. Wittig, J. Chem. Phys. **97**, 2536 (1992).
[17] K. Ravichandran and T. R. Fletcher, 209th ACS National Meeting, Anaheim, 1995.
[18] Q. Sun and J. M. Bowman, J. Chem. Phys. **92**, 5201 (1990).
[19] Q. Sun, D. L. Yang, N. S. Wang, J. M. Bowman, and M. C. Lin, J. Chem. Phys. **93**, 4730 (1990).
[20] A. N. Brooks and D. C. Clary, J. Chem. Phys. **92**, 4178 (1990).
[21] G. C. Schatz, J. Phys. Chem. **99**, 516 (1995).
[22] C. W. Eaker and G. C. Schatz, J. Chem. Phys. **81**, 2394 (1984).
[23] K. Kudla and G. C. Schatz, In: ***The Chemical Dynamics and Kinetics of Small Radicals*** edited by K. Liu and A. F. Wagner (World Scientific, 1995), in press.
[24] M. Simonson, K. S. Bradley and G. C. Schatz, Chem. Phys. Lett., in press (1995).
[25] M. ter Horst, G. C. Schatz and L. B. Harding, to be published.
[26] J. M. Bowman, B. Gazdy, J. A. Bentley, T. J. Lee, and C. E. Dateo, J. Chem. Phys. **99**, 308 (1993).
[27] H. M. Lambert, T. Carrington, S. V. Filseth and C. M. Sadowski, J. Phys. Chem. **97**, 128 (1993).
[28] F. F. Crim, private communication
[29] K. S. Bradley and G. C. Schatz, to be published.

Dynamics of Chemical Reactions Induced by Cluster Impact

T. Raz and R. D. Levine
The Fritz Haber Research Center for Molecular Dynamics
The Hebrew University, Jerusalem 91904, Israel

Abstract
A new area in dynamics, characterized by collision times shorter than hitherto studied, is discussed with special reference to four-center reactions. Under the unusual combination of conditions made possible within an impact heated cluster, the nominally four-center reactions can be made to proceed (on the computer) via the four-center mechanism as suggested by Bodenstein. It is even possible to achieve multi-center (> 4) reactions. The unique features of the new dynamical regime are discussed and illustrated by a variety of high-barrier processes, including the four center $H_2 + I_2 \rightarrow 2HI$ and $N_2 + O_2 \rightarrow 2NO$ reactions and a multi-center "burning of air" process.

1 Introduction

Cluster impact-induced chemistry (CIC, [1-3]) has the characteristics of both isolated elementary processes as in gas phase dynamics and of reactions in solution. Yet it is not a bridge between the two limits but a quite distinct regime in dynamics which provides for many unique possibilities and for a degree of control not readily available otherwise. Here we discuss CIC as studied on the computer. The present result only show what can happen. It remains for our colleagues to tell us what actually does happen.

The role of the environment in CIC is to provide for a high density of atoms and of energy. It is a super-hot and super-dense medium [4,5], where the nearest (peaceful) equivalents are reactions in supercritical solvents [6] or in front of a comet or of a space vehicle during entry into the atmosphere. The three unique features are the considerable variation that can be achieved under realistic laboratory conditions (e.g., temperatures in the range from 10^2 to $>10^5$ K [7]), the ultrafast time scale over which activation can be achieved (and it is for this reason that super -critical conditions can be achieved [7]), and the very brief duration (typically $\leq$1ps) during which these extreme conditions prevail. What the latter means is that the chemical process inside the cluster (even a large one [8]) is very rapidly quenched. This, of course, means that endothermic reactions (= high energy products) are possible because the reaction products are kinetically stable.

Figure 1 is an illustration of the high density of reactive atoms during cluster impact. The initial conditions are 7 N_2 and 7 O_2 molecules inside a cold Ne cluster. The rare gas atoms are not shown in the figure. The entire cluster, including the embedded molecules is first equilibrated (on the computer, at 30 K). It is then given an initial velocity in the direction of the surface. The box outlined is the smallest enclosure which can be drawn around the reactive atoms before the cluster impacts the surface at 12 km/sec. The figure shows the location of the 14 reactive atoms 70 fs after the impact. The atoms now occupy 75% of the original volume and the maximal density of reactive atoms is higher by almost 50% compared to the initial density (which is that of the glassy solid, $\approx$ 0.9 g/cm^3). At the peak compression there are

Springer Series in Chemical Physics, Volume 61
Gas Phase Chemical Reaction Systems
Eds.: J. Wolfrum, H.-R. Volpp, R. Rannacher, and J. Warnatz

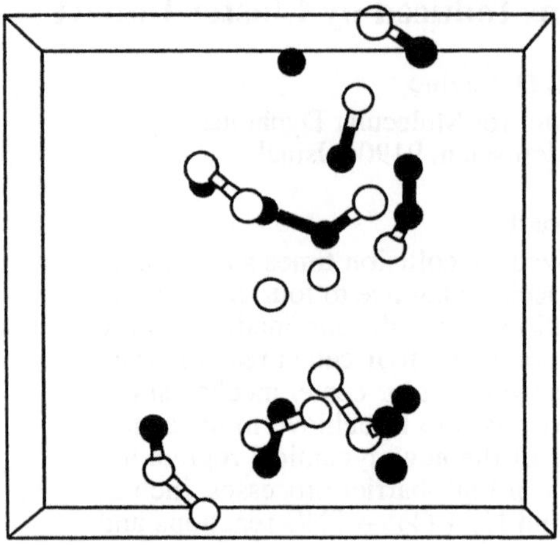

Figure 1: The spatial distribution of 14 nitrogen atoms (black) and oxygen atoms (white) inside a Ne cluster (not shown), 70 fs after surface impact at 12 km/sec. A tube connects those atoms which are within their equilibrium bond distance. The box is the smallest hypothetical enclosure drawn about all the reactive atoms in the cold cluster, prior to impact. Note the considerable compression and the chemical implication: there are several examples where three or four atoms are within the range of the chemical forces.

$\approx 4.8 \times 10^{22}$ reactive atoms/cm^3 or more than three orders of magnitude higher than for an ideal gas at STP (Loschmidt's number).

What is not immediately expected is that under such high densities, gas phase dynamics can provide useful insights. That however is very much the case [2] and the reason is the rather short interaction times that are involved. At such high velocities, the duration of a collision is short compared to a typical vibrational period even for such molecules as N_2. The reason is the essentially hard sphere nature of the short range atom-atom repulsion. Figures 2–4 illustrate several points regarding the dynamics of the reactive collision.

Figure 2 is a bond distance *vs.* time plot for the two old bonds and the two new bonds during the $N_2 + O_2 \rightarrow 2NO$ reactive collision in a cluster of 216 Ne atoms for an impact at 10.5 km/sec. The initial conditions are identical in both panels and the difference is in the mechanics of the surface at which the collision occurs. The left panel is for an idealized 'hard' and 'flat' surface which simply reverses the direction of the momentum of the atoms that hit it. Such a surface causes no dissipation of energy and so the available energy is higher than for the surface used in the other panel which is of a hard cube type [9] and adjusted [10] so that about 55% of the energy of an impacting atom is dissipated. The energy loss is clearly evident in the collision being slower and this is the case also in other examples, not only with respect to the role of surface dissipation but also say in the role of the mass of the rare gas cluster atoms etc. [10]. The role of the environment is in a large part in determining the energy available to the molecules just before the reactive collision.

Examination of Fig. 2 shows that the reactive atoms are compressed at the instant of impact. Not shown in the figure are those, two, atom-atom distances which are unbound both before and after the reaction. These contract just like the two new bond distances and expand just like the old bond distances. The bond switching oc-

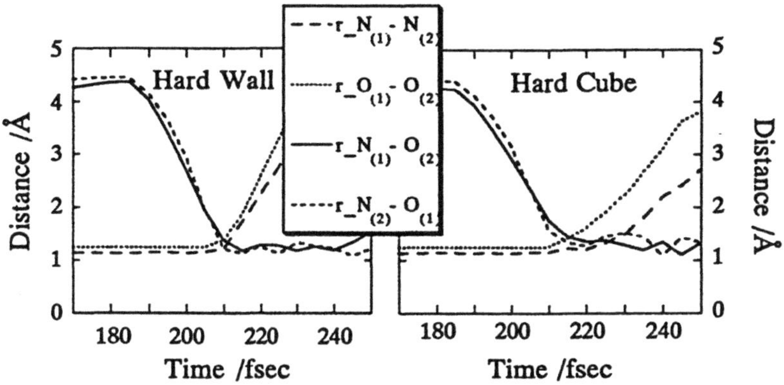

Figure 2: The bond distance vs. time (in fs) plot for the two old bonds and the two new bonds during the $N_2 + O_2 \rightarrow 2NO$ reactive collision in a cluster of 216 Ne atoms for an impact at 10.5 km/sec. The initial conditions of the cluster are identical in both panels. Left panel: a reflecting surface. Right panel: a surface dissipating 55% of the energy of impact. Note that the reaction is concerted four center and that the nascent NO molecules are vibrationally excited, as expected on the basis of the kinematic constraint on a four center reaction [11]. The reactive collision shown in this figure is a homogeneous one, occurring inside the cluster and away from the surface

curs in a concerted fashion in less than 10 fs and in physical units it is clearly faster than the vibrational period of the newly formed NO bonds.

The very short durations of both the reactive event and of the activation of the reactants before the collision are very clear in Fig. 2. This is seen in the distance vs. time plots, as the action of a force is evident as a change of slope (= velocity) in the plot. There are two clear implications. The first is that activation (or reaction) occurs on a time scale short compared even to fast vibrational periods. This is a new regime not previously well studied except in ultrashort laser pumping experiments [12] and, much earlier, in studies of hot atom chemistry [13]. The other point to which we return in Fig. 4 below is that the reactants are, de facto, isolated during the bond switching event.

The compression during reaction is also shown in Fig. 3 which is the weighted generalized hyperdistance [2] of the four reactive atoms vs. time for the right panel of Fig. 2. Also shown are the position of all atoms that are relevant to the dynamics of that reactive collision.

Four-center reactions have two kinds of barriers: an inherent one due to conservation of orbital symmetry [14] and an additional one, kinematic in origin [11] which requires vibrational excitation of the reactants (and leads to vibrationally excited products). For a homogeneous reaction, the energy required to overcome both barriers needs to be provided by the atoms of the cluster. Since the reaction shown in Fig. 2 occurs quite soon after impact, the energy is provided by those rare gas atoms which have already rebounded from the surface and are moving towards the reactants, which are still moving forward. Figure 4 identifies the key rare gas atoms that are involved.

The impression from Fig. 4 is quite general. At a given time, one cluster atom (or, at most, a few atoms) exercise a required force. The role of the rare gas cluster is very much binary in nature. Moreover, once the needed impulse was

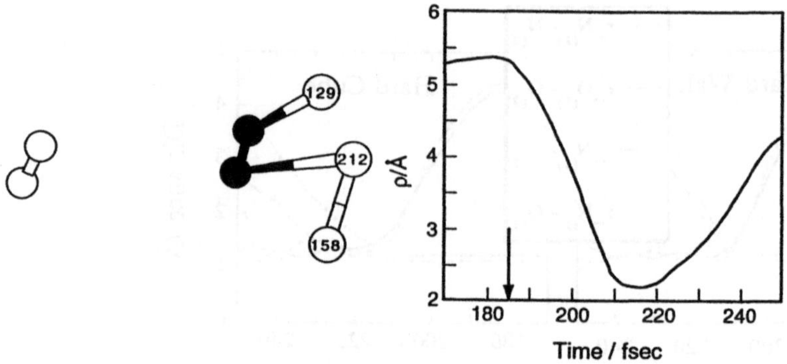

Figure 3: The hyperradius for the four atoms during the $N_2 + O_2 \rightarrow 2NO$ reactive collision *vs.* time (in fs), for the bottom panel of Fig. 2. The time 185 fs is indicated. The position of the four reactive atoms and of three cluster atoms (cf. Fig. 4 for an identification of the serial number) at that time is shown.

delivered, the reactive collision occurs as if the participating reactive atoms are isolated.

Figures 2–4 show another aspect whereby the cluster favors reaction. In the gas phase [11] most $N_2 + O_2$ collisions are nonreactive due to steric considerations. The cluster favors a nearly coplanar configuration of the reactants, near to one another. The result is a rather high yield of four-center events. However, since the kinematic constraints favor high vibrational excitation of the NO products, many molecules undergo collisional dissociation. The fragmentation of the cluster prevents more molecules from undergoing secondary processes.

Figures 2–4 were for a cluster containing just one N_2–O_2 pair. If more molecules are present, multi-atom (> 4) transition states are possible, see Fig. 5.

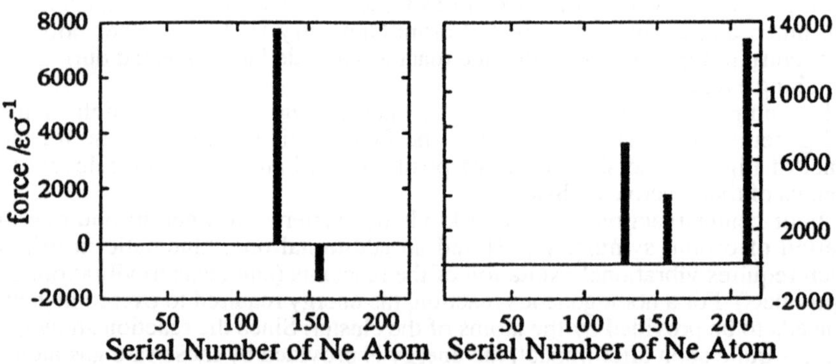

Figure 4: The force applied by each rare gas atom *vs.* its (arbitrary) serial number, at 185 fs. Left: the force applied on the N-N bond so as to cause vibrational excitation. Right: the force applied along the N_2–O_2 approach coordinate. The three rare gas atoms that are doing the work are identified in Fig. 3.

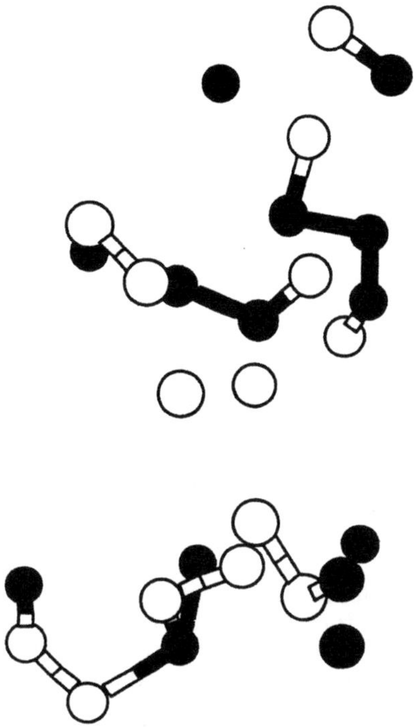

Figure 5: The location of the reactive atoms, (nitrogen: black, oxygen: white) shortly after a $(O_2)_7(N_2)_7Ne_{97}$ cluster impacted a surface at 12 km/sec. A tube connects those atoms that are chemically interacting. Note the 5 and 7 atom configurations.

2 Status Report

The simplest high barrier process that can be induced by cluster impact is that of bond dissociation [8,10,15,16], Fig. 6. One can think of the sequential hard Cl-Ar collisions seen in Fig. 6 as due to successive 'layers' of rare gas atoms that rebound from the surface [10]. This 'micro' shock wave is also seen in Fig. 7 which is for the four-center $N_2 + O_2$ reaction in a $(O_2)_7(N_2)_7Ne_{97}$ cluster impact. It shows the number of atom-atom bonds and the oscillations are due to successive compressions of the reactants by the cluster.

The 'classical' four-center reaction studied by Bodenstein was that of $H_2 + I_2 \rightarrow 2HI$ [17]. While in the bulk, under ordinary thermal conditions, the reaction may well proceed by a different mechanism ([18] and references therein), there is no question that in a cluster, a concerted four center reaction is the dominant mechanism [2], see Fig. 8. The novel aspect seen in Figure 8 is the rather different time scale for the H_2 and I_2 motion. The much faster moving H atoms scede from one another before the iodine molecule has moved much. Consequently, the newly formed HI molecules are typically rotationally excited. This is seen in the oscillation of the H-other I-distance *vs.* time.

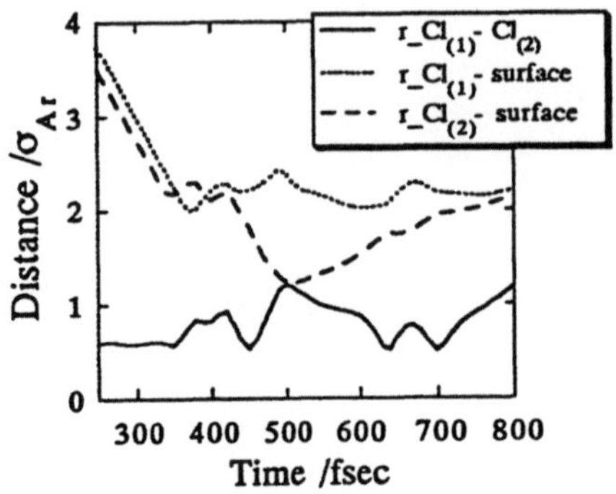

Figure 6: Distances vs. time plot for a Cl_2Ar_{500} cluster impact at 5 km/sec. The distance is measured in units of the Cl-Ar Lennard-Jones parameter σ. The Cl-surface distance establishes that the process occurs when there are two layers of Ar atoms between the molecule and the surface. The Cl-Cl bond distance shows that there are several sequential Cl-Ar collisions before the molecule actually dissociates. There is therefore a range of possible processes [8].

So far we have discussed concerted four-center reactions. An alternative is a sequential mechanism [12,19] in which one new bond forms before the other. Such a mechanism is unlikely for a diatom-diatom reaction in a cluster but is possible for a unimolecular reaction. A discussion of how the cluster can govern the relative importance of the two mechanisms can be found elsewhere [3].

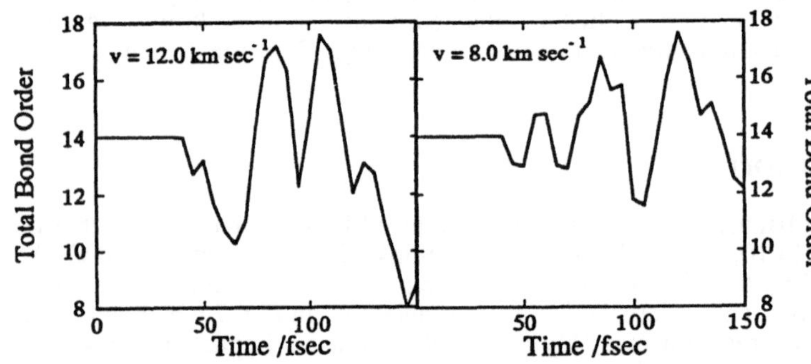

Figure 7: The bond order between all reactive atoms during the compression stage in a $(O_2)_7(N_2)_7Ne_{97}$ cluster impact at two velocities of impact: 12 km/sec (left panel) and 8 km/sec (right panel). There are 14 bonds to begin with and during the cluster compression many atoms acquire additional neighbours, cf. Fig. 5. Of course, some molecules dissociate and this makes a negative contribution to the total bond order. The oscillations are due to the 'micro shock wave' generated by successive layers of Ne atoms rebounding from the surface. The period of the oscillation (25 and 35 fs for the two panels) is given as the radius of a Ne atom over the velocity of impact. When more extensive energy dissipation at the surface is allowed, the oscillations are quenched due to the dispersion in both the magnitude and the direction of the velocities of the Ne atoms.

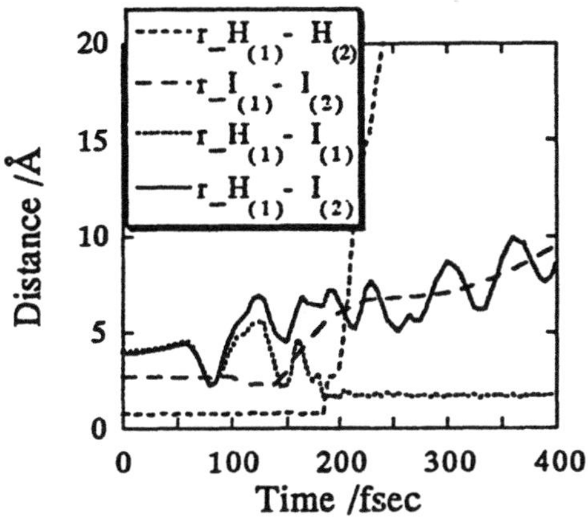

Figure 8: The bond distance *vs.* time (in fs) plot for the $H_2 + I_2 \rightarrow 2HI$ reactive collision in a cluster of 125 Xe atoms for an impact at 6 km/sec. Shown are the two old bonds, one new bond and one H - other I distance The numbering of the atoms is given in the inset.

3 The Burning of Air

A cluster of N_2/O_2 or several N_2/O_2 molecules embedded in a cluster can exhibit the long studied [20] burning of air. The primary issue for such a study is the nature of the many-atom potential which is used to describe a possibly multi-center reaction. While one can generalize the valence bond type [21] approximation due to London and known nowadays as LEPS, we chose a simpler route. The potential between the reactive atoms is one that has been used before in studies of many-atom systems [22]. It is of a form that allows for a weakening of a bond between a pair of atoms when one or more other reactive atoms are nearby. It also includes a long range physical interaction which describes the packing in a molecular liquid. With *i,j* etc. being indices of atoms, the *N* atom potential is given as a sum over all bonds:

$$V = \sum_{i=1}^{N} \sum_{j>1}^{N} V_{ij} \qquad (1).$$

$$V_{ij} = V_R(r_{ij}) - \tilde{b}_{ij} V_A(r_{ij}) + V_W(r_{ij})$$

Here V_{ij} is the bond potential which is the sum of a repulsive short range potential V_R and a corresponding longer range attractive potential V_A. When $\bar{b}=1$, these two terms specify the chemical part of the atom-atom potential. V_W is the long range van der Waals potential. The many body character of the potential (which means that V_{ij} can be a function not only of r_{ij} but also of the location of all the other reactive atoms) enters through b_{ij} which is a coordination number which serves to reduce the strength of the attractive chemical potential between atoms *i* and *j* when other reactive atoms are nearby:

$$\tilde{b}_{ij} = (b_{ij} + b_{ji})/2 \quad , \quad \overline{B} \equiv \sum_{i} \sum_{j \neq 1} \overline{b}_{ij}$$

$$b_{ij} = 1 \Big/ \left(1 + \sum_{k \neq i,j} f(r_{ik}) g(\theta_{ijk}) \exp((r_{ij} - r_{ij}^{o})/a) \right) \tag{2}$$

and the superscipt *0* designates an equilibrium value. $\overline{B}$ is the total coordination number. $f(r_{ik})$ is a cutoff function which turns off the influence of atom *k* when it is too far from atom *i* :

$$f(r) = \left(1 - \tanh((r - r^{o})/a')\right)/2 \tag{3}$$

$g(\theta_{ijk})$ is similarly an angular cutoff function. These definitions insure that $\overline{b}_{ij}$ is a monotonically decreasing function of the coordination number of atoms *i* and *j*. If there are no neighbors near the *i,j* pair, $\overline{b}_{ij} = 1$.

The first question is whether such a potential is realistic for our problem. Figure 9 compares it to the LEPS potential [2,11] for the four-center case, while Fig. 10 shows the potential for a six-center atomic configuration. It is to be expected [14,23] that a six-center bond switching will have a significantly lower barrier than the four-center case.

A note of caution, relevant to the entire discussion, should however be sounded. So far, the computational studies have not allowed for the participation of electronically excited states. The only available experimental *vs.* theoretical study [7] suggests that there is a velocity regime, say below 12 km/sec, where electronic excitation is indeed minor. More work on this point is clearly called for.

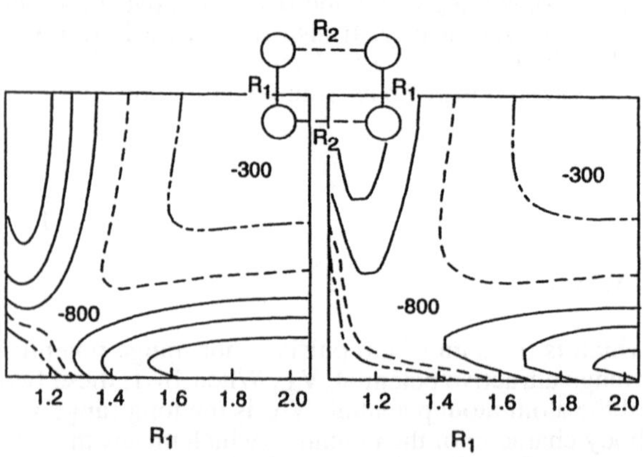

Figure 9: An equipotential contour plot for the potential energy (in kJ/mol) specified by Eqs. (1-3) and the LEPS potential [11] for a rectangular configuration of atoms in the four-center $N_2 + N_2 \rightarrow$ 2NN reaction. The axes are the old and new bond distances (in Å) which specify a rectangular configuration.

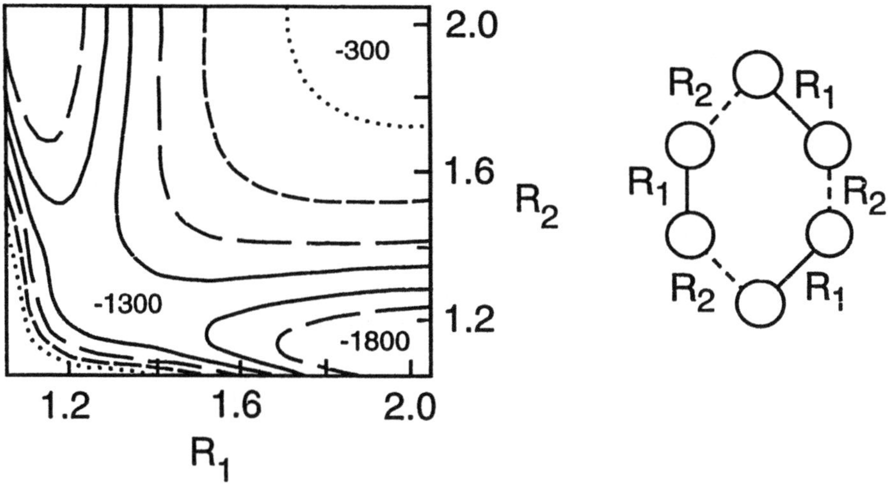

Figure 10: An equipotential contour plot for the potential energy (in kJ/mol) specified by Eqs. (1-3) for a hexagonal configuration of atoms in a six center exchange, as shown. The distances (in Å) are identified in the figure.

4 Concluding Remarks

Cluster impact chemistry provides a new dynamical regime, hitherto not accessible to systematic exploration. High barrier chemical reactions can be studied under *de facto* isolated gas phase collsion conditions. On the other hand, ternary (and even higher order) encounters, which were of special interest to Bodenstein [24], can be explored. The unique features of the scheme are the high material and energy density conditions which can be established very rapidly, leading to super-heating (which has already been experimentally demonstrated [7]) on a time scale shorter than even typical intramolecular events.

5 Acknowledgements

We thank R. Bersohn, W. Christen, U. Even, E. Hendell, J. Jortner, K.L. Kompa and I. Schek for discussions and for their interest. This work was supported by the Air Force Office of Scientific research, AFOSR. Computer time was provided by the Sonderforschungsbereich 377.

6 References

[1] T. Raz and R.D. Levine, J. Am. Chem. Soc. **116**, 11167 (1994).
[2] T. Raz and R.D. Levine, J. Phys. Chem. **99**, 7495 (1995).
[3] T. Raz and R.D. Levine, J. Phys. Chem., **99**, 13713 (1995).
[4] R.J. Beuhler and L. Friedman, Chem. Rev. **86**, 521 (1986).
[5] U. Even, I. Schek and J. Jortner, Chem. Phys. Lett. **202**, 303 (1993).

[6] R.W. Shaw et al., C&E News, December 23, Vol. **69**(51), 26 (1991).
[7] T. Raz, R.D. Levine and U. Even, J. Chem. Phys., **103**, 5394 (1995); E. Hendell, U. Even, T. Raz and R.D. Levine, Phys. Rev. Lett. **75**, 2670 (1995).
[8] I. Schek, J. Jortner, T. Raz and R.D. Levine, Chem. Phys. Lett. in press.
[9] C.W. Muhlhausen, L.R. Williams and J.C. Tully, J. Chem. Phys. **83**, 2594 (1985).
[10] T. Raz, I. Schek, M. Ben-Nun, U. Even, J. Jortner and R.D. Levine, J. Chem. Phys. **101**, 8606 (1994).
[11] T. Raz and R.D. Levine, Chem. Phys. Lett. **226**, 47 (1994).
[12] S. Pedersen, J. L. Herek, A. H. Zewail, Science **266**, 1359 (1994).
[13] R. Wolfgang, Acc. Chem. Res. **2**, 148 (1969).
[14] R.B. Woodward and R. Hoffmann, ***The Conservation of Orbital Symmetry;*** (Verlag Chemie, 1971).
[15] I. Schek, T. Raz, R.D. Levine and J. Jortner, J. Chem. Phys. **101**, 8596 (1994).
[16] T. Raz, I. Schek, J. Jortner and R.D. Levine, Proc. Yamada Cluster Conference (1995).
[17] M. Bodenstein, Z. phys. Chem. **13**, 56 (1893), *ibid.* **29**, 295 (1899).
[18] J.B. Anderson, J. Chem. Phys. **100**, 4253 (1994).
[19] J.A. Berson, Science **266**, 1338 (1994); R. Hoffmann, S. Swaminathan, B.G. Odell, R. Gleiter, J. Am. Chem. Soc. **92**, 7091 (1970).
[20] Priestley (1779); Cavendish (1785).
[21] J.H. Van Vleck and A. Sherman, Rev. Mod. Phys. **7**, 167 (1935).
[22] D.W. Brenner, D.H. Robertson, M.L. Ebert and C.T. Wight, Phys. Rev. Lett. **70**, 2174 (1993); J. Tersoff, Phys. Rev. B **37**, 6991 (1988).
[23] D.R. Herschbach, Pure & Appl. Chem. **47**, 61 (1976).
[24] M. Bodenstein, Z. phys. Chem. **100**, 118 (1922).

Part III

Investigations of Thermal Kinetics: Experiment

The Hydrogen-Iodine Reactions: 100 Years Later

J.B. Anderson
Department of Chemistry, The Pennsylvania State University
University Park, Pennsylvania 16802, U.S.A.

Abstract

Max Bodenstein opened the field of gas phase chemical kinetics in 1894 with his report of experimental studies of the hydrogen-iodine reaction $H_2 + I_2 \rightarrow HI + HI$ and its reverse $HI + HI \rightarrow H_2 + I_2$. Bodenstein measured the rates of the forward and reverse reactions, their equilibria, and their temperature dependence. He found second order kinetic expressions and an Arrhenius temperature dependence for the rate constants. He suggested several mechanisms for these reactions.

Modern theoretical and experimental studies have revealed additional details of the reaction kinetics. The results of these studies suggest that the low temperature thermal reaction proceeds by both the direct bimolecular reaction of hydrogen molecules with vibrationally excited iodine molecules $H_2 + I_2(\text{hi v}) \rightarrow HI + HI$ and the termolecular reaction of hydrogen molecules with iodine atoms $H_2 + I + I \rightarrow HI + HI$. The direct bimolecular reaction mechanism and the termolecular reaction mechanism were among those suggested by Bodenstein one hundred years ago.

1 Introduction

Max Bodenstein opened the field of gas phase chemical kinetics at Heidelberg one hundred years ago with the publication of his landmark paper [1] reporting experimental measurements of the rates of the hydrogen-iodine reaction, $H_2 + I_2 \rightarrow HI + HI$, and its reverse, $HI + HI \rightarrow H_2 + I_2$. In this first systematic study of the kinetics of chemical reactions in the gas phase he determined the effects of reactant concentrations and temperature for both reactions. He found that the reactions follow overall second-order kinetic expressions, that their rate constants have an Arrhenius temperature dependence, and that the equilibrium constant is given by the ratio of forward to reverse rate constants.

Bodenstein considered several possible mechanisms for the reactions including the direct bimolecular reactions $H_2 + I_2 \rightarrow HI + HI$ and $HI + HI \rightarrow H_2 + I_2$ as well as combinations of reactions involving intermediate steps such as $H + I_2 \rightarrow HI + I$ and $I + H_2 \rightarrow HI + H$. He also noted [2] that the termolecular reaction involving "the combination of existing iodine atoms with hydrogen would have to be looked (at): $2J + H_2 \rightarrow 2\,HJ$ ...". In modern terms this is $I_2 + M \leftrightarrow I + I + M$ and $H_2 + I + I \rightarrow HI + HI$.

Despite Bodenstein's considerations of alternative mechanisms the forward and reverse reactions were generally assumed to occur as direct bimolecular reactions and these reactions soon became "textbook examples" of bimolecular reactions. Lewis [3] used Bodenstein's data to demonstrate the applicability of Arrhenius' concept of active molecules and the Arrhenius rate expression. Hinshelwood [4] cited the hydrogen-iodine reaction and its reverse as evidence for

Springer Series in Chemical Physics, Volume 61
Gas Phase Chemical Reaction Systems
Eds.: J. Wolfrum, H.-R. Volpp, R. Rannacher, and J. Warnatz

the principle of activation by collision. Langevin and Rery [5] found the measured rates to be in agreement with a simple collision theory of reactions. Both Hinshelwood [4] and Kistiakowsky [6] raised the question of the nature of the activation energy and which forms of molecular excitation lead to reaction. With the discovery of deuterium the isotope effects on the reactions were examined [7,8]. One of the early applications of transition state theory was the prediction by Wheeler, Topley, and Eyring [9], with apparent success, of the rates of the hydrogen-iodine reactions.

In the 1950's the possibility of a chain reaction was investigated in detail by Benson and Srinivasan [10] and by Sullivan [11] who found the chain mechanism to contribute to reaction above 600 K and to be dominant above 750-800 K. New and accurate measurements of the rates of reaction at lower temperatures carried out by Sullivan [11,12] with more modern techniques confirmed the accuracy of the second-order rate constants determined by Bodenstein 60 years earlier.

At about the same time the possibility of a termolecular mechanism for the reaction of H_2 with I_2 was reintroduced by Semenov and others [13]. Since then the question of the bimolecular mechanism *versus* the termolecular mechanism has been investigated in several studies, both experimental and theoretical, but it has not been completely resolved. Nevertheless, it is now clear that the two mechanisms are really very similar. It is likely that both mechanisms contribute to the overall reaction.

The reaction of H_2 with I_2 to produce HI and HI is exothermic by about 3.0 kcal/mol. The measured activation energy for the low-temperature reaction of H_2 with I_2 is 40.6 kcal/mol and that for HI with HI is 43.6 kcal/mol. The dissociation energy of I_2 is 36.2 kcal/mol, 4.4 kcal/mol below the activation energy for the overall reaction of H_2 with I_2.

In Table 1 are summarized the energetically allowed mechanisms which have been proposed over the years for the overall reaction of H_2 with I_2. For the reverse reaction of HI with HI the mechanisms listed may simply be reversed. The direct reaction mechanism was listed explicitly by Bodenstein and, except for the one with the H_2I intermediate, the others were at least anticipated by Bodenstein. For a thermal equilibrium among the possible reactant species the mechanisms of direct bimolecular reaction, termolecular reaction, and that with the H_2I intermediate have rates proportional to the product $[H_2][I_2]$ as shown. They cannot be distinguished in simple rate experiments.

2 Sullivan's Photochemical Experiments

In 1967 Sullivan [14] reported experimental measurements of the rate of the overall reaction $H_2 + I + I \rightarrow HI + HI$. Using a low-temperature photochemical source to produce I atoms Sullivan measured their reaction rate with H_2 and determined the (apparent) rate constant for the termolecular reaction and its temperature dependence. Extrapolation of the rate constant to the higher temperature range of the thermal reaction data showed that the former could account for the entire thermal rate. It was thus shown that the dominant mechanism for the thermal reaction of H_2 with I_2 at temperatures below about 700 K is either the termolecular reaction $H_2 + I + I \rightarrow HI + HI$ or another mechanism which is kinetically indistinguishable from it.

Sullivan pointed out that the behavior observed in his experiments could be explained by a mechanism involving a loosely bound intermediate H_2I in rapid

Table 1. Possible mechanisms for the reaction $H_2 + I_2 \rightarrow HI + HI$

Direct bimolecular reaction

$H_2 + I_2 \rightarrow HI + HI$ slow

rate = k $[H_2]$ $[I_2]$

a subset

$I_2 + M \rightleftharpoons I_2(\text{hi v}) + M$ fast

$H_2 + I_2(\text{hi v}) \rightarrow HI + HI$ slow

rate = k $[H_2(\text{hi v})]$ $[I_2]$ = k K $[H_2]$ $[I_2]$

Termolecular reaction involving I atoms

$I_2 + M \rightleftharpoons I + I + M$ fast

$H_2 + I + I \rightarrow HI + HI$ slow

rate = k $[H_2]$ [I] [I] = k K $[H_2]$ $[I_2]$

via H_2I intermediate

$I_2 + M \rightleftharpoons I + I + M$ fast

$H_2 + I + M \rightleftharpoons H_2I + M$ fast

$H_2I + I \rightarrow HI + HI$ slow

rate = k $[H_2I]$ [I] = k K $[H_2]$ [I] [I] = k K K $[H_2]$ $[I_2]$

Chain reaction

$I_2 + M \rightarrow I + I + M$ (and others) initiation

$I + H_2 \rightarrow HI + H$ propagation

$H + I_2 \rightarrow HI + I$ propagation

$I + I + M \rightarrow I_2 + M$ (and others) termination

reversible equilibrium with reactants H_2 and I as for the H_2I mechanism listed in Table 2. This mechanism is kinetically indistinguishable from the termolecular reaction in Sullivan's photochemical experiments. Another possibility was suggested by classical trajectory calculations and pointed out by Jaffe *et al.* [15] and by Anderson [16]. The observed behavior could be explained by a mechanism involving vibrationally excited iodine molecules $I_2(\text{hi v})$ in rapid reversible equilibrium with reactants I + I as for the $I_2(\text{hi v})$ mechanism listed in Table 2. This mechanism is also kinetically indistinguishable from the termolecular reaction in Sullivan's photochemical experiments.

Table 2. Possible mechanisms for the reaction $H_2 + I + I \rightarrow HI + HI$

Termolecular		
	$H_2 + I + I \rightarrow HI + HI$	slow
	rate = k $[H_2]$ [I] [I]	
via $I_2(\text{hi v})$ intermediate		
	$I + I + M \rightleftharpoons I_2(\text{hi v}) + M$	fast
	$H_2 + I_2(\text{hi v}) \rightarrow HI + HI$	slow
	rate = k $[H_2]$ $[I_2(\text{hi v})]$ = k K $[H_2]$ [I] [I]	
via H_2I intermediate		
	$H_2 + I + M \rightleftharpoons H_2I + M$	fast
	$H_2I + I \rightarrow HI + HI$	slow
	rate = k $[H_2I]$ [I] = k K $[H_2]$ [I] [I]	

Sullivan's photochemical experiments eliminate the direct bimolecular reaction of H_2 with I_2 in low vibrational states as the mechanism for thermal reaction, but they leave open the possibility of the direct bimolecular reaction of H_2 with I_2 in high vibrational states.

3 Potential Energy Surfaces

The earliest calculations of the potential energy surface for the H_2-I_2 system were semiempirical London-Eyring-Polanyi (LEP) calculations by Wheeler, Topley, and Eyring [9] which predicted reaction of H_2 with I_2 to occur through a trapezoidal transition state. Much later, Hoffmann [17] gave arguments based on orbital symmetry suggesting the trapezoidal configuration to be energetically unlikely, and Cusachs, Krieger, and McCurdy [18] reported 16-electron semiempirical molecular orbital calculations suggesting the reaction barrier to be lower for the collinear configuration I-H-H-I than for a trapezoidal configuration.

Raff *et al.* [19] used a semiempirical four-electron valence bond formalism to obtain a complete potential energy expression of the London-Eyring-Polanyi-Sato (LEPS) type. Their surface allows reaction through both the trapezoidal configuration (barrier 42.0 kcal/mol above the minimum for separated H_2 and I_2) and the collinear configuration (barrier 45.6 kcal/mol). In similar four-electron valence-bond calculations Minn and Hanratty [20] found the trapezoidal barrier slightly higher than the collinear, but both barriers were about 50 percent higher than the experimental activation energy of 40.6 kcal/mol. We made an *ad hoc* modification [16] of the LEPS surface of Raff *et al.* [19] for use in classical trajectory calculations by lowering the barrier in the collinear region (to 39.6 kcal/mol) and raising it in the trapezoidal region (to 43.5 kcal/mol) to produce a surface favoring collinear reaction. More recent electronic structure calculations

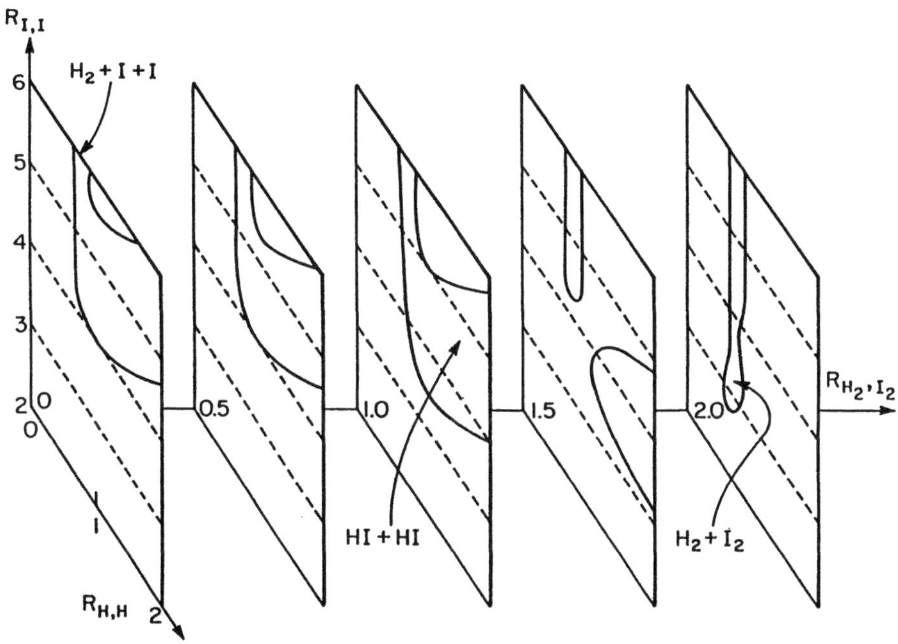

Figure 1: Calculated potential energy surface in $(R_{H,H}, R_{I,I}, R_{H_2,I_2})$ space. The curves are potential energy contours 5.0 kcal/mol above the energy at the saddle point. From Anderson [21].

indicate a potential surface favoring the collinear reaction path. We recently [21] carried out analytic variational calculations with single- and double-substitution configuration interactions for the 16 valence shell electrons. An effective core potential was used to replace the 46 electrons corresponding to the $\mathrm{Kr}(4d)^{10}$ inner-shell electrons of each iodine atom. The basis set and effective potentials were those of Hay and Wadt [22] incorporated into standard Gaussian 92 programs. The resulting potential energy surface is illustrated in Fig. 1. The minimum in the barrier occurs for a collinear configuration 35.1 kcal/mol above the minimum for separated H_2 and I_2. The barrier rises with distortion toward a trapezoidal configuration and no saddle point for a trapezoidal configuration is observed. The surface is qualitatively similar to the modified LEPS surface favoring the collinear path mentioned above. Since the potential energy is a function of six coordinates

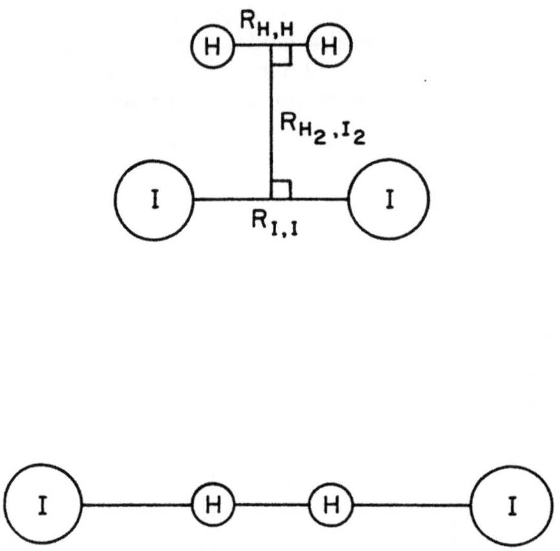

Figure 2: Symmetric trapezoidal and collinear configuations lying on prospective pathways for the hydrogen-iodine reaction. From Anderson [21].

the surface cannot be described fully by a two- or three-dimensional contour plot. However, since any deviation from the planar, right-angle symmetry of the configurations shown in Fig. 2 results in an increase in the potential energy, the low energy pathways for reaction may be traced in a reduced configuration space limited to that symmetry. In Fig. 1 our analytic variational [21] potential energy surface is mapped as a function of the remaining three coordinates R(H-H), the distance between the H nuclei, R(I-I) between the I nuclei, and R(H_2-I_2) between the centers of masses of H_2 and I_2. The minimum in the barrier to reaction from either $H_2 + I_2$ or $H_2 + I + I$ to HI + HI occurs for R(H_2-I_2) = 0, i.e., in the first plane from the left in Fig. 1. The contour lines indicated are those for a potential energy 5.0 kcal/mol above the minimum in the barrier. Thus, the paths of reactive and non-reactive trajectories with energies less than 5.0 kcal/mol above the minimum for reaction are constrained between the contours. It may be seen that the minimum in the barrier is accessible to $H_2 + I_2$, $H_2 + I + I$, and HI + HI.

The reliability of such calculations is not completely known and the resulting surface cannot be considered as final. Nevertheless, the same calculations gave potential energy curves for the diatomics H_2, I_2, and HI which are qualitatively correct and barrier heights for the three-body reactions $H + I_2 \rightarrow HI + I$ and $I + H_2 \rightarrow HI + H$ which are qualitatively correct. Thus, we may expect the predictions for the H_2-I_2 reaction to be qualitatively correct.

4 Classical Trajectory Studies

Classical trajectory calculations for the reaction $H_2 + I_2 \rightarrow HI + HI$ and its reverse have been carried out for two potential energy surfaces. Such calculations are not easy to perform because of the large number of possible states of reactant species, the mismatch of the masses of H and I atoms, and the low probability of reaction. The light H atoms require small time steps to avoid time-step error and the heavy I atoms require a long time for movement. The probability of reaction of two HI molecules, randomly selected from pairs at 700 K with sufficient total energy to react, is about 1 in 10^6 [15]. In a thermal system collisions of H_2 with I_2 (or 2I) which result in reaction to form HI + HI are indeed rare events.

The first set of calculations was reported in 1970 by Raff *et al.* [23] for the LEPS surface [19] favoring reaction through the trapezoidal configuration. Raff *et al.* [23] followed trajectories starting from the separated reactants $H_2 + I_2$ in a variety of combinations of rotational, vibrational, and translational energies, but excessive computational requirements prevented them from discovering the primary reaction pathway for that surface.

A second set of calculations was reported by Jaffe, Henry, and Anderson [15] for the same LEPS surface. Using the combined phase-space/trajectory approach devised earlier by Keck [24] and Jaffe *et al.* [15] were able to gain a factor of about 10^6 in calculation efficiency relative to that of conventional calculations and they obtained a sample of 693 reactive trajectories representative of reaction at 700 K for that surface. The gain in efficiency which allows such rare events to be investigated results from sampling trajectories crossing a dividing surface in the region of the barrier and following them forward and backward to determine their complete paths. The method has since been applied in dynamics problems ranging from simple reactions in the gas phase to protein folding in solution to binary star formation in space. It has come to be known as the "rare event" method.

Reaction on the LEPS surface was observed to occur through both the trapezoidal and the collinear saddle point regions. For trajectories passing through the trapezoidal region the reactants were H_2 molecules and I_2 molecules with vibrational energies less than half the dissociation energy of I_2. For trajectories passing through the collinear region the reactants were H_2 molecules and vibrationally excited bound and quasibound I_2 molecules with energies near the dissociation energy along with some separated I atom pairs. In either case the product HI molecules were found to have substantial vibrational excitation.

A third set of trajectory calculations [16] was carried out with the modified LEPS surface favoring collinear reaction. On this surface reaction was observed almost exclusively through the region of the collinear saddle point. Reactant energy was found concentrated primarily in I_2 vibration and dissociation and, to a lesser extent, in H_2 translation. The product energy was found primarily in HI vibration. The calculated reactant distribution was

$H_2 + I_2$(bound, hi v) 42 percent,
$H_2 + I_2$(quasibound) 13 percent,
$H_2 + I + I$ 45 percent.

The calculated rate constant for the overall reaction of H_2 with I_2 at 700 K was 11.6 $cm^3\ mol^{-1}\ sec^{-1}$, which may be compared with the corresponding experimental value [12] of 29.3 $cm^3\ mol^{-1}\ sec^{-1}$.

Examination of individual trajectories for the modified LEPS surface reveals reaction of vibrationally excited bound or quasibound I_2 with H_2 through the nearly collinear configuration I-H-H-I. With the slowly moving I_2 molecule stretched to an internuclear distance of about 5 Å, the H_2 molecule approaches from the side and inserts between the two atoms. The H atoms separate and as the I atoms move apart one H atom goes with each. The reaction of I + I with H_2 is revealed to be almost identical. As the slowly moving I atoms pass by each other at a distance of about 5 Å, the H_2 molecule approaches from the side and inserts between the two atoms. The H atoms separate and as the I atoms move apart one H atom goes with each. Thus, the bimolecular reaction $H_2 + I_2$(hi v) $\rightarrow$ HI + HI is not much different from the termolecular reaction $H_2 + I + I \rightarrow$ HI + HI.

5 Experiments in Molecular Dynamics

Experiments to determine the energy requirements for the reaction HI + HI $\rightarrow$ $H_2 + I_2$ (or 2I) were reported by Jaffe and Anderson [25] in 1968. In these DI was substituted for one HI and the experiments were designed to determine cross sections for the formation of HD in energetic collisions of HI with DI. Molecular beams of HI, accelerated to energies of 40 to 215 kcal/mole by expansion from nozzles in mixtures with hydrogen or helium, were passed through a reaction chamber containing DI at low density. In the range of collison energies investigated, at one-half to two and one-half times the activation energy of 43.6 kcal/mol, no HD attributable to reaction was detected. The upper limit to the reaction cross section was determined to be 0.04 $Å^2$. The results indicate that translational energy is not effective in promoting the reaction and that vibrational or rotational excitation of one or both of the collision partners is required for an appreciable probability of reaction.

An investigation of the effects of vibrational excitation of HI in promoting the reaction HI + HI $\rightarrow H_2 + I_2$ was reported in 1981 by Horiguchi and Tsuchiya [26]. In these experiments the rates of the reaction were measured in mixtures with carbon dioxide which was excited vibrationally by irradiation with light from a cw laser. An enhancement of reaction rate by a factor of about 2.5 was observed for mixtures with carbon dioxide compared to those without carbon dioxide when both were irradiated. The enhancement was attributed to vibrational excitation of HI through collisional transfers of energy from laser-excited carbon dioxide. The results indicate that vibrational excitation of one or both of the colliding HI molecules promotes the reaction.

6 Discussion

We have argued previously [16] that if vibrationally excited I_2(hi v) formed in recombination of I atoms in Sullivan's photochemical experiments is slow, the bimolecular and termolecular mechanisms would be kinetically indistinguishable in Sullivan's photochemical experiments. Recent measurements of the vibrational relaxation of $I_2(v = 43)$ by Nowlin and Heaven [27] have shown the rate to be very slow. This gives additional support to the arguments for indistinguishability in Sullivan's experiments.

However, very slow vibrational relaxation of I_2 leads to additional questions pointed out by Truhlar [28]. If the upward relaxation is sufficiently slow that it becomes the slow step or bottleneck in mechanisms involving reaction of H_2 with I_2(hi v) or I + I, then these species would be depleted by reaction and the mechanisms would be kinetically distinguishable from reaction involving unexcited iodine molecules. Perhaps the relative rates of relaxation and reaction are similar and the effect of slow relaxation has not yet been detected.

Our recent electronic structure calculations yield a potential energy surface adequate to explain, at least qualitatively and within the uncertainties due to an incomplete knowledge of relaxation rates, the available experimental observations for the hydrogen-iodine reaction. The rate expressions, the rate constants, their temperature dependence, the vibrational excitation of HI products, the excitation and/or dissociation of reactant I_2, the photochemical rates - all are compatible with the recent ab initio potential energy surface and with the classical trajectory calculations carried out with a similar surface. And all are compatible with either the bimolecular or termolecular mechanisms. It appears most likely that both mechanisms contribute, but the matter is not resolved as yet.

7 Acknowledgments

Support of this work by the National Science Foundation (Grant No. CHE-8714613) and the Office of Naval Research (Grant N00014-92-J-1340) is gratefully acknowledged. The author is indebted also to the Humboldt Foundation for the opportunity to carry out this work.

8 References

[1] M. Bodenstein, Z. phys. Chem. **13**, 56 (1894).
[2] Ref. 1, p. 122.
[3] W. C. McC. Lewis, J. Chem. Soc. **113**, 471 (1918).
[4] C. N. Hinshelwood, ***The Kinetics of Chemical Change in Gaseous Systems*** (Oxford University Press, Oxford, England 1926) p. 52.
[5] P. Langevin and J.-J. Rery, Radium (Paris) **10**, 142 (1913).
[6] G. B. Kistiakowsky, J. Am. Chem. Soc. **50**, 2315 (1928).
[7] J. C. L. Blagg and G. M. Murphy, J. Chem. Phys. **4**, 631 (1936).
[8] K. H. Geib and A. Lendle, Z. phys. Chem. (Abt. B) **32**, 463 (1936).
[9] A. Wheeler, B. Topley, and H. Eyring, J. Chem. Phys. **4**, 178 (1936).
[10] S. W. Benson and R. Srinivasan, J. Chem. Phys. **23**, 200 (1955).
[11] J. H. Sullivan, J. Chem. Phys. **30**, 1291 (1959).
[12] J. H. Sullivan, J. Chem. Phys. **30**, 1577 (1959); *ibid.* **36**, 1925 (1962); *ibid.* **39**, 300 (1963).
[13] N. N. Semenov, ***Some Problems in Chemical Kinetics and Reactivity,*** Vol. **2** (Princeton University Press, Princeton, New Jersey 1959) pp 73 and 74.
[14] J. H. Sullivan, J. Chem. Phys. **46**, 73 (1967).

[15] R. L. Jaffe, J. M. Henry, and J. B. Anderson, J. Am. Chem. Soc. **98**, 1140 (1976); J. M. Henry, J. B. Anderson, and R. L. Jaffe, Chem. Phys. Lett. **20**, 138 (1973); J. B. Anderson, J. M. Henry, and R. L. Jaffe, J. Chem. Phys. **60**, 3725 (1974).
[16] J. B. Anderson, J. Chem. Phys. **61**, 3390 (1974).
[17] R. Hoffmann, J. Chem. Phys. **49**, 3739 (1968).
[18] L. Cusachs, M. Krieger, and C. W. McCurdy, J. Chem. Phys. **49**, 3740 (1968).
[19] L. M. Raff, L. Stivers, R. N. Porter, D. L. Thompson, and L. B. Sims, J. Chem. Phys. **52**, 3449 (1970).
[20] F. L. Minn and A. B. Hanratty, J. Chem. Phys. **53**, 2543 (1970); Theor. Chim. Acta **19**, 390 (1970).
[21] J. B. Anderson, J. Chem. Phys. **100**, 4253 (1994).
[22] P. J. Hay and W. R. Wadt, J. Chem. Phys. **82**, 299 (1985).
[23] L. M. Raff, D. L. Thompson, L. B. Sims, and R. N. Porter, J. Chem. Phys. **56**, 5998 (1972).
[24] J. C. Keck, Disc. Faraday Soc. **33**, 173 (1962).
[25] S. B. Jaffe and J. B. Anderson, J. Chem. Phys. **49**, 2859 (1968); *ibid.* **51**, 1059 (1969).
[26] H. Horiguchi and S. Tsuchiya, Int. J. Chem. Kin. **13**, 1085 (1981).
[27] M. L. Nowlin and M. C. Heaven, J. Chem. Phys. **99**, 5654 (1993).
[28] D. G. Truhlar, discussion (1995).

Recent Advances in the Measurement of High-Temperature Bimolecular Rate Constants

J. V. Michael
Chemistry Division, Argonne National Laboratory
Argonne, IL 60439 USA

Abstract

Recent advances in the measurement of high-temperature reaction rate constants are discussed. The studies carried out by shock tube methods are particularly considered because these results are important not only in theoretical chemical kinetics but also in practical applications. The work on five chemical reactions are reviewed in detail. These are: D + H_2, Cl + H_2, H + O_2, CH_3 + CH_3, and H + NO_2.

1 Introduction

The field of thermal gas phase chemical kinetics and, in particular, bimolecular reactions, has continually expanded since the first such study by Bodenstein about 100 years ago [1]. There are two main reasons for this continuing interest. First, most of these reactions occur with activation barriers that can be accurately determined from the thermal experiment. These barriers can then be compared to theoretical potential energy surfaces from modern *ab initio* electronic structure calculations that in turn can then serve as the basis for theoretically estimating the rate behavior with modern dynamical theories. Hence, the first reason for this type of study is to support theoretical chemistry and vice versa. However, there is a second and equally important motivation that is practical. A number of naturally occurring and anthropogenically altered environmental systems are now known to be influenced by the thermal rates of chemical processes. These include the earth's mesosphere, stratosphere, and troposphere; the various planetary atmospheres; interstellar clouds; air pollution; and high temperature incineration and, in general, combustion chemistry. Most gas phase kinetics investigations are currently being supported for use in developing mechanistic descriptions for these practical systems.

In this paper, current work in the field of high temperature bimolecular rate processes is considered. There have been two significant experimental developments in the past ~20 years, namely, the high temperature fast-flow reactor (HTFFR) method with the related high temperature photochemistry (HTP) option [2] and the flash or laser photolysis-shock tube (F/LP-ST) technique [3]. Both can be used for temperatures in excess of 1000 K; however, in favorable cases, the latter can be extended to ~2500 K. Hence, the combination of the F/LP-ST and the classic thermal shock tube experiment [4] has proven to be particularly powerful in providing rate constant data over very large temperature ranges. There are two recent reviews of high temperature bimolecular reactions studied in shock tubes that are particularly relevant [5,6].

In the remainder of the paper, selected examples of shock tube studies from our laboratory will be considered. In particular, the experimental and theoretical results on the simplest and most important reaction of all, namely H + H_2 (and deuterated analogues), will be discussed in order to illustrate the first motivation above. Subsequently, results on another important "theoretical test case" reaction, Cl + H_2 (and D_2) will be reviewed. In an attempt to illustrate the second motivation,

Springer Series in Chemical Physics, Volume 61
Gas Phase Chemical Reaction Systems
Eds.: J. Wolfrum, H.-R. Volpp, R. Rannacher, and J. Warnatz

recent results on the most important reaction in all combustion systems, $H + O_2 \rightarrow OH + O$, will be discussed. A vibrationally hot species is initially formed followed by a forward dissociation process that is higher lying than reactants. High temperature results will then be presented for two barrierless bimolecular rate processes that are also chemical activation cases, namely, $2CH_3 \rightarrow C_2H_5 + H$ and $H + NO_2 \rightarrow OH + NO$.

2 Experimental

The F/LP-ST technique has been described previously [3,7]. Figure 1 shows a schematic diagram of the shock tube apparatus. He is used as the driver gas, and the test gas is mostly Ar (or Kr) with small quantities of added source and reactant molecules. The source molecule on photolysis gives the transient species that is spectroscopically measured as it reacts with the reactant molecules. Two different molecules are generally used requiring accurate premixture determinations with capacitance manometers.

Experiments are performed behind reflected shock waves where the hot gas is effectively stagnant and not flowing. Flash or laser photolysis occurs after the reflected shock wave has traversed the spectroscopic observation station. Transient species are observed radially across the shock tube. Reflected shock pressure and temperature are kept low so that thermal decomposition is minimized. The initial transient species concentration is initiated by photolysis, and its decay is then totally determined by bimolecular reaction. Diffusion out of the viewing zone is negligibly slow on the experimental time scale. This experiment is then an adaptation of the static kinetic spectroscopy experiment with the reflected shock serving as a source of high temperature and density; i. e., shock heating is equivalent to a pulsed furnace.

Similar thermally induced experiments [4] can be carried out if a molecule exists that thermally decomposes to transients on a shorter time scale than that for transient decay through bimolecular reaction. Additionally, the reactant molecule must be thermally stable under thermodynamic conditions where the source has completely dissociated. In this type of experiment, the zero time is set by shock wave passage past the observation station. In both types of experiments, pressure transducers, mounted equidistant along the shock tube, are used to measure the incident shock wave velocity. Temperature and density in the reflected shock wave regime are calculated from incident shock velocities through relations and correction procedures [8] that account for boundary layer perturbations. Since the initial composition is known, the thermodynamic state of the system is fully determined.

In the atomic resonance absorption spectrometric (ARAS) adaptation of the methods, atomic species are spectroscopically monitored as a function of time. H- [7,9], D- [7,10], O- [7,11], N- [12], Cl- [13] and I-atom [14] reactions have been studied. Beer's law holds if absorbance, (ABS), is kept low, and then $(ABS) \equiv -\ln(I/I_0)$ (I and I_0 are transmitted and incident intensities of the resonance light, respectively) is proportional to the atomic concentration. If the decay of atom A is controlled by a bimolecular reaction, A + R, where R is the stable reactant molecule, then the decay rate is pseudo-first-order provided [R]>>[A]. Because (ABS) is proportional to [A], observation of $(ABS)_t$ is sufficient to determine the decay constant. Values for k_{bim} for each experiment are then determined by dividing the decay constant by [R]. The results from many experiments are usually displayed as Arrhenius plots. If a reaction is pressure dependent, experiments can also be carried by varying total density. Termolecular reactions can therefore be studied. In certain cases, chemical isolation is not possible, and numerical chemical simulations of the

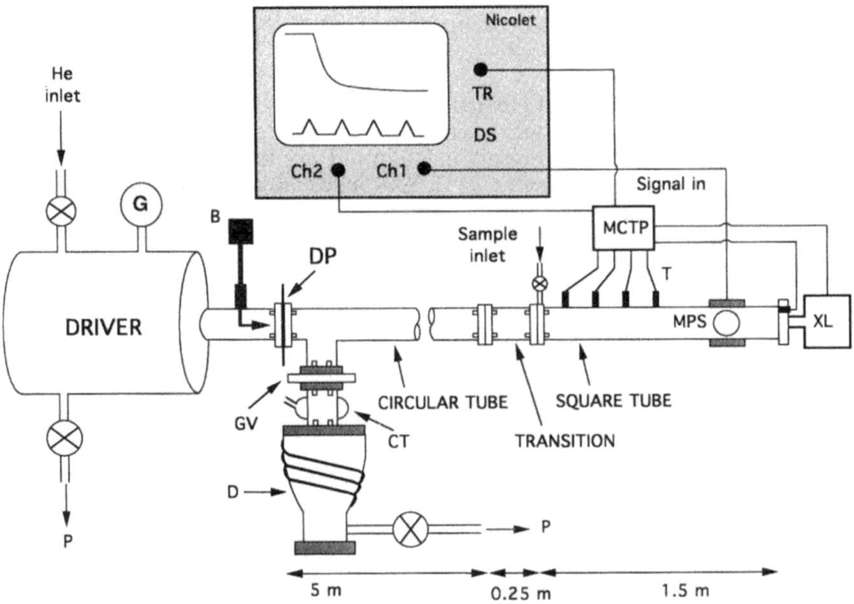

Figure 1: Schematic diagram of the shock tube apparatus

important chemical reactions then become necessary. In these cases, concentration profiles have to be derived, and this means that the atomic absorption curve-of-growth has to be carefully determined.

Lastly, we have recently modified the apparatus so that multi-pass optics can be incorporated into the photometric system [15]. The optical path length is 105 cm, and this technique has now been successfully applied to detect, CF_2 [16] and OH [17] radicals in two kinetics investigations.

3 Results and Discussion

In the sections to follow, rate data for 5 bimolecular reactions are reviewed. The first two will be compared to theoretical calculations that use modern methods for both potential energy evaluation and dynamical estimation for thermal rate constants. The latter three involve the formation of vibrationally excited species that subsequently forward dissociate into products.

3.1 D + H_2 → HD + H and H + D_2 → HD + D: FP-ST experiments have been carried out on these reactions over the temperature range, ~650 to 2200 K [9,10], and the absolute rate constant results have already been reviewed [6,7,18]. Because of the historical importance of these reactions in gas phase chemical kinetics, there are a large number of earlier lower temperature studies primarily due to Le Roy and co-workers [19], Westenberg and de Haas [20], and Jayaweera and Pacey [21]. All sets of data for D + H_2 have been combined to give the Arrhenius plot shown in Fig. 2. Both reactions have been evaluated giving analytic expressions for the rate behaviour over quite large temperature ranges [9,10].

Because these are the simplest reactions, they have served from the beginnings of gas phase chemical kinetics as a test case for theory. The theoretical potential energy surface for H + H_2 is known with sufficient accuracy that adjustments of vibrational frequencies and/or total electronic binding energy are not allowed; i. e., electronic structure and subsequent dynamics calculations can no longer be parameterized. *Ab initio* potential energy calculations are from Liu [22], Siegbahn and Liu [23], Truhlar and Horowitz [24], Varandas et al. [25], Diedrich and Anderson [26], and Boothroyd et al. [27] giving the LSTH [22-24], DMBE [25], and BKMP [27] surfaces. The LSTH surface has then been used in variational transition state theory (VTST) [28] and also in reduced dimensionality collinear exact quantum bend (CEQB) [29] calculations to estimate the thermal rate behaviour for D + H_2 and H + D_2, and the results are in fairly good agreement with the experimental evaluations. Garrett et al. [30] calculated VTST rate constants for D + H_2 and Michael et al. [18] calculated CEQB rate constants for both reactions with the DMBE potential energy surface. These results show improvement over the original LSTH calculations particularly in the low temperature region where tunnelling is dominate.

In addition to the thermal studies, researchers have also investigated the two processes using well established experimental dynamical methods [31]. These experiments involve hot hydrogen atoms and therefore sample the potential energy surface well above the potential barrier height. Several calculations using the LSTH surface have been carried out with modern quantum scattering theory, and one of the

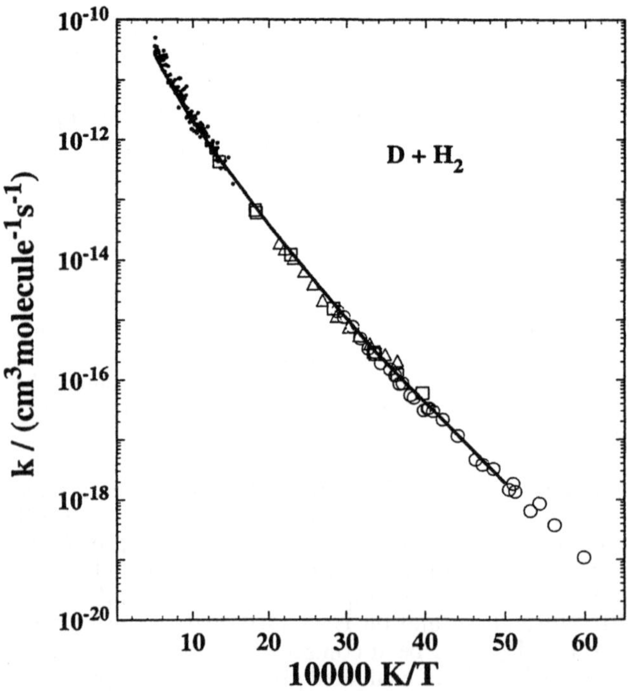

Figure 2: Data points from Ref. 10 and Ref. 19-21. Line is Theoretical from Ref. 35.

most notable explains the differential cross section data at 1.29 eV relative energy quite well if the geometric phase effect is taken into account [32]. This agreement suggests that the LSTH surface is relatively accurate above the barrier height.

Since the thermal results sample the potential energy surface below the barrier, we suggested [18] that quantum scattering theory should also be used to rationalize these data. Subsequently, studies have recently appeared by Park and Light [33] and Mielke et al. [34], and, in latter case, excellent agreement with the D + H_2 experiments is obtained if the DMBE surface is used. The newer BKMP surface gives worse agreement with experiment particularly in the lower temperature tunnelling region. Mielke et al. also compare their cumulative reaction probabilities with Wang and Bowman's [35] modified J-shifted CEQB method on the DMBE surface. This theory-to-theory comparison is exceptionally good. Hence, the thermal rate behaviour predicted by Wang and Bowman agrees well with Mielke et al. and also with experiment as shown in Fig. 2. It should be noted that the new CEQB results are almost identical to the earlier calculation on the DMBE surface [18] and also to DMBE surface based simple conventional transition state theory calculations with Wigner tunnelling (CTST/W) from 300 to 2200 K [9,10].

The agreement between theory and experiment shown in Fig. 2 is impressive. If additional quantum scattering calculations were to be carried out on H + D_2 the CEQB-DMBE results [18] would undoubtedly be reproduced, and therefore, such calculations are probably not necessary. However, Fleming and co-workers have measured the T-dependence of Mu + H_2 and D_2 [36]. The Mu-atom has 1/9 the mass of H, and therefore, tunnelling is still quite significant at higher temperatures than in H + D_2 and D + H_2. Even though VTST [37] and quantum scattering [38] calculations on the LSTH surface have already been reported, a rigorous quantum scattering calculation using the DMBE surface would be particularly interesting for rationalizing these data.

It is true that the theoretical H + H_2 work to date represents the most significant theoretical development in dynamics and kinetics since the first experiments by Bodenstein [1]. For nearly 70 years, countless theoreticians have been involved in this effort. The H + H_2 reaction served as the inspiration for transition state theory and for developing more accurate electronic structure calculational methods. To many experimentalists, the credibility of all other theoretical advances in dynamics and kinetics has really rested on the solution to the H + H_2 problem. In view of the fact that good agreement has been obtained between the above mentioned experiments and theory, one is tempted to declare that the H + H_2 reaction is now solved. Clearly theoreticians should be congratulated on this major accomplishment.

3.2 Cl + H_2 → HCl + H and Cl + D_2 → DCl + D: Since the work of Nernst [39], this reactive system has also served as one of the test cases for thermal bimolecular reaction rate theory [40-42]. With the Cl-atom ARAS method bimolecular rate constants have been measured between 296 and 3000 K using both thermally and photochemically (LP-ST) produced Cl-atoms [43]. Equilibrium constants were also measured and were found to be in good agreement with JANAF values [44]. When combined with several lower temperature studies, the evaluated experimental results can be expressed by,

$$k_{Cl,H_2} = 2.52 \times 10^{-11} \exp(-2214\ \mathrm{K}/T) \text{ for } 199 \le T \le 354\ \mathrm{K}, \quad (1)$$

$$k_{Cl,H_2} = 1.57 \times 10^{-16}\, T^{1.72} \exp(-1544\ K/T) \text{ for } 354 \le T \le 2939\ K, \text{ and} \quad (2)$$

$$k_{Cl,D_2} = 2.77 \times 10^{-16}\, T^{1.62} \exp(-2162\ K/T) \text{ for } 255 \le T \le 3020\ K, \quad (3)$$

all in molecular units. Arrhenius plots of these evaluations are shown in Fig. 3.

Tucker et al. [42] considered 11 potential energy surfaces, some of which had been previously derived using semi-empirical methods, in their improved canonical variational transition state theory with least action ground-state transmission coefficient (ICVT-LAG) calculations. These rate constant predictions were compared to the lower temperature experimental data obtained before 1985. In our work [43], we continued the comparison into the high temperature region (i. e., with Eq. (1)-(3)) using the CTST/W method. At this level of theory, none of the 11 surfaces from Tucker et al. can explain the data. Harding has performed new high level *ab initio* electronic structure calculations [45], and the CTST/W comparison to experiment is still unsatisfactory. CTST calculations using Eckart tunnelling, CTST/E, were also carried out, and there was no significant improvement in predictions. The data summarized by Eq. (1)-(3) and plotted in Fig. 3 exhibits too little curvature, suggesting that the barrier height and/or bending or imaginary vibrations are in error for all 12 of the surfaces. Accordingly, these quantities are parametrically varied until good agreement is found for $Cl + H_2$ at the CTST/E level of calculation. For $Cl + H_2$, $V^{\ddagger} = 6.77$ kcal mole^{-1}, $\nu_b = 782$ cm^{-1}, and $\nu_i = 1000i$ cm^{-1} gives the CTST/E

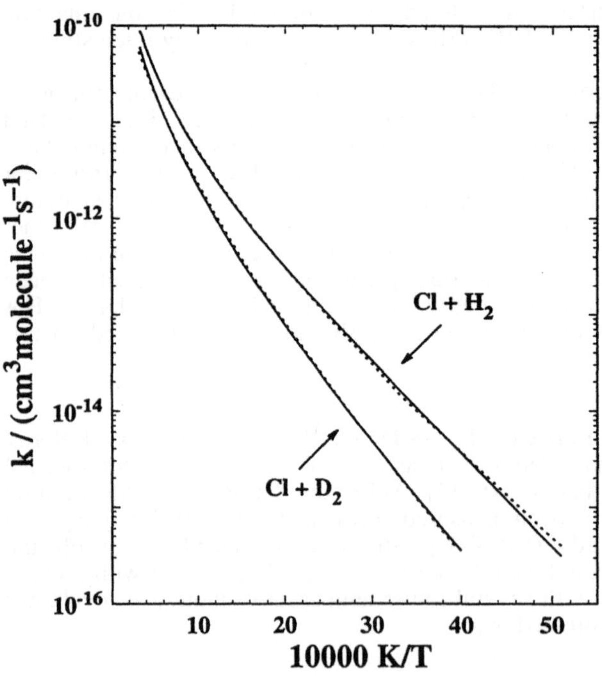

Figure 3: Solid lines are experimental evaluations (Ref. 43). Dashed lines are from CTST/E theory (see text).

theoretical prediction shown in Fig. 3. These saddle point properties are consistently transformed for Cl + D_2, and this prediction is also shown in Fig. 3. Even though the comparisons are excellent, the derived saddle point properties are not unique. Similar agreement can be obtained with other combinations of $V^{\ddagger}$, ν_b, and ν_i at the CTST/E level. Clearly, quite different conclusions about the saddle point and the reaction path might be found if these data and the new hot atom dynamical data of Alagia et al. [46] were to be considered using modern dynamical theory in exactly the same way as in the H + H_2 reaction. Such an approach might reconcile the experiments with the newest and most accurate potential energy surface by Harding [45].

3.3 H + O_2 → OH + O and D + O_2 → OD + O: This reaction is the most important reaction in combustion. It is the principal branching reaction in the oxidation of H_2 but also occurs in the oxidation mechanisms of all hydrocarbons [47]. This branching reaction has the highest enthalpy of all reactions that significantly consume O_2 with most other O_2 consuming processes being linear propagation reactions. Because of this it is usually the most important "slow step" in combustion mechanisms and is therefore often rate controlling. Since it is so important, the rate data have been reviewed many times [48]. There are a number of notable experimental studies on H + O_2 dating from 1967 to the present [49,50]; however, there has been only one direct study [49m] reported for D + O_2. As is to be expected, there is a substantial theoretical effort on these reactions [51].

The data and theoretical calculations up to 1992 have already been evaluated and reviewed [48d]; however, in the past three years there have been additional experimental work and developments. Ryu et al. [50] have made new measurements, have included new measurements by Yang et al. [49o], and have incorporated a new interpretation of earlier results by Yu et al. [49p] in an evaluation that takes into account all of the earlier direct studies [49]. This grand evaluation yields the result in molecular units,

$$k_{H,O_2} = 1.30 \times 10^{-10} \exp(-7105\ K/T) \text{ for } 960 \leq T \leq 5330\ K, \qquad (4)$$

Schott's experiments [49c] and Miller's [51a-c] theory both suggested a negative T-dependence in the A-factor, and this catalyzed the new interest starting in ~1981. Quasiclassical trajectory calculations (QCT) [51a-c] on the Melius-Blint potential energy surface [52] were used to rationalize the negative T-dependence as being due to increased re-crossing as temperature increases; however, it has subsequently been shown that this surface is in error. More recently, Varandas and co-workers have used their DMBE IV surface in QCT, QCT-quantum mechanical threshold (labeled QMT), and QCT-internal energy quantum mechanical threshold (labeled IEQMT) calculations [51i,j], and their results at 5 temperatures ranging from 1000-3000 K indicate that re-crossing exists but has no major effect because their implied A-factor is not strongly T-dependent in agreement with Eq. (4). These workers have also theoretically examined the rate behaviour for D + O_2 and find an isotope effect that is near to unity in agreement with experiment [49m].

Lastly, this new evaluation, Eq. (4), is only ~20-40% higher than the recommendation of Baulch et al. [48a]. This 1972 evaluation was based primarily on induction period and explosion limit measurements in the H_2/O_2 system showing that these early measurements were of high quality.

3.4 $CH_3 + CH_3 \rightarrow C_2H_5 + H$: This reaction was studied by observing H-atom formation rates using the ARAS technique in the thermal dissociation of CH_3I [53]. CH_3I dissociation is fast compared to the subsequent bimolecular dissociative recombination. In the analysis, all of the absolute $[H]_t$ results were fitted using numerical simulations on a mechanism that consisted of several reactions most of which were known. The main complication came from thermal cracking of CH_3 to give either CH_2 + H or CH + H_2. These processes dominated above ~2100 K, and therefore, the useful temperature range for determining the rate behaviour was limited to 1224-1938 K. Because the initially formed C_2H_5-radical is thermally unstable above 1224 K, it rapidly decomposes to C_2H_4 + H on a time scale that is short compared to the title reaction. Hence, the system effectively gives C_2H_4 + 2H at the dissociative recombination rate. The rate constants can be expressed in Arrhenius form by,

$$k_{CH_3+CH_3} = (5.25 \pm 1.19) \times 10^{-11} \exp(-7384 \pm 312 \text{ K/T}), \qquad (5)$$

in molecular units for $1224 \leq T \leq 1938$ K. As can be seen in Fig. 4, Eq. (5) is only 15-25% lower than earlier results by Frank and Braun-Unkhoff [54] (1320-2300 K) and is therefore in excellent agreement.

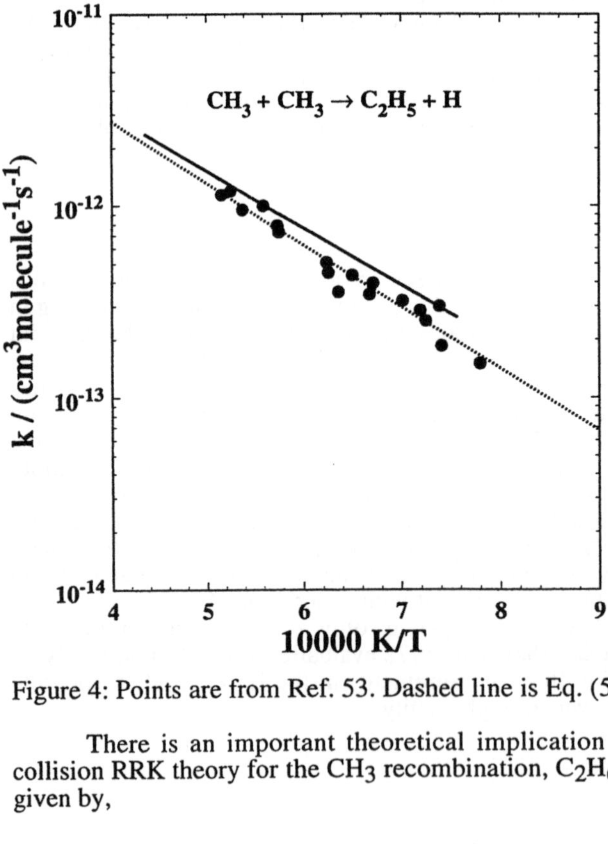

Figure 4: Points are from Ref. 53. Dashed line is Eq. (5). Solid line is from Ref. 54.

There is an important theoretical implication from Eq. (5). Using strong collision RRK theory for the CH_3 recombination, C_2H_6 formation rate constants are given by,

$$k_r = k_{add} \int_{\varepsilon_0}^{\infty} \frac{\beta\omega f(\varepsilon) d\varepsilon}{(k_{f\varepsilon} + k_{b\varepsilon} + \beta\omega)}, \quad (6)$$

where k_{add}, $k_{f\varepsilon}$, $k_{b\varepsilon}$, β, ω, and $f(\varepsilon)$ refer to (a) the high pressure rate constant for CH_3 recombination, (b) the RRK (or RRKM) rate constant for forward dissociation, (c) the RRK (or RRKM) rate constant for backward dissociation at the threshold energy, ε_O, (d) the collisional deactivation efficiency, (e) the collision rate constant, and (f) the RRK (or RRKM) normalized distribution function for a given temperature, respectively. Eq. (6) shows that $k_\infty = k_{add}$. Rate constants based on H formation from $2CH_3 \rightarrow C_2H_5 + H$ are,

$$k^{bi} = k_{add} \int_{\varepsilon_0}^{\infty} \frac{k_{f\varepsilon} f(\varepsilon) d\varepsilon}{(k_{f\varepsilon} + k_{b\varepsilon} + \beta\omega)}, \quad (7)$$

implying that at high temperature and low pressure, $k^{bi}_{\lim P \rightarrow 0} \leq k_{add}$. The A-factor from Eq. (5) is 5.25 x 10^{-11} cm^3 molecule^{-1} s^{-1} which is already the collision rate for CH_3 + CH_3 corrected for the electronic degeneracy ratio. In contrast, recombination experiments and analysis, based on Eq. (6), by Walter et al. [55] suggest decreasing k_{add} with increasing temperature implying a value of 1.7 x 10^{-11} cm^3 molecule^{-1} s^{-1} at 2000 K. This disagreement will require a continuing theoretical assessment on this important reaction.

3.5 $H + NO_2 \rightarrow OH + NO$: Rate constants were measured over the temperature range, 1100 to 2000 K, in reflected shock wave experiments using two different methods of analysis [17]. In both methods, the source of H-atoms was from C_2H_5I decomposition followed by the faster subsequent dissociation of C_2H_5. In the first method, absolute $[H]_t$ was measured with the ARAS technique. Experiments were performed under such low $[C_2H_5I]_0$ that the decay of H-atoms was strictly first-order. The second method utilized a multi-pass optical system [15] for observing the product OH-radical. A resonance lamp was used as the absorption source. Extensive calibration was required giving a modified Beer's law description of the curve-of-growth for use in converting absorption data to absolute $[OH]_t$ from which rate constants were determined. The combined results from the two methods give,

$$k_{H,NO_2} = (1.61 \pm 0.41) \times 10^{-10} \text{ for } 1100 \leq T \leq 2000 \text{ K}, \quad (8)$$

in molecular units. These results have been compared to earlier work at lower temperatures [56], and the combined database yields the temperature independent value k_{H,NO_2} = (1.45 ± 0.30) x 10^{-10} cm^3 molecule^{-1} s^{-1} for 195≤T≤2000 K. The combined results are shown in Fig. 5 along with this T-independent value. We are currently carrying out a theoretical analysis using *ab initio* electronic structure calculations combined with modern dynamical theory in order to rationalize the surprising T-invariance [17].

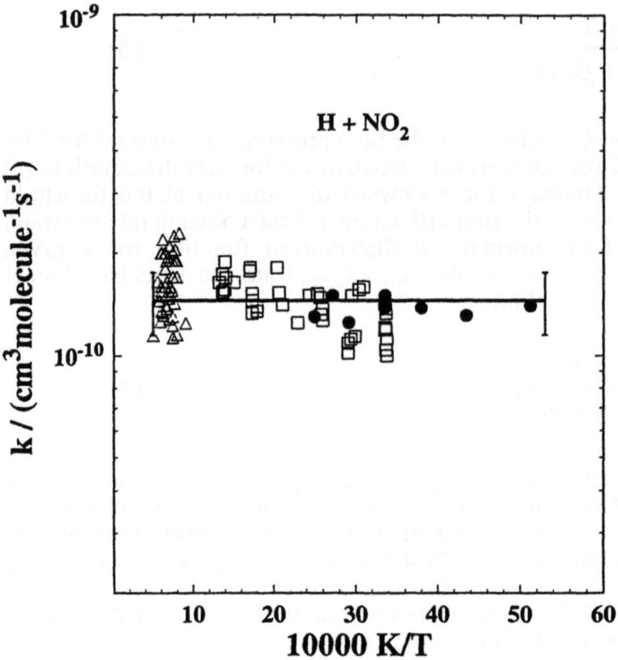

Figure 5: Data points from Ref. 17 and Refs. 56g-i. Line is the T-independent evaluation.

4 Conclusions

The field of chemical research started by Bodenstein [1] 100 years ago, gas phase chemical kinetics, is healthy. Experimental methods continue to mature, but the exciting field of theoretical kinetics and dynamics is still being formulated and tested against experimental results. Because these tests are always going to be necessary to assess success, chemical kinetics will remain a vibrant field in chemistry even though it is continually being de-emphasized in academic chemistry departments at least in the United States. Fortunately for theoretical chemistry, the need to model complex chemical systems in the practical disciplines has supplied intellectual support for continued work, and therefore, much of the current work is being carried out in departments of engineering in private and state supported laboratories and universities throughout the world. This trend will probably continue into the next century, and it is undoubtedly true that "advances in experimental chemical kinetics" will still be a topic of interest 200 years after Max Bodenstein.

5 Acknowledgements

The author wants to thank former colleagues, J.R. Fisher and K.S. Shin, and present colleagues, Drs. K.P. Lim, S.S. Kumaran, M.-C. Su, and A.F. Wagner for thorough readings of the manuscript. This work was supported by the U. S. Department of Energy, Office of Basic Energy Sciences, Division of Chemical Sciences, under Contract No. W-31-109-Eng-38.

6 References

[1] M. Bodenstein, Z. phys. Chem. **13**, 56 (1894); **22**, 1 (1897); **29**, 295 (1899).

[2] (a) A. Fontijn and S. C. Kurzius, Chem. Phys. Lett. **13**, 507 (1972); (b)W. Felder, A. Fontijn, H. N. Volltrauer, and D. R. Voorhees, Rev. Sci. Inst. **51**, 195 (1980).

[3] (a) G. Burns and D. F. Hornig, Can. J. Chem. **38**, 1702 (1960); (b) J. Ernst, H. Gg. Wagner, and R. Zellner, Ber. Bunsenges.. Phys. Chem. **82**, 409 (1978); (c) K. J. Niemitz, H. Gg. Wagner, and R. Zellner, Z. phys. Chem. **124**, 155 (1981); (d) J. V. Michael, J. W. Sutherland and R. B. Klemm, Int. J. Chem. Kin. **17**, 315 (1985).

[4] J. N. Bradley, ***Shock Waves in Chemistry and Physics***, Wiley, New York, NY, 1962.

[5] W. Tsang and A. Lifshitz, Annu. Rev. Phys. Chem. **41**, 559 (1990).

[6] J. V. Michael and K. P. Lim, Annu. Rev. Phys. Chem. **44**, 429 (1993).

[7] (a) J. V. Michael, ***Advances in Chemical Kinetics and Dynamics, Vol. 1***, J. R. Barker, Ed., JAI Press, Greenwich, CT, 1992, p. 47. (b) J. V. Michael, ***Isotope Effects in Gas-Phase Chemistry***, ACS Symp. Ser. 502, J. A. Kaye, Ed., American Chemical Society, Washington, 1992, p. 80.

[8] (a) J. V. Michael and J. W. Sutherland, Int. J. Chem. Kinet. **18**, 409 (1986). (b) J. V. Michael and J. R. Fisher, ***Seventeenth International Symposium on Shock Waves and Shock Tubes;*** Y. K. Kim, Ed.; American Institute of Physics, New York, NY, 1989, p. 210.

[9] J. V. Michael, J. Chem. Phys. **92**, 3394 (1990).

[10] J. V. Michael and J. R. Fisher, J. Phys. Chem. **94**, 3318 (1990).

[11] (a) J. V. Michael, J. Chem. Phys. **90**, 189 (1989). (b) M.-C. Su, K. P. Lim, J. V. Michael, J. Hranisavljevic, Y. M. Xun, and A. Fontijn, J. Phys. Chem. **98**, 8411 (1994).

[12] (a) D. F. Davidson and R. K. Hanson, Int. J. Chem. Kinet. **22**, 843 (1990). (b) M. Koshi, M. Yoshimura, K. Fukuda, H. Matsui, K. Saito, M. Watanabe, A. Imamura, and C. Chen, J. Chem. Phys. **93**, 8703 (1990). (c) J. V. Michael and K. P. Lim, J. Chem. Phys. **97**, 3228 (1992).

[13] S. S. Kumaran, K. P. Lim, J. V. Michael, and A. F. Wagner, J. Phys. Chem. **99**, 8673 (1995).

[14] S. S. Kumaran, M.-C. Su, K. P. Lim, and J. V. Michael, Chem. Phys. Lett., in press.

[15] M.-C. Su, S. S. Kumaran, K. P. Lim, and J. V. Michael, Rev. Sci. Inst., in press.

[16] S. S. Kumaran, M.-C. Su, K. P. Lim, J. V. Michael, and A. F. Wagner, J. Phys. Chem., submitted.

[17] M.-C. Su, S. S. Kumaran, K. P. Lim, A. F. Wagner, L. B. Harding, and J. V. Michael, in preparation.

[18] J. V. Michael, J. R. Fisher, J. M. Bowman, and Q. Sun, Science **249**, 269 (1990).

[19] (a) B. A. Ridley, W. R. Schulz, and D. J. Le Roy, J. Chem. Phys. **44**, 3344 (1966). (b) D. N. Mitchell and D. J. Le Roy, J. Chem. Phys. **58**, 3449 (1973). (c) W. R. Schulz and D. J. Le Roy, Can. J. Chem. **42**, 2480 (1964). (d) W. R. Schulz and D. J. Le Roy, J. Chem. Phys. **42**, 3869 (1965).
[20] A. A. Westenberg and N. de Haas, J. Chem. Phys. **47**,1393 (1967).
[21] I. S. Jayaweera and P. D. Pacey, J. Phys. Chem. **94**, 3614 (1990).
[22] B. Liu, J. Chem. Phys. **58**, 1925 (1973).
[23] P. Siegbahn and B. Liu, J. Chem. Phys. **68**, 2457 (1978).
[24] D. G. Truhlar and C. J. Horowitz, J. Chem. Phys. **68**, 2466 (1978).
[25] A. J. C. Varandas, F. B. Brown, C. A. Mead, D. G. Truhlar, and N. C. Blais, J. Chem. Phys. **86**, 6258 (1987).
[26] D. L. Diedrich and J. B. Anderson, Science **258**, 786 (1992).
[27] A. I. Boothroyd, W. J. Keogh, P. G. Martin, and M. R. Peterson, J. Chem. Phys. **95**, 4343 (1991).
[28] B. C. Garrett and D. G. Truhlar, Proc. Natl. Acad. Sci. **76**, 4755 (1979). (b) ibid., J. Chem. Phys. **72,** 3460 (1980).
[29] Q. Sun and J. M. Bowman, J. Phys. Chem. **94**, 718 (1990).
[30] B. C. Garrett, D. G. Truhlar, A. J. C. Varandas, and N. C. Blais, Int. J. Chem. Kinet. **18**, 1065 (1986).
[31] T. N. Kitsopoulos, M. A. Buntine, D. P. Baldwin, R. N. Zare, and D. W. Chandler, Science **260**, 1605 (1993), and references therein.
[32] Y.-S. M. Wu and A. Kuppermann, Chem. Phys. Lett. **235**, 105 (1995).
[33] T. J. Park and J. C. Light, J. Chem. Phys. **94**, 2946 (1991); **96**, 8853 (1992).
[34] S. L. Mielke, G. C. Lynch, D. G. Truhlar, and D. W. Schwenke, J. Phys. Chem. **98**, 8000 (1994).
[35] D. Wang and J. M. Bowman, J. Phys. Chem. **98**, 7994 (1994).
[36] I. D. Reid, D. M. Garner, L. Y. Lee, M. Senba, D. J. Arseneau, and D. G. Fleming, J. Chem. Phys. **86**, 5578 (1987).
[37] D. K. Biondi, D. C. Clary, J. N. L. Connor, B. C. Garrett, and D. G. Truhlar, J. Chem. Phys. **76**, 4986 (1982).
[38] (a) G. C. Schatz, J. Chem. Phys. **83**, 3441 (1985); K. Tsuda, K. Moribayashi, and H. Nakamura, Chem. Phys. Lett. **231**, 439 (1994).
[39] W. Nernst, Z. Elektrochem. **24**, 335 (1918).
[40] A. Wheeler, B. Topley, and H. Eyring, J. Chem. Phys. **4,** 178 (1936).
[41] R. E. Weston, Jr., J. Phys. Chem. **83**, 61 (1979).
[42] For a review see, S. C. Tucker, D. G. Truhlar, B. C. Garrett, and A. D. Isaacson, J. Chem. Phys. **82**, 4102 (1985).
[43] S. S. Kumaran, K. P. Lim, and J. V. Michael, J. Phys. Chem. **101**, 9587 (1994).
[44] M. W. Chase, Jr., C. A. Davies, J. R. Downey, Jr., D. J. Frurip, R. A. McDonald, and A. N. Syverud, J. Phys. Chem. Ref. Data **14**, Suppl. 1 (1985).
[45] L. B. Harding, private communication.
[46] M. Alagia, N. Balucani, P. Casavecchia, E. H. van Kleef, and G. G. Volpi, 13th International Symposium on Gas Kinetics, paper A4, Dublin, Sept., 1994.
[47] (a) C. K. Westbrook and F. L. Dryer, Prog. Energy Combust. Sci. **10**, 1 (1984). (b) J. A. Miller and C. T. Bowman, ibid. **15**, 287 (1989), and references therein.
[48] (a) D. L. Baulch, D. D. Drysdale, D. G. Horne, and A. C. Lloyd, ***Evaluated Data for High Temperature Reactions, Vol. 1***, Butterworths, London (1972). (b) N. Cohen, and K. R. Westberg, J. Phys. Chem. Ref. Data **12**, 531 (1983). (c) J. Warnatz, ***Combustion Chemistry***, W. C. Gardiner, Jr., Ed.,

Springer-Verlag, New York, 1984. (d) J. V. Michael, Prog. Energy Combust. Sci. **18**, 327 (1992).
[49] (a) D. Gutman and G. L. Schott, J. Chem. Phys. **46** 4576 (1967). (b) D. Gutman, E. A. Hardwidge, F. A. Dougherty, and R. W. Lutz, J. Chem. Phys. **47**, 4400 (1967). (c) G. L. Schott, Symp. (Int.) Combust., [Proc.] **12**, 569 (1968). (d) G. L. Schott, Combust. Flame **13**, 357 (1973). (e) K. M. Pamidimukkala and G. B. Skinner, ***Thirteenth International Symposium on Shock Waves and Shock Tubes***, SUNY Press, Albany, 1981, p. 585. (g) P. Frank and Th. Just, Ber. Bunsenges.. Phys. Chem. **89**, 181 (1985). (h) N. Fujii and K. Shin, Chem. Phys. Lett. **151**, 461 (1988). (i) N. Fujii, T. Sato, H. Miyama, K. S. Shin, and W. C. Gardiner, Jr., ***Current Topics in Shock Waves, Seventeenth International Symposium on Shock Waves and Shock Tubes***, Y. W. Kim, Ed., American Institute of Physics, New York, 1989, p. 456. (j) A. N. Pirraglia, J. V. Michael, J. W. Sutherland, and R. B. Klemm, J. Phys. Chem. **93**, 282 (1989). (k) D. A. Masten, R. K. Hanson, and C. T. Bowman, J. Phys. Chem. **94**, 7119 (1990). (l) T. Yuan, C. Wang, C.-L. Yu, M. Frenklach, and M. J. Rabinowitz, J. Phys. Chem. **95**, 1258 (1991). (m) K. S. Shin and J. V. Michael, J. Chem. Phys. **95**, 262 (1991). (n) H. Du and J. P. Hessler, J. Chem. Phys. **96**, 1077 (1992). (o) H. Yang, W. C. Gardiner, Jr., K. S. Shin, and N. Fujii, Chem. Phys. Lett. **231**, 449 (1994). (p) C.-L. Yu, M. Frenklach, D. A. Masten, R. K. Hanson, and C. T. Bowman, J. Phys. Chem. **98**, 4770 (1994).
[50] S.-O. Ryu, S. M. Hwang, and M. J. Rabinowitz, J. Phys. Chem., in press.
[51] (a) J. A. Miller, J. Chem. Phys. **74**, 5120 (1981). (b) ibid. **75**, 5349 (1981). (c) ibid. **84**, 6170 (1986). (d) S. N. Rai and D. G. Truhlar, J. Chem. Phys. **79**, 6046 (1983). (e) C. J. Cobos, H. Hippler, and J. Troe, J. Phys. Chem. **89**, 342 (1985). (f) J. Troe, ibid. **90**, 3485 (1986). (g) D. C. Clary and H. J. Werner, Chem. Phys. Lett. **112**, 346 (1984). (h) M. M. Graff and A. F. Wagner, J. Chem. Phys. **92**, 2423 (1990). (i) M. R. Pastrana, L. A. M. Quintales, J. Brand⁄ao, and A. J. C. Varandas, J. Phys. Chem. **94**, 8073 (1990). (j) A. J. C. Varandas, J. Brand⁄ao, and M. R. Pastrana, J. Chem. Phys. **96**, 5137 (1992).
[52] C. F. Melius and R. J. Blint, Chem. Phys. Lett. **64**, 183 (1979).
[53] K. P. Lim and J. V. Michael, Symp. (Int.) Combust., [Proc.] **25**, 713 (1994).
[54] P. Frank and M. Braun-Unkhoff, ***Proceedings of the Sixteenth Symposium on Shock Tubes and Waves***, VCH, Weinheim, 1988, p. 379.
[55] D. Walter, H.-H. Grotheer, J. W. Davies, M. J. Pilling, and A. F. Wagner, Symp. (Int.) Combust., [Proc.] **23**, 107 (1990).
[56] (a) W. A. Rosser and H. Wise, J. Phys. Chem. **85**, 532, 2227 (1961). (b) P. G. Ashmore and B. J. Tyler, Trans. Faraday Soc. **58**, 1108 (1962). (c) L. F. Phillips and H. I. Schiff, J. Chem. Phys. **37**, 1233 (1962). (d) H. Gg. Wagner, U. Welzbacher, and R. Zellner, Ber. Bunsenges. Phys. Chem. **80**, 1023 (1976). (e) M. A. A. Clyne and P. B. Monkhouse, J. Chem. Soc. Faraday Trans. II **73**, 298 (1977). (f) P. P. Bemand and M. A. A. Clyne , J. Chem. Soc. Faraday Trans. II **73**, 394 (1977). (g) J. V. Michael, D. F. Nava, W. A. Payne, and L. J. Stief, J. Phys. Chem. **83**, 2818 (1979). (h) S. J. Wategaonkar and D. W. Setser, J. Chem. Phys. **90**, 251 (1989). (i) T. Ko and A. Fontijn, J. Phys. Chem. **95**, 3984 (1991).

Kinetics at Ultra-low Temperatures: Non-Arrhenius Behaviour and Applications to the Chemistry of Interstellar Clouds

I.W.M. Smith,[a] B.R. Rowe[b], and I.R. Sims[a]
[a]School of Chemistry, The University of Birmingham, Edgbaston, Birmingham B15 2TT, U.K.
[b]Département de Physique Atomique et Moléculaire, U.A. 1203 du C.N.R.S., Campus de Beaulieu, Université de Rennes I, 35042 Rennes Cedex, France

Abtract
A brief review is presented of the techniques used to measure the rate constants of elementary gas-phase reactions at low (298 ≥ (T/K) ≥ 77) and ultralow ((T/K) < 77) temperatures and the results obtained by the applications of these methods. A variety of reactions remain rapid at temperatures as low as 13 K. Methods are suggested for estimating rate constants at the temperatures of interstellar clouds.

1 Introduction

Ever since the rates of chemical reactions were first systematically measured, chemical kineticists have sought to measure, characterise, and understand the effect of temperature on these rates. One of the earliest and most celebrated efforts in this direction was the proposal by Svante Arrhenius [1] that the temperature-dependence of reaction rate constants can be expressed in terms of what we now know as the Arrhenius equation:

$$k(T) = A \exp(-E_a/RT) \qquad (1)$$

where, implicitly, E_a, the activation energy, and A, the A- or pre-exponential factor, are independent of temperature.

Arrhenius first formulated his proposal on the basis of parallels with the van't Hoff equation. In modern times, it has become more usual [2] to seek relationships between E_a and other quantities characteristic of an elementary reaction:

- E^o: the threshold or critical energy, which is the minimum reagent energy for which reaction is possible;
- V^*: the potential energy barrier on the reaction path leading from reagents to products; and
- $\Delta E^{\ddagger}{}_0$: the difference between the lowest quantised states in the transition state for the reaction and in the reagent(s).

As the quality and range of kinetic data have improved, it has become evident that the simple Arrhenius expression (1) cannot adequately describe how the value of the rate constant for a particular reaction varies over a wide range of temperature. For *bimolecular* reactions, for example between a free radical and a saturated molecule, which exhibits a moderately strong positive temperature dependence, the rate constant can be measured over a broad range of temperature and plots of *ln* k *versus* (1/T) or (1/RT) frequently exhibit an upward curvature; i.e., E_a as defined by

Springer Series in Chemical Physics, Volume 61
Gas Phase Chemical Reaction Systems
Eds.: J. Wolfrum, H.-R. Volpp, R. Rannacher, and J. Warnatz

$$E_a = - d\, ln\, k(T) / d\, (1/RT) \quad (2)$$

increases with T. An example of such behaviour is shown in Fig. 1 for the reaction [3]:

$$CN + H_2 \rightarrow HCN + H; \Delta_r H^o{}_{298} = - 82 \text{ kJ mol}^{-1} \quad (3)$$

In contrast, the great majority of the many *unimolecular* reactions which have been studied experimentally [4] have rates which increase strongly with temperature. As a consequence, their kinetics can only be studied over a limited range of temperature and any non-Arrhenius behaviour is impossible to discern. An exception is when a single molecule dissociates to two free radicals. The kinetics can then be investigated over a wide temperature range by observing the rates of *dissociation* at high temperature and of the reverse radical *association* at low temperatures, with the two sets of data being linked via the principle of detailed balance [2]. For radical association, there is generally no barrier on the minimum energy path for reaction and therefore no fundamental reason to expect the rate constants for such processes to obey the Arrhenius equation.

The qualitative and quantitative growth in kinetic data that has occurred over the past two decades has exposed the limitations of the simple Arrhenius equation. To provide a fit to kinetic data which yield curved 'Arrhenius plots' it has become usual to use a modified form of Eq. (1) which may be written as:

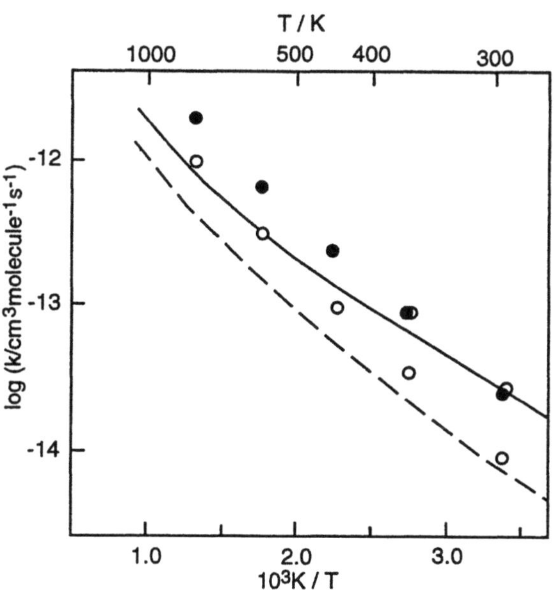

Figure 1: Rate constants for $CN + H_2 \rightarrow HCN + H$ and $CN + D_2 \rightarrow DCN + D$ showing curvature in the Arrhenius plot typical of bimolecular reactions showing a positive temperature dependence. The continuous curves are the results of transition state theory calculations

$$k(T) = A' (T/298)^n \exp(-\theta/T) \quad (4)$$

for which the activation energy depends on temperature according to the equation:

$$E_a = R (\theta + n T) \quad (5)$$

This modified form of the Arrhenius equation can be given some theoretical justification *via* transition state theory (TST) or, for bimolecular reactions, via collisional models. In the later case, the rate constant is a thermal average over the energy-dependent reaction cross section. (Strictly this may only be true for state-to-state rate constants with the thermal rate constant also involving a weighted sum over the internal states of the reagents.).

The simplest line-of-centres (LOC) collision model and the angle-dependent line-of-centres (ADLOC) collision model predict [5] expressions for the reaction cross section which are proportional to $(E-E^o)/E$ and $(E-E^o)^2/E$, respectively, yielding rate expressions like Eq. (4) for k(T) with n = 0.5 in the LOC case and n = 1.5 for the ADLOC model. In the TST expression for k(T), the pre-exponential part (A") can be written as

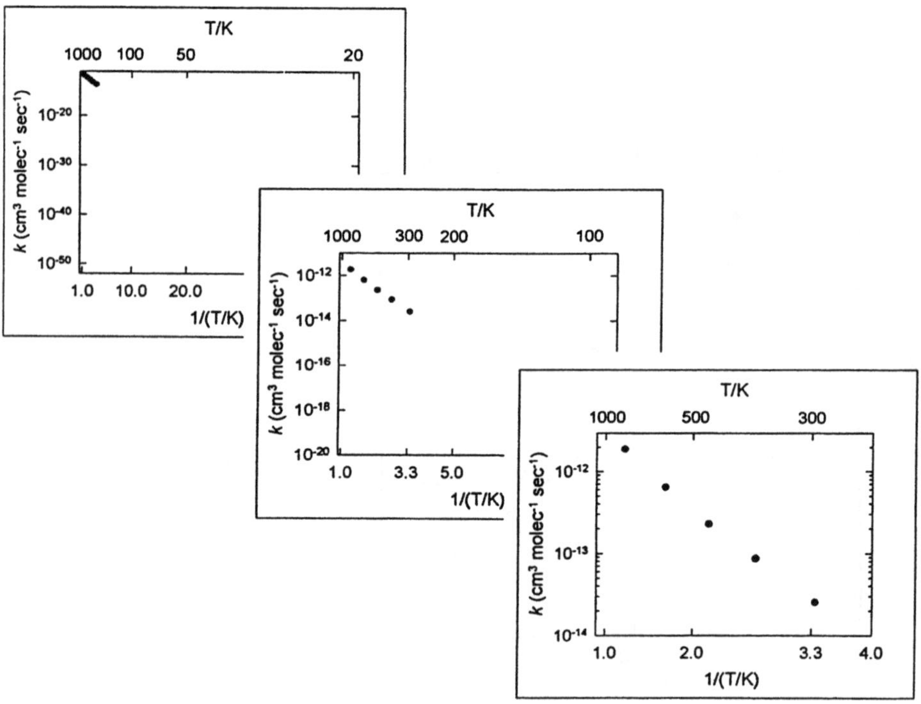

Figure 2: Arrhenius plots for $CN + H_2 \rightarrow HCN + H$ on three different scales showing the extent of 'Arrhenius space' which is explored in experiments at ultra-low temperautures.

$$A'' = (k_BT/h)\ (Q^{\ddagger}/Q_{reag}), \qquad (6)$$

where $Q^{\ddagger}$ and Q_{reag} are the 'per volume' partition functions for the transition state and reagent species, respectively. For a bimolecular reaction, in which the partition functions for the translational and rotational motions can be assumed to take their limiting classical values, the temperature-dependence of A" can be reduced. For example, for two diatomic species reacting through a non-linear transition state

$$A'' \propto (Q^{\ddagger}_{vib,app})/T. \qquad (7)$$

The partition functions for the '*conserved oscillators*' [6] have been cancelled with those of the reagents. The '*appearing oscillators*' [6], three in the example which we have chosen, are low frequency bending or rocking modes which will, in the limit, have partition functions proportional to T suggesting n must be ≤ 2.

Our chief concern here is with the kinetics of reactions at low ($298 \geq (T/K) \geq 77$) and ultra-low ($(T/K) < 77$) temperatures [7]. Two aspects of such measurements are illustrated in Fig. 2. The first is how, by measuring rate constants below room temperature, one can explore large regions of 'Arrhenius space' and hence examine Arrhenius behaviour to a degree not previously possible. However, the diagrams in Fig. 2 also show a difficulty in such measurements. If the reaction has a significant activation energy, then its rate will soon become immeasurably slow as the temperature is lowered.

The fact that many reactions between neutral species behave in the way illustrated in Fig. 2 has led to them being largely neglected in chemical models of dense interstellar clouds where the temperatures are typically 10–50 K. Such models have been dominated by processes involving charged species: especially, ion-molecule reactions and dissociative recombinations [8]. It has been appreciated for many years that such processes not only have large rate coefficients at room temperature but that they usually become even faster at the temperatures found in dense interstellar clouds. The fundamental reason for the rapidity of most ion-molecule reactions is the lack of a potential barrier restricting the probability of reaction in thermal collisions. The electrostatic attraction counteracts any increase in potential energy due to charge or bond reorganisation. In practice, a rich variety of kinetic behaviour is found in ion-molecule reactions [9] but, in many cases, the rate is apparently determined by 'adiabatic capture' by the long-range potential determined by ion-induced dipole and/or ion-dipole forces [10]. Partly as a result of the kinetic measurements which we review below, and partly as a result of theoretical efforts [10], it is now becoming widely recognised that the rates of many neutral-neutral reactions, like those of ion-molecule reactions, are controlled by capture rather than being limited by a potential energy barrier or well-defined transition state.

The most obvious candidates for such behaviour are reactions between pairs of free radicals. As mentioned above, these reactions often proceed by association e.g.

$$OH + NO\ (+ M) \rightarrow HONO\ (+ M) \qquad (8)$$

but they can also yield two new species in a metathesis which usually occurs via a transient complex; e.g.

$$CN + O_2 \rightarrow (NCOO) \rightarrow NCO + O. \qquad (9)$$

For such reactions no single well-defined transition state exists. In order to estimate a thermal rate constant it is necessary to adopt a microcanonical approach in which 'detailed' rate constants, k(E,J), are calculated for the formation of complexes of given total energy and total angular momentum. Values of k(E,J) are then averaged over E and J to yield k(T). This procedure lies at the heart of Troe's statistical adiabatic channel model [6].

Empirically, the rate constants for radical-radical reactions like that between CN and O_2, for radical-radical association in the limit of high pressure, and for the vibrational relaxation of one radical by another are frequently fitted to an expression like Eq. (4) but with θ set to zero; i.e.

$$k(T) = k_{298} (T/298)^n \qquad (10)$$

and n is found to have values between 0 and −1, at least over limited ranges of temperature. Such expressions have been used in models of interstellar clouds for reactions where it has been recognised there is no potential energy barrier. However, quite frequently n has been set equal to +0.5 (consistent with the hard-sphere collision model) which could lead to a considerable underestimate of the rate constants for these reactions below 50 K.

In what follows we review very briefly some of the results obtained in recent experiments at low and ultra-low temperatures. Examples are chosen to illustrate the diversity of non-Arrhenius behaviour which has been discovered in our kinetic experiments below room temperature. In the final section, we attempt to review what has been learnt in the limited number of experiments so far and to provide some cautious guidance to modellers of interstellar clouds as to how to estimate reasonable rate constant values for reactions between neutral species.

2 Experiments at Low and Ultra-low Temperatures

No attempt will be made here to provide a detailed description of the kinetic experiments at low and ultra-low temperatures. In both cases, pulsed laser photolysis (PLP) is used to generate free radicals (CN, OH, CH) from a suitable precursor (e.g. NCNO or ICN, HNO_3 or H_2O_2, $CHBr_3$) and laser-induced fluorescence (LIF) is used to observe them. The LIF signal decays as the time delay between the pulses from the photolysis and probe lasers is increased, providing information about the rate of loss of radicals by reaction.

This well-established technique is implemented either in cryogenically cooled cells, to reach temperatures down to that of liquid N_2 [11], or in a CRESU (Cinétique de Réaction en Ecoulement Supersonique Uniforme) apparatus [12]. In the latter case, ultra-low temperatures are achieved by expanding a gas mixture through a convergent-divergent Laval nozzle, producing a supersonic flow of relatively dense gas in which collisions occur and a definite temperature is established. By pre-cooling the gas reservoir upstream of the nozzle, temperatures as low as 13 K have been reached in the supersonic flow. In this case, He must be used as the carrier gas to prevent clustering during and after the expansion and the non-radical reagent must not condense within the reservoir or its concentration in the flowing gas will not be accurately know. In addition, care must be taken to ensure that this molecular species does not form dimers and higher oligomers at the ultra-low temperatures following the expansion. Effects due to O_2 clustering were observed, but could be allowed for, in experiments on reaction (9).

Because the radicals decay according to pseudo-first-order kinetics, condensation of the photochemical precursor for the radicals is unimportant in CRESU experiments, as in those in cooled cells, as long as sufficient precursor survives to provide a measurable concentration of radicals.

Unique as they are in accessing ultra-low temperatures, there are other limitations to the experiments performed in a CRESU apparatus. The fact that a separate nozzle is required for each combination of carrier gas, carrier gas density and temperature is no more than an experimental irritant, but the relatively limited range of total pressure or gas density does restrict the study of association reactions, although the rate of reaction (8) has been measured at 53 K at five different Ar densities, 0.51–8.2 x 10^{17} molecule cm^{-3} [13]. By contrast, experiments in Birmingham have been performed in cooled cells, for example on the reaction between CH radicals and H_2, at total pressures up to 400 Torr [14], although this is a rather modest value when compared with the pressures reached by Forster *et al.* [15] in their experiments on reactions of the OH radical.

The complementarity of cooled cell and CRESU experiments is further demonstrated by studies on the reactions of OH radicals with CO (down to 80 K) [11(b)] and $O(^3P)$ atoms (down to 158 K) [16].

$$OH + CO \rightarrow CO_2 + H \qquad (11)$$

$$O(^3P) + OH \rightarrow O_2 + H \qquad (12)$$

Reaction (11), which may also proceed via collisional stabilisation to a HOCO radical has been the subject of numerous kinetic and dynamics studies over the past 30 years. At room temperature its rate constant is $k_{298} = 1.5 \times 10^{-13}$ cm^3 $molecule^{-1}$ s^{-1} but below 500 K the activation energy is almost zero. Recent measurements [11(b)] show that the rate constant only falls slightly as the temperature declines to 80 K. Unfortunately, the reaction is too slow to measure at lower temperatures in the present generation of CRESU experiments. The linear flow is so fast, that the first-order rate constant for decay of the radical concentration must be $\geq 5 \times 10^3$ s^{-1} with the result that any bimolecular reactions for which the rate constant is $\leq 10^{-12}$ cm^3 $molecule^{-1}$ s^{-1} cannot be successfully studied.

Finally in this section, we note that the CRESU experiments so far (see Table 1 below) have only provided rate constants for reactions between free radicals generated photochemically and reagents which can be introduced as stable species into the gas flow. Reactions like (12), between two radicals which are highly reactive under normal laboratory conditions, have not been investigated due to the difficulties associated with generating a large enough, known concentration of any radical species. The rates of reaction (12) and those of N atoms with OH [16] and with CH have recently been measured at temperatures down to *ca.* 100 K in experiments in which discharge-flow techniques are used to create a known steady-state concentration of atomic radicals before a conventional PLP-LIF experiment is performed.

Developments aimed at removing two of the limitations of the CRESU method are currently under way in Rennes. Attempts are being made to couple a discharge to the gasflow in order to generate large concentrations of radical atoms, and new diffusers and nozzles are being designed in order to access temperatures close to and below 10 K without cooling the gas mixture prior to its expansion.

3 Results at Low and Ultra-low Temperatures

Table 1 shows the processes for which rate constants have been obtained by the PLP-LIF technique applied in the CRESU apparatus and the lowest temperature at which the rate of each such process has been measured. A rich variety of temperature dependences have been observed, although for all the processes identified in Table 1, the general trend is for the rate constant to increase as the temperature is lowered.

Log-log plots showing the variation of rate constant with temperature are shown for three representative cases in Fig. 3. Reaction (9) between CN and O_2

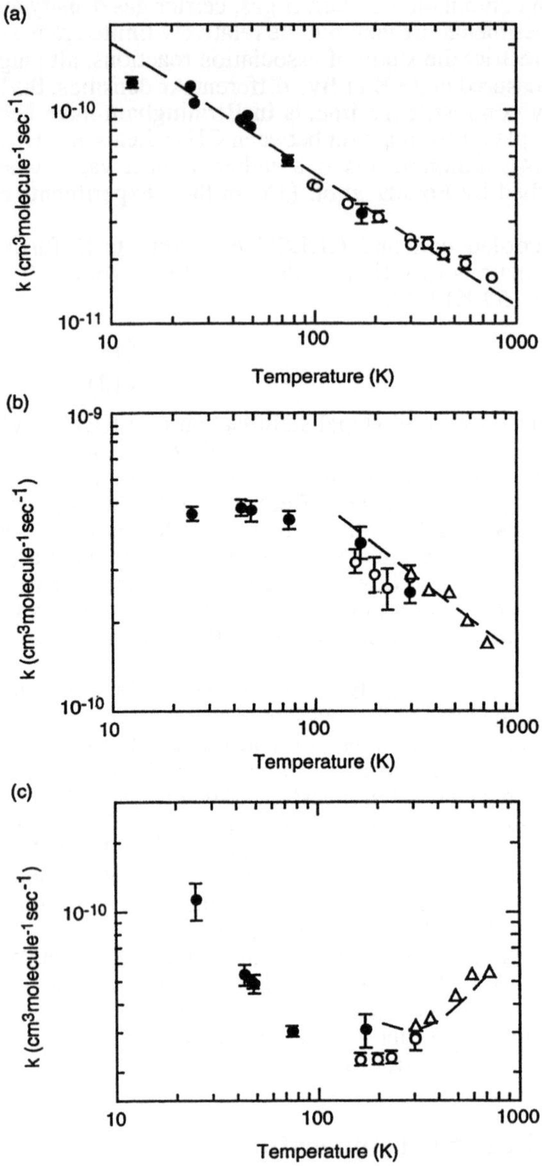

Figure 3: Log-log plots showing the variation of rate constant with temperature for three examples: (a) $CN + O_2$, (b) $CN + C_2H_2$, (c) $CN + C_2H_6$.

serves as an example of a radical-radical reaction proceeding across an entirely attractive surface through a minimum or potential 'well'. The rate constants for this reaction have now been measured over an enormous range 13–3800 K [17] and increase monotonically to lower temperatures. The observation that they rather accurately fit the form represented by equation (10) is almost certainly fortuitous. This form of behaviour, in which the rate constants exhibit a relatively mild increase as the temperature falls and at least approximately fit Eq. (10) with $0 \geq n \geq -1$ seems fairly typical of the (few) radical-radical processes which we have studied. With such processes, reaction (12) between $O(^3P)$ atoms and OH radicals is a good example [16], part of the temperature dependence arises because the electronic partition functions of the reagents, and therefore the fraction of collisions occurring on the attractive surface – or possibly surfaces – leading to products change with temperature. If allowance is made for this factor, it appears that the rate constant at the lowest temperatures at which experiments have been performed may correspond approximately with that for capture on the lowest surface.

At high temperatures, with adiabatic states of high angular momentum playing an increasing role, the crucial part of the potential between the reagents moves to smaller separations. Chemical forces start to act, the states associated with motions orthogonal to the reaction coordinate become more widely spaced, and the thermal rate constant falls. Kim and Klippenstein [18] have been successful in reproducing the temperature dependence for $CN + O_2$ by performing TST calculations based on an *ab initio* long-, or medium-range potential.

In some of the cases which we have studied the rate constants appear to reach a 'saturation value' at low temperatures; if the temperature is lowered still further, the rate constants increase no more and actually may decrease. The clearest examples of this behaviour are the reactions of CN with C_2H_4 and C_2H_2 (see Fig. 3(b)) [19] and of OH with butenes [20]. The involvement of unsaturated molecules where the dispersion and induction forces between them and polar free radicals are relatively large strongly suggests that the rates of these reactions are also determined by 'capture' but that now the range at which 'capture' is determined by long-range electrostatic forces extends to high temperatures. As with radical-radical reactions, the decrease in rate constants at high temperatures presumably reflects a general movement of the transition state 'bottleneck' to shorter inter-reagent separations where motions which are 'free' in the limit of separated reagents become restricted and the associated energy states become more widely separated.

The third example which is chosen to illustrate the diversity of behaviour found in our experiments at low and ultra-low temperatures [19] is the reaction (see Fig. 3(c)):

$$CN + C_2H_6 \rightarrow HCN + C_2H_5 \quad \Delta_rH^o{}_{298} = -\,92\ \text{kJ mol}^{-1} \qquad (13)$$

As an H-atom abstraction reaction, one could be forgiven for expecting that the rate constants would vary according to the simple or modified form of the Arrhenius equation. In practice, we have found the most surprising kinetics for any reaction that we have studied at room temperature and above, the rate constants are found to increase with temperature, albeit quite slightly. However, as the temperature is lowered the rate constants pass through a minimum and then rise quite steeply to attain their highest value at 25 K.

The results show clearly that there cannot be any maximum on the path of minimum electronic potential energy leading from reagents to products. The variation of k(T) with temperature must reflect subtle changes in the nature and location of the transition state bottleneck as E,J change. At the lowest temperatures, k(T) is approximately a third of the value that might be expected on the basis of a simple capture

Table 1: Minimum temperatures (in Kelvin) at which reactions of CN, OH, CH(v=0) and CH(v=1) have been studied in the CRESU apparatus.

Reagent	CN	OH	CH(v=0)	CH(v=1)
CH_4	-	-	23	-
C_2H_6	25	-	23	-
NH_3	25	-	23	23
H_2	-	-	53	23
D_2	-	-	-	-
HBr	-	23	-	-
N_2	-	-	-	23
CO	-	-	-	23
C_2H_4	25	-	23	-
C_2H_2	25	-	23	-
butenes	-	23	23	-
O_2	13	-	13	-
NO	-	23	13	-

model invoking the long-range electrostatic potential. The decrease in k(T) as T is raised from its smallest value is similar to that found for other reactions between pairs of radicals and between radicals and unsaturated molecules and presumably has a similar cause: a 'tightening' of the transition state. However, in this case, this trend reverses, and above ca. 200 K, k(T) starts to increase again. In this range it seems likely that the rate of reaction is determined by passage through a transition state associated with transfer of the H atom from C_2H_6 to the attacking CN radical. Although not associated with a maximum in the electronic potential energy, this configuration, through this range, is associated with a maximum in local free energy, so k(T) shows a positive dependence on temperature. Alternatively, one can view this positive dependence as arising from the partition function of the transition state depending more steeply on T than the product of the partition functions of the reagents.

4 Estimating Rate Constants at Ultra-low Temperatures

Although there is no substitute for accurate measurements, it is unlikely that experiments will provide rate constants for all the reactions that may be important in chemical models of interstellar clouds. Already, however, the measured values provide a fertile testing ground for theories of the kinds propounded by Clary [10] and Troe [6]. Until such time as either experiment or detailed theory provides a reliable kinetic data base, modellers are likely to appreciate guidelines as to how they might estimate approximate rate constants for neutral-neutral reactions between 10 and 50 K, since it is now clear that the rate of *any* neutral-neutral reaction which has a rate constant within an order of magnitude of the collisional value at room temperature may have a rate constant which is determined by 'capture' [21] and which *may* increase as the temperature is lowered.

Perhaps the most important class of neutral-neutral reaction in interstellar clouds is that between two small free radicals. Where measurements of the rate constants are available at higher temperatures [e.g. 16], one can estimate values at ultra-low temperatures by allowing for the temperature dependence of the reagent partition functions in order to find a rate constant associated only with those collisions occurring on the potential energy surface(s) leading adiabatically from reagents to products without a potential barrier. An extrapolation according to Eq. (10) seems likely to yield a reasonable approximation to the rate constants below 50 K. There is one caveat. If such a procedure yields values larger than estimates of the simple capture rate constant on the reactive surface(s) then the values should be 'capped' at those values [21]. Based on the very few examples where data are available it seems that the rate constants are likely to be close to their simple capture values at the temperatures of interstellar clouds.

Unsaturated molecules have relatively high polarisabilities and consequently strong long-range dispersion and, in collisions with polar species, induction forces [21]. Our results on the reactions of CN [19] and OH [20] with alkenes and alkynes for which the rate constants appear to reach a limiting value below *ca.* 50 K are strongly suggestive of the notion that, at low enough temperatures, the rate constants for neutral-neutral reactions which remain fast at ultra-low temperatures reach a value determined by adiabatic capture [10, 21]. The temperature at which this happens will be determined by the strength of the long-range forces between the two collision partners. Once again it may necessary to allow for any effects due to collisions occurring on more than one potential energy surface and the fact that the fraction colliding on the attractive reactive surface may depend on temperature [21].

Undoubtedly it is most difficult to predict the rate constants at ultra-low temperatures for reactions between radicals and saturated molecules. The kinetic behaviour of reaction (13) and of the reaction between OH radicals and HBr [22] was unexpected. The message seems to be that, even for reactions of this type, the rate constants may approach the value determined by capture at ultra-low temperatures if the room temperature value is itself within a factor of about ten of the collisional value.

Finally, we note that Herbst *et al.* [23] have examined the effect of including various fast neutral-neutral reactions into models of the chemistry of quiescent, dark interstellar clouds. The calculated abundances of many molecules were greatly changed from previous values. It seems clear that the inclusion of faster neutral-neutral chemistry is likely to have a profound effect on our understanding of the chemistry in interstellar clouds over the next few years.

5 Acknowledgements

It is a pleasure to acknowledge the support that we have received from several agencies for our work on kinetics at low and ultra-low temperatures: SERC, EPSRC, the CEC under the Science Plan, the British Council under the Alliance Programme, and GDR's "Physicochemie des Molécules Interstellaires" and "Dynamique des Réactions Moléculaires" programmes.

6 References

[1] (a) S. Arrhenius, Z. phys. Chem. **4**, 226 (1889); (b) K.J. Laidler and M.C. King, J. Chem. Educ. **61**, 494 (1984).

[2] I.W.M. Smith, ***Kinetics and Dynamics of Elementary Gas Reactions*** (Butterworths, London, 1980).
[3] I.R. Sims and I.W.M. Smith, Chem. Phys. Lett. **149**, 565 (1988).
[4] P.J. Robinson and K.A. Holbrook, ***Unimolecular Reactions*** (Wiley, New York, 1972).
[5] (a) ref. [3] p. 89; (b) I.W.M. Smith, J. Chem. Educ. **59**, 9 (1982).
[6] (a) J. Troe, J. Chem. Phys. **75**, 226 (1981); (b) J. Troe, J. Chem. Phys. **79**, 6017 (1983).
[7] I.R. Sims and I.W.M. Smith, Ann. Rev. Phys. Chem. **41**, 109 (1995).
[8] for example, S.S. Prasad, W.T., Huntress Jr., Astrophys. J. Suppl. Ser. **43**, 1 (1980).
[9] M.A. Smith, in ***Unimolecular and Bimolecular Reaction Dynamics***, ed. C.Y. Ng, T. Baer, I. Powis (Wiley, New York, 1994).
[10] D.C. Clary, Ann. Rev. Phys. Chem. **41**, 61 (1991).
[11] (a) I.R. Sims and I.W.M. Smith, Chem. Phys. Lett. **151**, 481 (1988); (b) M.J. Frost, P. Sharkey and I.W.M. Smith, J. Phys. Chem. **89**, 12254.
[12] I.R. Sims, J.-L. Queffelec, A. Defrance, C. Rebrion-Rowe, D. Travers, P. Bocherel, B.R. Rowe and I.W.M. Smith, J. Chem. Phys. **100**, 4229 (1994).
[13] P. Sharkey, I.R. Sims, I.W.M. Smith, P. Bocherel and B.R. Rowe, J. Chem. Soc. Faraday Trans. **90**, 3609 (1994).
[14] R.A. Brownsword, I.W.M. Smith and D.W.A. Stewart, to be published.
[15] R. Forster, M. Frost, D. Fulle, H.F. Hamann, H. Hippler, A. Schlepegrell and J. Troe, J. Chem. Phys., in press (1995).
[16] I.W.M. Smith and D.W.A. Stewart, J. Chem. Soc. Faraday Trans. **90**, 3221 (1994).
[17] I.W.M. Smith in ***Chemical Kinetics and Dynamics of Small Radicals,*** ed. K. Liu and A.F. Wagner (World Scientific, Signapore, 1995).
[18] S.J. Klippenstein and Y.-W. Kim, J. Chem. Phys. **99**, 5790 (1993).
[19] I.R. Sims, J.-L. Queffelec, D. Travers, B.R. Rowe, L.B. Herbert, J. Karthauser and I.W.M. Smith, Chem. Phys. Lett. **211**, 461 (1993).
[20] I.R. Sims, P. Bocherel, A. Defrance, D. Travers, B.R. Rowe and I.W.M. Smith, J. Chem. Soc. Faraday Trans. **90**, 1473 (1994).
[21] I.W.M. Smith, Int. J. Mass. Spectrom. Ion Proc. in press (1995).
[22] I.R. Sims, I.W.M. Smith, D.C. Clary, P. Bocherel and B.R. Rowe, J. Chem. Phys. **101**, 1748 (1994).
[23] E. Herbst, H.-H. Lee, D.A. Howe and T.J. Millar, Mon. Not. Roy. Astronom. Soc., **268**, 335 (1994).

Part IV

State Specific and Thermal Rates: Theory

The Influence of Hindered Rotations on Recombination/Dissociation Kinetics

A.F. Wagner,[a] L.B. Harding,[a] S.H. Robertson,[b] and D.M. Wardlaw[c]
[a]Chemistry Division, Argonne National Laboratory
Argonne, IL 60439 USA
[b]School of Chemistry, University of Leeds
Leeds LS2 9JT, UK
[c]Department of Chemistry, Queen's University
Kingston, ON K7L 3N6, CANADA

Abstract.
A simple formula for the canonical flexible transition state theory expression for the thermal reaction rate constant is derived that is *exact* in the limit of the reaction path being well approximated by the distance between the centers of mass of the reactants. This formula evaluates classically the contribution to the rate constant from transitional degrees of freedom (those that evolve from free rotations in the limit of infinite separation of the reactants). Three applications of this theory are carried out: D + CH_3, H + CH_2, and F + CH_3. The last reaction involves the influence of surface crossings on the reaction kinetics.

1 Introduction

Recently considerable theoretical and experimental attention has been given to reactions, either dissociative or associative, whose transition states or "bottlenecks" are characterized by large amplitude motion. This kind of motion is typical when there is no pronounced potential barrier for the formation of a parent molecule from a pair of constituent molecular fragments. Such "barrier-less" reactions are commonplace in chemistry and are particularly important in combustion, atmospheric, and interstellar processes. Despite ongoing advances in computer technology, classical trajectories, and quantum dynamics, the most widespread theories for modeling reaction rate coefficients are statistical theories. One such class of theories specifically designed to address large amplitude motion is flexible transition state theory (FTST) [1,2]. This class can be used to get the high pressure limiting form of the recombination rate constant.

FTST is a variational theory that minimizes the rate constant by varying the bottleneck location along the reaction path, thus optimizing bottleneck location. FTST also treats the bottleneck or transition state as "floppy" with large amplitude transitional modes which are coupled to each other and to overall rotation of the molecular system. Typically transitional modes correspond to free rotations of the separated fragments that evolve into vibrational motions of the parent molecule. In FTST, these modes are treated by classical mechanics which makes feasible the inclusion of the full transitional mode potential (or, the entire potential energy surface (PES), if available). However, as originally developed, intensive numerical

Springer Series in Chemical Physics, Volume 61
Gas Phase Chemical Reaction Systems
Eds.: J. Wolfrum, H.-R. Volpp, R. Rannacher, and J. Warnatz

calculations, with consequent loss of physical insight, were required to incorporate complex potentials into the calculation of the rate constant.

Recently, Smith [3], Klippenstein *et al.* [4,5], Aubanel, Robertson, and Wardlaw [6] and others have developed versions or approximations to the original FTST which are easier to implement and are physically more transparent. In a recent paper by the authors [7], hereafter called paper I, a particularly simple but exact canonical version of FTST was developed. This paper will review this canonical theory and apply it to three different recombination reactions: $D+CH_3$, $H+CH_2$, and $F+CH_3$.

2 CFTST Model

In this section, the basic components of a canonical FTST theory are reviewed. A more detailed derivation can be found in paper I. The canonical variational transition state theory expression for the high pressure limiting rate constant is (with $\beta = 1/kT$)

$$k(T) = \frac{g_e}{\beta h}\frac{\sigma}{\sigma^\ddagger}\frac{Q^\ddagger_{TS}(T)}{Q_{react}(T)}e^{-\beta V^\ddagger} \tag{1}$$

The quantity $Q^\ddagger_{TS} = Q_{TS}(R^\ddagger,T)$ is the pseudo-partition function of the system at the transition state, the transition state location $R^\ddagger$ being that value of the reaction coordinate R which, at a given T, minimizes $Q_{TS}(R,T)$. This partition function has its zero of energy on the reaction path where the potential is $V^\ddagger = V_{rxnpath}(R^\ddagger)$. The form of Q_{TS} will be described in detail below. Q_{react} is the partition function of the reactants; $\sigma/\sigma^\ddagger$ is the ratio of reactant and transition state symmetry factors and g_e is the ratio of electronic degeneracy factors for the reactants and transition state. The incorporation of large amplitude transition state motion is through Q_{TS}.

There are three major assumptions of FTST which determine Q_{TS}. First, the overall rotation of the system and the large amplitude transitional modes (as a group) are separable from the small amplitude conserved modes (as a group):

$$Q_{TS}(R,T) = Q_c(R,T)Q_{tr}(R,T) \tag{2}$$

where Q_c and Q_{tr} are the conserved mode and transitional/external-rotational mode partition functions, respectively. The conserved modes are typically those degrees of freedom present in the recombined product and the isolated reactants. Second, Q_c is evaluated quantum mechanically (almost always as independent harmonic oscillators from normal modes orthogonal to the reaction coordinate) while Q_{tr} is treated classically through the expression

$$Q_{tr} = 1/h^n \int\cdots\int e^{-\beta H}\, \mathbf{dp}\, \mathbf{dq} \tag{3}$$

where H is the classical Hamiltonian for the n transitional/external-rotation modes (excluding of course the reaction coordinate) and **q** and **p** are generalized coordinates and their conjugate momenta **p**. H can be written as the sum of a transitional mode kinetic energy T_{tr} and a potential energy V_{tr}. Third and last, in describing T_{tr} and V_{tr}, the fragments will be assumed to be rigid bodies whose shapes are those at the given value of R along the reaction path. At any point on the reaction path, each fragment adjusts its optimized internal geometry in response to the presence of the other fragment. When either fragment departs from the reaction path in transitional mode directions, its internal geometry will be presumed fixed and only its orientation relative to the other fragment will change.

This last assumption means that each fragment, no matter how complicated, can be treated as either an atom or a linear, spherical, symmetric, or asymmetric top for the purposes of transitional motion. Principal moment of inertia analysis of each fragment at position R fixes the locations of the atoms of the fragment with respect to the principal axes. The relative orientation of the fragments to each other, i.e., the transitional modes, are then determined by only the angles that govern the orientation of the principal axes of each top. When decomposed from the full PES, V_{tr} can ultimately be formulated as a function of only those orientation angles and R.

With the above assumptions, large amplitude motion is contained exclusively in Q_{tr} which can be drastically simplified. First, since V_{tr} cannot depend on **p**, the integration over d**p** in Eq. (3) can be done, as it turns out, analytically:

$$Q_{tr} = (2\pi/\beta h^2)^{n/2} \int\cdots\int |\mathbf{A}|^{1/2} e^{-\beta V_{tr}} d\mathbf{q} \tag{4}$$

where $|\mathbf{A}|$ is the determinant of the matrix whose elements define T_{tr} by

$$2T_{tr} = \sum_{i,j=1}^{n} A_{ij} \dot{\mathbf{q}}_i \dot{\mathbf{q}}_j \tag{5}$$

The explicit expression for $|\mathbf{A}|$ depends on the definition of the reaction path because **A** defines kinetic motion off the reaction path during which R is constrained to be fixed. For many reaction path definitions, such constrained motion implies a complicated expression for $|\mathbf{A}|$. However, if R is defined as the separation of the centers of mass of the two fragments, a general expression for $|\mathbf{A}|$ has been found (see paper I):

$$|\mathbf{A}| = (\mu R^2)^2 I_{1a} I_{1b} I_{1c} I_{2a} I_{2b} I_{2c} \sin^2\theta_1 \sin^2\theta_2 \sin^2\Theta \tag{6}$$

where μ is the reduced collisional mass of the system, (I_{ia}, I_{ib}, I_{ic}) are the principal moments of inertia of fragment i, θ_i is the spherical angle measuring the departure of the principal axis of fragment i from a z axis coincident with **R**, and Θ is one of three "external" angles (Ψ, Θ, Φ) needed to orient the body-fixed collision system relative to a space-fixed system. For convenience, the R dependence of (I_{ia}, I_{ib}, I_{ic}) and hence of $|\mathbf{A}|$ has been suppressed. Equation (6) is written for two asymmetric tops. However, by the appropriate adjustment of moments of inertia and the value of n in Eq. (4), all cases involving spherical tops, linear tops, and atoms can be accommodated.

Substitution of Eq. (6) into Eq. (4) allows the analytic integration over Ψ, Θ, and Φ which are external coordinates and therefore leave V_{tr} in the integrand unchanged. For two symmetric tops, the final result is:

$$Q_{tr}(R,T) = \Gamma(R,T)\, Q_{pd}(R,T)\, Q_{fr,1}(R,T)\, Q_{fr,2}(R,T) \tag{7}$$

where

$$Q_{pd}(R,T) = \frac{2\mu R^2}{\beta\hbar^2} \tag{8}$$

$$Q_{fr,i}(R,T) = \left(\frac{2\pi I_{ib}}{\beta\hbar^2}\right)^{1/2}\left(\frac{2 I_{ia}}{\beta\hbar^2}\right) \tag{9}$$

$$\Gamma(R,T) = \frac{1}{2^5\pi^3}\int_0^{\pi}d\theta_1 \sin\theta_1 \int_0^{\pi}d\theta_2 \sin\theta_2 \int_0^{2\pi}d\phi_1 \int_0^{2\pi}d\phi_2 \int_0^{2\pi}d\chi\, e^{-\beta V_{tr}} \tag{10}$$

Here χ is the difference in the azimuthal angles associated with θ_1 and θ_2, and ϕ_i is the third Euler angle needed to orient the ith rigid fragment with respect to **R**. $Q_{pd}(R,T)$ is the partition function for the pseudo-diatomic formed from the centers of mass (CM) of the two fragments. $Q_{fr,i}$ is the partition for the free rotation of symmetric top i. $\Gamma(R,T)$ is a hindering function in the form of a normalized configuration integral. Equation (9) can be modified for any other combination of tops or atoms.

The above expression for Q_{tr} has two factors: a kinematics factor ($Q_{pd}Q_{fr,1},Q_{fr,2}$) dependent only on masses and fragment shape changes, and a steric factor (Γ) exclusively dependent on V_{tr}. Note that when $V_{tr}=0$ (i.e., no inter-fragment interaction), $\Gamma(R,T) = 1$. By definition, V_{tr} is zero at the reaction path geometry. If the reaction path geometry has the most attractive geometry at a given value of R (as is generally the case), then at that value of R, V_{tr} will by definition never be less than zero and consequently the hindering function will always fall between 0 and 1. Thus Q_{tr} is generally a potential-dependent fraction times a shape and mass-dependent kinematic factor.

Substitution of the above expressions into Eq. (1) followed by minimization with respect to R yields

$$k(T) = \frac{g_e}{\beta h}\frac{\sigma}{\sigma^\ddagger}\frac{e^{-\beta V^\ddagger}}{Q_{trans}(T)}\left[\frac{Q_c^\ddagger(T)}{Q_{vib,1}(T)Q_{vib,2}(T)}\right]\left[\frac{Q_{pd}^\ddagger(T)Q_{fr,1}^\ddagger(T)Q_{fr,2}^\ddagger(T)}{Q_{fr,1}(T)Q_{fr,2}(T)}\right]\Gamma^\ddagger(T) \tag{11}$$

where Q_{react} in Eq. (1) has been explicitly decomposed into $Q_{trans}Q_{vib,1}Q_{vib,2}$ $Q_{fr,1}Q_{fr,2}$. Of the last three factors in Eq. (11), the first contains all the vibrational information, the second all the shape information, and the last all the transitional mode potential information. If it is assumed that the fragments do not significantly alter their shape or frequencies in the kinetically important region, then Eq. (12) simplifies to

$$k(T) = \frac{g_e}{\beta h}\frac{\sigma}{\sigma^\ddagger}\frac{e^{-\beta V^\ddagger}}{Q_{trans}(T)}Q_{pd}^\ddagger(T)\Gamma^\ddagger(T). \tag{12}$$

In these expressions, Γ can equally well accommodate any representation of V_{tr}, even if it does not involve large amplitude motion. The only requirement is that the degree of freedom be treated classically. Thus these expressions offer a seamless way to follow the evolution of transitional modes from free rotors, to hindered rotors, to harmonic oscillators and allow a variational search of the reaction bottleneck to locate the transition state in whatever regime is important for the given temperature.

3 Results and Discussion

In this section, the above canonical rate constant theory will be applied to three reactions.

3.1 D + CH_3 → CDH_3: The recombination reaction of CH_3 with H(D) has frequently been studied by FTST approaches [1,6,4,8] and paper I. In all these publications, the *ab initio* fully dimensional CH_4 potential energy surface of Hase *et al.* [9] has been used. These studies have shown that canonical FTST overestimates microcanonical FTST by only 5% to 9% for temperatures from 300 K to 2000 K. Thus a canonical theory is largely reliable in comparisons to experiment. In paper I, rate constant calculations for H+CH_3 *via* Eq. (11) reproduced previous canonical FTST studies with full numerical phase space integrations. However, the only reliable measurements of the high pressure recombination rate constant for this system are for the D+CH_3 isotopic variant [10]. Here we apply Eq. (11) to this variant.

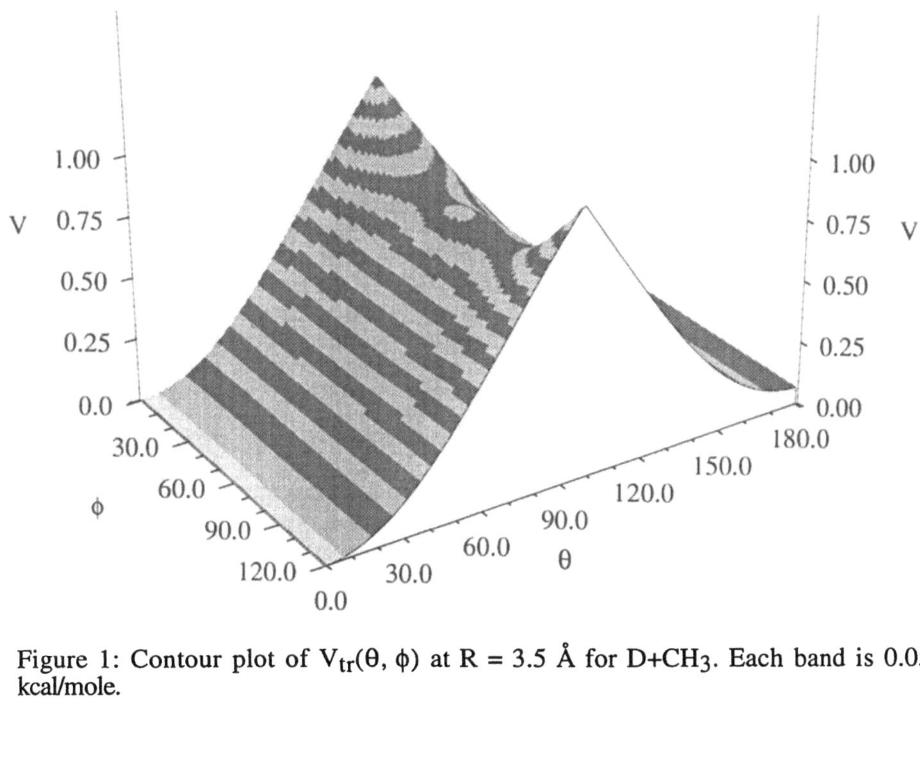

Figure 1: Contour plot of $V_{tr}(\theta, \phi)$ at R = 3.5 Å for D+CH_3. Each band is 0.05 kcal/mole.

The last factor in Eq. (11) contains the effect of $V_{tr}(\theta,\phi)$ where the spherical angle θ and the azimuthal angle ϕ locate the D with respect to the C_{3v} axis of CH_3 whose origin is at the CM of CH_3. A perspective contour plot of V_{tr} is displayed in Fig. 1 at R = 3.5Å where R is measured from the CM of the slightly splayed CH_3. This distance is between the locations of the canonical FTST transition states for 300K and 400 K. In the figure, the value of R and the splayed shape of CH_3 on the reaction path at R are held constant while the spherical angle θ is varied from 0° to 180° and the azimuthal angle ϕ varies from 0° to 120° (the potential has a periodicity in this coordinate of 120°).

In Fig. 1, V_{tr} clearly has a plane of symmetry about $\phi = 60°$, as it must. V_{tr} peaks at $\theta \approx 90°$, i.e., where D is closest to the H atoms on the CH_3 fragment. At this value of θ, V_{tr} is lowest for $\phi = 60°$ where D passes between two C-H bonds on the CH_3 fragment. V_{tr} is highest at $\phi = 0°$ or 120°, where D passes right over a C-H bond. This dependence gives the hindering potential an overall saddle appearance. At $\theta = 0°$ or 180°, V_{tr} is essentially independent of ϕ. This is as required for a 2D harmonic oscillator which is what the motion of the attacking H atom becomes in the small amplitude, normal mode limit (i.e., a doubly degenerate rock). The figure also shows that V_{tr} at $\theta = 0°$ is lower than V_{tr} at $\theta = 180°$ This is due to the splaying of CH_3 at R which distinguishes a frontside (θ = 0°) and backside (θ = 180°) approach for the attacking atom. Consequently, V_{tr} is not symmetric about $\theta = 90°$. As a consequence, the integral in $\Gamma^{\ddagger}$ cannot be rigorously carried out over half the range of θ and the result multiplied by two. This factor of two is precisely the reaction path degeneracy that is normally thought of in reactions of this type and that has been invoked in previous non-FTST studies of this reaction.

In Fig. 2, the resulting canonical FTST rate constant determined from Eq. (11) is labelled HR for hindered rotor. As required by its derivation, this result is essentially identical to the published [8] canonical FTST rate constant for the same PES. This earlier calculation involved time consuming numerical evaluations of multi-dimensional integrals, many of which in the canonical limit are done analytically in the derivations above.

As was done in paper I for $H+CH_3$, the free rotor and harmonic oscillator versions of V_{tr} have been constructed and the resulting rates for $D+CH_3$ are included in the figure under the labels FR and HO, respectively. In paper I, these rates were shown to be very similar in value to those from "traditional" VTST calculations. The HO rate in the figure was constructed with a reaction path degeneracy of two and a quadratic form of V_{tr} accurate as θ approaches the $\theta = 0°$ line in Fig. 1. There is of course another limiting quadratic form of V_{tr} that arises as θ approaches the θ = 180° line in Fig. 1. This is *not* included in the V_{tr} used in Eq. (10) for $\Gamma(R,T)$ but approximately accounted for by the reaction path degeneracy of two. Implicitly, all previous studies of this reaction have used this approach. However as discussed in paper I, for $H+CH_3$ up to 10% overestimations of the rate can occur.

The results in Fig. 2 indicate that the HO, HR, and FR models are all noticeably different from each other over the 300 K to 2000 K temperature range. Also in Fig. 2 is the error box containing the experimentally measured rates [10]. Generally the calculated HR rate is quite close to the measured values at the lowest experimental temperature and about 15% higher at the higher experimental temperature. The fact that the computed rate constant is a canonical value leads in $H+CH_3$ to a 5% to 10% overestimation of the more rigorous microcanonical rate. Overall, the agreement between theory and experiment is quite satisfactory.

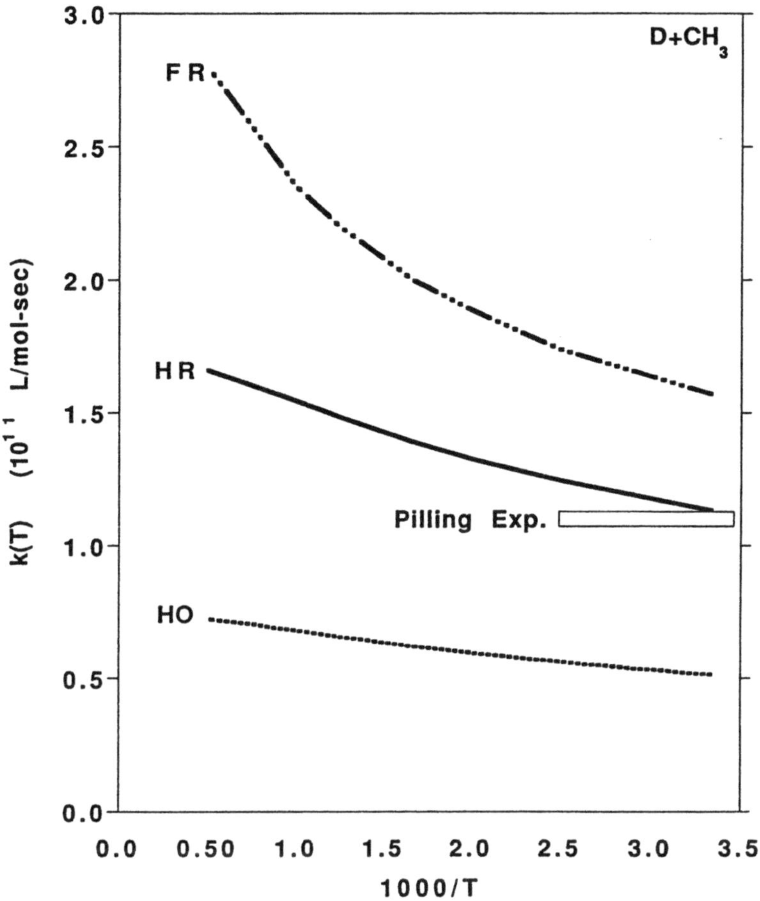

Figure 2: Calculated rate constant k(T) *vs.* 1000/T for $D+CH_3$ (see text). Experimental values are from Ref. [10].

3.2 $H + CH_2 \rightarrow CH_3$: This reaction has been the subject of several experimental [11] and theoretical studies [12]. The published theoretical work calculated portions of the hindered rotation barriers for different values of $R_{H\text{-}C}$, the distance of the attacking H from CH_2. This calculation employed a tzp basis set [13] with an multi-reference singles and doubles configuration interaction (MRSDCI) wavefunction and was carried out with the COLUMBUS program system [14]. This work has now been supplemented by a much more extensive mapping of the hindered rotational potential with a less expensive dzp basis set [13]. Tests show that V_{tr} is not sensitive to dzp or tzp basis set selection.

In the dzp calculations reported here, the potential energy was obtained at approximately 500 different geometries on a grid of values for $R_{H\text{-}C}$ and for two spherical angles describing the attacking H with respect to the C_{2v} axis originating on the C atom. The CH_2 was kept frozen at the equilibrium position for these calculations. The SURVIB program system [15] was used to generate a fit of the potential

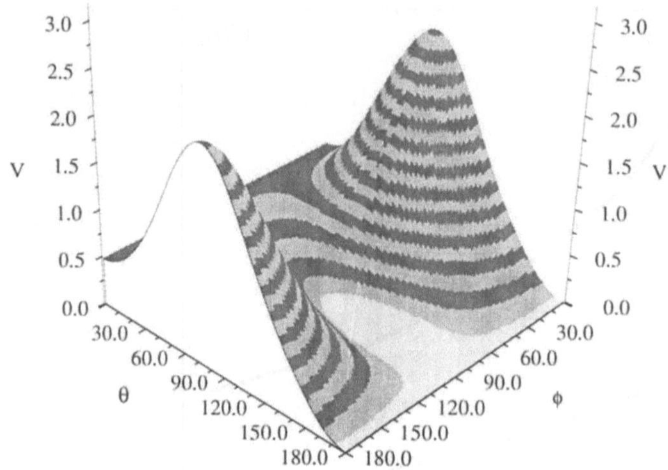

Figure 3: Contour plot of $V_{tr}(\theta, \phi)$ at $R_{H\text{-}CH2} = 3.15$ Å for H+CH_2. Each band is 0.1 kcal/mole.

relative to the C_{2v} reaction path in inverse power expansions of the distance and sinusoidal terms in the angles. Ultimately this fit was converted into $V_{tr}(\theta,\phi)$, the change in potential from the C_{2v} axis when $R_{H\text{-}CH2}$, the distance of H from the center of mass of the CH_2, is held fixed. θ and ϕ are the spherical angles describing the location of the attacking H atom with respect to the C_{2v} axis and with an origin on the CH_2 center of mass.

In Fig. 3, $V_{tr}(\theta,\phi)$ is displayed for $R_{H\text{-}CH2} = 3.15$Å, a distance that is approximately at the reaction bottleneck for 1500 K. The range of ϕ is 0° to 180°, reflecting the C_{2v} symmetry of CH_2, while the range of ϕ in Fig. 1 is 0° to 120° reflecting the C_{3v} symmetry of CH_3. As a consequence of the broader azimuthal angular space, the valley between the two hills in the vicinity of $\theta \approx 90°$ is very much more pronounced in CH_2 than in CH_3 (see Fig. 1). Here, it is easier for the attacking H to fit between two adjacent C-H bonds in CH_2 than in CH_3. However, when θ approaches 180°, i.e., the backside approach, it is more difficult for the H to fit between the more highly splayed CH_2 than the more nearly planar CH_3. Consequently, the difference in energies between the backside and frontside approach is more pronounced for H attack of CH_2 than for H attack of CH_3 (see Fig. 1).

The valley floor in Fig. 3 has an interesting feature. The lowest contour encompassing the $\theta = 0°$ line, i.e., the supposed reaction path geometry, extends well into the valley. In effect, the true reaction path at this value of $R_{H\text{-}CH2}$ is *not* along the C_{2v} axis but at an angle of approximately 60° with this axis. In fact, the calculations show that at larger distances θ approaches 90° before eventually arriving at its asymptotic value of 0° determined by long-range electrostatic forces. The kinetics theory of the preceding section does not require a correct identification of the reaction path. However, the hindering function $\Gamma(R,T)$ in the theory can only be guaranteed to be bounded by unity if the path to which $V_{tr}(\theta,\phi)$ refers is in fact the lowest energy path.

Since no conserved frequencies or shape changes have yet been studied, Eq. (12) for k(T) applies. The final computed high pressure limiting rate constants, labelled HR, along with experimental rate constants at finite pressures, are shown in Fig. 4. Calculated rate constants with the free rotor (FR) and harmonic (HO) models are also shown. Unlike the $H+CH_3$ case, the hindered and free rotor models are very similar to one another and, in a relative sense, noticeably higher than the harmonic model. The calculated rate constants for the high pressure limit are very much higher, especially at high temperatures, than the experimental values at finite pressure. In the earlier theoretical studies [12], the effect of finite pressure was found to be minor. The resolution of experiment-theory differences remains a subject for future work.

3.3 $F + CH_3 \rightarrow CH_3F$: The ^2P character of F gives rise to three distinct potential energy surfaces in its recombination with the methyl radical. The $F + CH_3$ system is a simple version of a general pattern in the recombination of radicals with non-zero electronic angular momentum. The multiple potential energy surfaces can separately display attractive electrostatic, hydrogen-bonding, and chemical-bonding

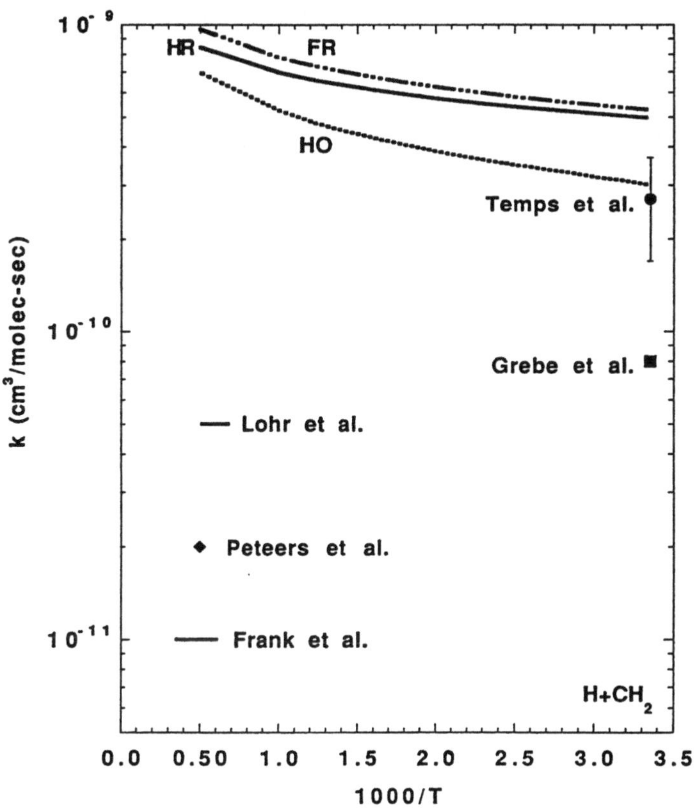

Figure 4: Calculated rate constant k(T) *vs.* 1000/T for $H+CH_2$ (see text). Experimental values are from Ref. [11].

interactions that occur at different distances and angles of approach leading to complex surface crossings during hindered rotations. Thus even though there are as yet no experimental studies of recombination or dissociation kinetics, this relatively simple extension of the H + CH_3 kinetics provides an opportunity to study the influence of angularly dependent surface crossings on the kinetics.

Electronic structure calculations have been carried out for F approaching CH_3 frozen in its equilibrium configuration. Only three directions of approach were examined: (1) along the CH_3 C_{3v} axis, (2) in the 4 atom plane approaching the H end of a C_{2v} C-H bond of CH_3, and (3) as previous but approaching the C end of a C_{2v} C-H bond. As in the H+CH_2 study, a dzp basis set [13] was employed in an MRSDCI calculation with the COLUMBUS program system [14]. All three surfaces were characterized along each of the three directions of approach.

Qualitatively, the three potential energy surfaces can be correlated with the orientation of the radical p orbital of F with the direction of approach. For the C_{3v} direction of approach, the orientation of the radical orbital along that axis leads to bond formation and is very strongly favoured. The placement of either doubly occupied orbital along that axis produces identical potential energies (by symmetry) that are generally repulsive in the kinetically important regions. However, when the doubly occupied orbital is aligned along the C_{2v} direction of approach against the H end of the CH_3, a favourable hydrogen-bonding type interaction occurs at large distances. The degeneracy is split but either curve is much more attractive than that for the radical orbital oriented along the C_{2v} axis. Similar but weaker interactions are also seen when the C end of the C-H bond is approached along the C_{2v} axis. Of course, all three potentials become repulsive for either C_{2v} approach at short distances because the radical orbitals are not aligned properly to form a new bond.

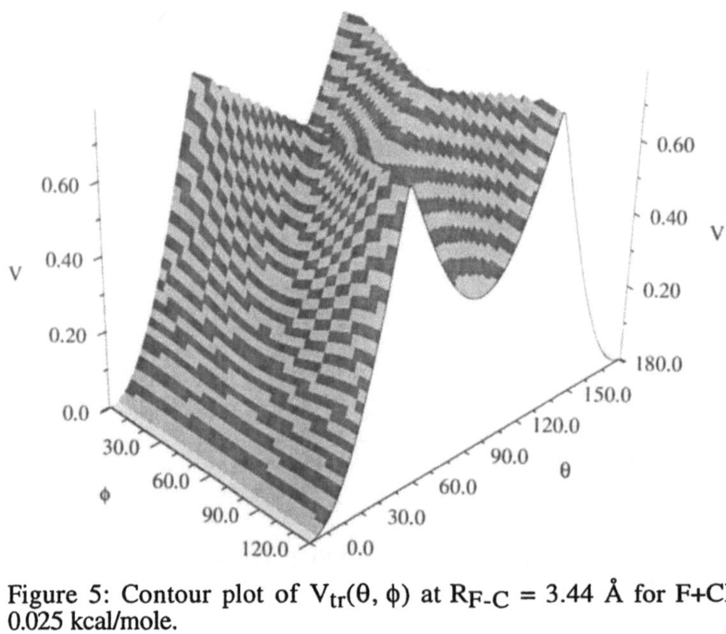

Figure 5: Contour plot of $V_{tr}(\theta, \phi)$ at $R_{F\text{-}C}$ = 3.44 Å for F+CH_3. Each band is 0.025 kcal/mole.

If the direction of approach is varied but the orientation of the radical orbital to the direction of approach is preserved, the available calculations allow three, qualitative *diabatic* potentials $V_{tr,L}(\theta,\phi)$ to be constructed for spherical angles θ and ϕ with respect to the C_{3v} axis. Here L denotes the diabatic surface according to L = 1 (radical orbital aligned along the direction of approach), 2 (radical orbital perpendicular to the direction of approach and in the plane for 4 atom planar geometries), and 3 (radical orbital perpendicular to axis of approach but perpendicular to the plane for 4 atom planar geometries). $V_{tr,L}(\theta,\phi)$ is the change in energy from the reaction path. If the reaction path geometry is assumed to be along the C_{3v} axis, then there are three potential curves for that path which are designated $V_{rxnpath,L}$. Of these, the lowest is the one whose radical orbital is along the C_{3v} axis, i.e., $V_{rxnpath,1}$. Then $V_{tr,L}(\theta,\phi)$ can be assumed to have the form:

$$V_{tr,L}(\theta,\phi) = V_{rxnpath,L}(0,0) - V_{rxnpath1}(0,0) + a(R_{F\text{-}C})\sin 2\theta[1 + b(R_{F\text{-}C})\cos 3\phi \sin\theta]$$

where constants a and b are determined from the computed energies on the two C_{2v} angles of approach. This fit does display the appropriate symmetry about the C_{3v} and C_{2v} axes and has a minimal number of sinusoidal terms in the spherical angles. The computed variation of $V_{rxnpath,1}$ determines $V^{\ddagger}$ and completes the description of that portion of the PES required by the kinetics.

Since no conserved frequencies or shape changes have yet been studied, Eq. (12) for k(T) applies. However, two extreme models of $V_{tr}(\theta,\phi)$ can be constructed from $V_{tr,L}(\theta,\phi)$ for L = 1, 2, 3:

$$V_{tr}(\theta,\phi) = V_{tr,1}(\theta,\phi) \quad \text{(diabatic model)}$$
$$= \min[\,V_{tr,L}(\theta,\phi)]_L \quad \text{(adiabatic model)}$$

In the diabatic model, the hindered rotations pass through every avoided crossing diabatically and consequently changes in electronic character are not used to minimize hindered rotation barriers. In the adiabatic model, hindered rotations stay on the adiabatic surface as it makes sharp changes in electronic character and consequently there is a minimization of hindered rotation barriers. The actual coupling of electronic and rotational motion will range between these two extremes and the variation in the rate constant for these two models is a measure of the maximum kinetics impact this coupling can have.

In Fig. 5, the adiabatic model of $V_{tr}(\theta,\phi)$ is displayed for $R_{F\text{-}C}$ = 3.44Å, the approximate location of the transition state at 600 K. The lower part of the potential has the same qualitative shape as that for $H+CH_3$ (see Fig. 1). However, about the top one third of the diabatic hindered rotation barrier has been eliminated by a surface crossing. The L = 2 diabatic surface at θ = 90° is the lower surface at all azimuthal angles. If the dynamics stay on the adiabatic surface, the barrier to hindered rotation, as seen from the figure, is about 0.7 kcal/mole. The full diabatic barrier at this distance is about 0.9 kcal/mole.

In Fig. 6, the calculated adiabatic and diabatic variational canonical rate constant is displayed as a function of temperature. The differences between the two models are largest at low temperatures where the reaction bottleneck occurs at large distances that favour the hydrogen-bonding configuration in the 4 atom plane over

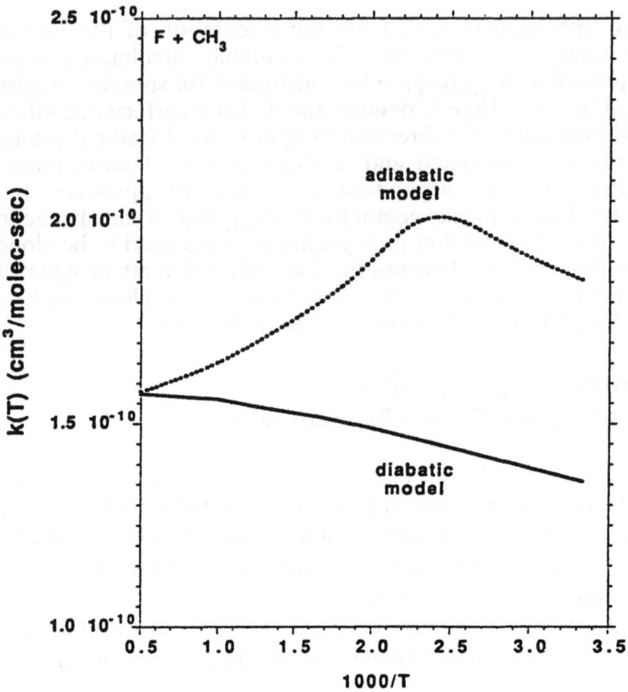

Figure 6: Calculated rate constant k(T) *vs.* 1000/T for F+CH_3 (see text).

the chemical-bonding configuration in the C_{3v} approach. This leads to severe surface crossings. At high temperature the differences vanish. This is due to the fact that at higher temperatures the reaction bottleneck occurs at smaller distances where all surfaces are similarly repulsive except the L=1 surface in the vicinity of the C_{3v} axis. Consequently, surface crossings are minor or non-existent. The two different rate constants are never more than ±15% different over the entire temperature range of 300 K to 2000 K. Thus the surface crossing effects are not expected to be pronounced.

4 Conclusions

This paper has reviewed a simple variational canonical rate constant theory and successfully applied it to three different recombination reactions. The theory classically accommodates large amplitude motion from any given point on the selected reaction path if such motion is constrained to leave the distance between the centers of mass of the reactants unchanged. The resulting theory is rigorously equivalent to FTST in the canonical limit The three applications, D+CH_3, H+CH_2, and F+CH_3, display different degrees of hindrance during large amplitude motion and one case, F+CH_3, exhibits surface crossings. Future developments of this approach will remove the constraint that large amplitude angular motion keeps the separation of the centers of mass of the fragments fixed.

5 Acknowledgements

This work was supported by the U.S. Department of Energy, Office of Basic Energy Sciences, Division of Chemical Sciences, under Contract No. W-31-109-Eng-38 (AW, LH), by the U.K. Engineering and Physical Sciences Research Council (SR), and by an operating grant from the Natural Sciences and Engineering Research Council of Canada (DW).

6 References

[1] D.M. Wardlaw and R.A. Marcus, Chem. Phys. Lett. **110**, 230 (1984); J. Chem. Phys. **83**, 3462 (1985).

[2] W.L. Hase and D.M. Wardlaw, in ***Bimolecular Collisions,*** eds. M.N.R. Ashfold and J.E Baggott (Chem. Soc., London, 1989); D.M. Wardlaw and R.A. Marcus, Adv. Chem. Phys. **70**, 231 (1988), part 1.

[3] (a) S.C. Smith, J. Chem. Phys. **95**, 3404 (1991); (b) *ibid.* **97**, 2406 (1992); (c) J. Phys. Chem. **97**, 7034 (1993).

[4] (a) S.J. Klippenstein and R.A. Marcus, J. Chem. Phys. **87**, 3410 (1987); (b) J. Phys. Chem. **92**, 3105 (1988).

[5] (a) S.J. Klippenstein, Chem. Phys. Lett. **170**, 71 (1990); J. Chem. Phys. **94**, 6469 (1991); *ibid.* **96**, 367 (1992); (b) S.J. Klippenstein, J. Phy. Chem. **98**, 11459 (1994); (c) S.J. Klippenstein, Chem. Phys. Lett. **214**, 418 (1993).

[6] E.E. Aubanel, S.H. Robertson and D.M. Wardlaw, J. Chem. Soc. Faraday Trans. **87**, 2291 (1991).

[7] S.H. Robertson, A.F. Wagner, and D.M. Wardlaw, J. Chem. Phys. **103**, 2917 (1995).

[8] E.E. Aubanel and D.M. Wardlaw, J. Phys. Chem. **93**, 3117 (1989).

[9] W.L. Hase, S.L. Mondro, R.J. Duchovic, and D.M. Hirst, J. Am. Chem. Soc. **109**, 2916 (1987).

[10] M. Brouard and M.J. Pilling, Chem. Phys. Lett. **129**, 439 (1986)

[11] P. Frank, K.A. Bhaskaran and Th. Just, J. Phys. Chem. **90**, 226 (1986); T. Bohland and F. Temps, Ber. Bunsenges. Phys. Chem. **88**, 459 (1984); J. Grebe and K.H. Homann, *ibid.* **86**, 581 (1982); R. Lohr and P. Roth, *ibid.* **85**, 153 (1981); J. Peeters and C. Vinckier, *15th Symposium (Int.) on Combustion,* The Combustion Institute, Pittsburgh, PA, 1975, p. 969.

[12] A.F. Wagner and L.B. Harding, ACS Symp. Series **502**, 48 (1992); M. Aoyagi, R. Shepard, A.F. Wagner, Intl. J. Supercomp. Appl. **5**, 72 (1991); M. Aoyagi, R. Shepard, A.F. Wagner, T.H. Dunning, Jr., J. Phys. Chem. **94**, 3236 (1990); T. H. Dunning, Jr., L.B. Harding, R.A. Bair, R.A. Eades, and R.L. Shepard, *ibid.* **90**, 344 (1986); B.R. Brooks and H.F. Schaefer, III, *ibid.* **67**, 5146, (1977); A. Merkel and L. Zulicke, Mol. Phys. **60**, 1379 (1987);

[13] T.H. Dunning, Jr., J. Chem. Phys. **90**, 1007 (1989), R.A. Kendall, T.H. Dunning, Jr., and R.J. Harrison, *ibid.***96**, 6796 (1992), D.E. Woon and T.H. Dunning, Jr., *ibid.* **98**, 1358 (1993).

[14] Shepard, R.; I. Shavitt, R.M. Pitzer, D.C. Comeau, M. Pepper, H. Lischka, P.G. Szalay, R. Ahlrichs, F.B. Brown, and J.-G. Zhao, Int. J. Quantum Chem. **S22**, 149 (1988).

[15] L.B. Harding, and W.C. Ermler, J. Comp. Chem. **6**, 13 (1985); W.C. Ermler, H.C. Hsieh, and L.B. Harding, Comp. Phys. Comm. **51**, 257 (1988).

State-to-State Reaction Dynamics of Polyatomic Molecules

D.C. Clary
Department of Chemistry, University of Cambridge
Cambridge CB2 1EW, UK

Abstract

Quantum reactive scattering calculations on the state-to-state dynamics of gas phase chemical reactions involving four or more atoms can now be carried out. This article describes how the Rotating Bond Approximation (RBA) can be applied to the four-atom reactions OH + HCl → H_2O + Cl and OH + HBr → H_2O + Br. This form of the RBA is state-selective in the local OH-stretching and bending vibrations of H_2O, and the rotational states of OH. The results give insight into the effect of rotational states on the vibrational product distributions of these atmospherically important reactions and also provide a new explanation of the strong negative temperature dependence observed in the rate constant for the OH+HBr reaction.

1 Introduction

Chemical reactions in the gas phase that involve four or more atoms are of considerable current interest from both the experimental and theoretical points of view[1-6]. Of particular concern is how the initial vibrational and rotational states of AB and CD affect the reaction cross sections and product ABC ro-vibrational distributions in AB + CD → ABC + D. Effects such as bond and mode selectivity and correlations between the quantum states of reactants and products are of special interest.

The reaction dynamics of the internal quantum states of molecules is also relevant to understanding the kinetics of reactions such as the temperature dependence of rate constants. Furthermore, reactions such as those considered here between diatomic free radicals often have very large rate coefficients with unusual temperature dependences that are not explained using simple theories that do not treat the full reaction dynamics of the chemical reaction [7].

The OH + HCl → H_2O + Cl and OH + HBr → H_2O + Br reactions are very important in atmospheric chemistry as they produce halogen atoms that react with ozone. The room temperature rate constant [8] for the OH + HCl reaction is of the order 10^{-13} $cm^3 s^{-1} molec^{-1}$ but is not dependent on temperature below 300K. The OH+HBr reaction is much faster [9] (10^{-11} $cm^3 s^{-1} molec^{-1}$) and the rate constant has a very strong negative temperature dependence, with the rate constant decreasing by a factor of ten as the temperature is decreased from 300 to 25K.

The first quantum reactive scattering calculations on a realistic four-atom reaction were done on the H_2 + CN → H + HCN reaction with linear geometry [10,11]. The first quantum reactive scattering calculation on a state-selected four-atom reaction with non-linear geometry was reported in 1991 [12] in which the reaction OH + H_2 → H_2O + H and the reverse reaction H + H_2O → OH + H_2 was

Springer Series in Chemical Physics, Volume 61
Gas Phase Chemical Reaction Systems
Eds.: J. Wolfrum, H.-R. Volpp, R. Rannacher, and J. Warnatz

treated. In the fullest form of the theory used in these calculations, which has since been called the Rotating Bond Approximation (RBA) [13], all three vibrations of H_2O are treated explicitly by a close-coupling expansion, as are the vibrations of the OH and H_2, and the rotation of OH. However, the rotational motion of H_2 and H_2O are approximated by using adiabatic bend corrections.

The RBA is a general theory that has the particular advantage that it can treat explicitly both the bending mode of triatomic ABC and rotational states of AB in the reaction ABC + D → AB + CD. The method has also been applied to the reactions OH + CO → CO_2 + H [14] and H + HCN → H_2 + CN [15]. It has also been extended to polyatomic reactions such as OH + CH_4 → H_2O + CH_3 [16] and, very recently, the Walden inversion reaction Cl^- + CH_3Br → CH_3Cl + Br^- [17]. In these latter calculations the spectator bonds not taking part directly in the reaction were held fixed, an approximation which is expected to be reliable on the basis of our previous computations.

In the remainder of this paper we first of all describe how the RBA is applied to the reactions of OH with HCl and HBr and then go on to describe some results.

2 The Rotating Bond Approximation

We present a brief description of the Rotating Bond Approximation applied to the OH(j) + HX(v) → H_2O(m,n) + X reaction, where X is Cl or Br. Most of the details of the theory are presented in Ref. 12. The vector joining the centres of mass of the OH and HX molecules is denoted $\mathbf{R_1}$ and the vector between the H and O atoms of OH is $\mathbf{R_2}$. The vector $\mathbf{R_3}$ that joins the H and X atoms of HX has an angle γ with respect to $\mathbf{R_1}$. The angle between $\mathbf{R_1}$ and $\mathbf{R_2}$ is θ and the torsional angle between $\mathbf{R_2}$ and $\mathbf{R_3}$ is ϕ. In the RBA, $\mathbf{R_3}$ is aligned along $\mathbf{R_1}$ so that γ and ϕ are set to zero. However, the zero-point energies associated with the γ and ϕ motions are calculated and these adiabatic energies are added to the potential energy surface as described below. The bond length R_2 is fixed during the calculation.

Transforming R_1 and R_3 to the hyperspherical coordinates ρ and δ gives the Hamiltonian

$$\begin{aligned} H = &-\frac{\hbar^2}{2\mu}\frac{\partial^2}{\partial\rho^2} - \frac{\hbar^2}{2\mu\rho^2}\frac{\partial^2}{\partial\delta^2} + \left(\frac{\hbar^2}{2MR^2}\right)(j_2^2 - K^2) + Bj_2^2 \\ &+ \frac{\hbar^2}{2\mu\rho^2}\left(J(J+1) - K^2 + \frac{3}{4}\right) + V(\rho,\delta,\theta) \end{aligned} \tag{1}$$

where $M = m_H\,(m_O+m_H)/(2m_H + m_O)$, V is the potential energy surface, B is the rotor constant of OH and R is the distance from the bond-breaking H atom of H_2O to the centre of mass of OH. Also $j_2{}^2$ is the operator associated with the rotation of OH, J is the total angular momentum and K, which is assumed to be conserved, is the projection of both $\mathbf{j_2}$ and $\mathbf{J}$ along the intermolecular axis $\mathbf{R_1}$.

The wavefunction for initial quantum state labelled by k' is expanded in the coupled-channel form

$$\Psi_{k'}(\rho,\delta,\theta;\rho_i) = \sum_{k}^{N} f_{kk'}(\rho;\rho_i)\psi_k(\delta,\theta;\rho_i) \tag{2}$$

The $\{\psi_k(\delta,\theta;\rho_i)\}$ functions are computed by diagonalising the Hamiltonian Eq. (1) for $\rho = \rho_i$ with

$$\psi_k(\delta,\theta;\rho_i) = \sum_{k_1}^{N_\delta} \sum_{k_2}^{N_\theta} c^k_{k_1,k_2}(\rho_i)\, \psi_{k_1}(\delta;\rho_i)\, \psi_{k_2}(\theta;\rho_i) \qquad (3)$$

The energies $\{E_k(\rho_i)\}$ obtained from this diagonalisation give the so-called hyperspherical adiabats when plotted versus ρ. These hyperspherical adiabats correlate with different quantum states of reactants or products for large ρ and avoided crossing between them can indicate large reaction probabilities.

The functions $\{ \psi_{k_1}(\delta;\rho_i) \}$ of Eq. (3) are calculated by diagonalising the Hamiltonian

$$H = -\frac{\hbar^2}{2\mu\rho_i^2}\frac{\partial^2}{\partial\delta^2} + V_o(\delta;\rho_i) \qquad (4)$$

with a basis set of n_δ equally spaced distributed Gaussian functions. An appropriate potential $V_o(\delta;\rho_i)$ is obtained by using the equilibrium bond angle θ_{eq} of H_2O in $V(\rho_i,\delta,\theta_{eq})$. The functions $\{ \psi_{k_2}(\theta;\rho_i)\}$ are expanded as linear combination of n_θ spherical harmonics $\{Y_j^K(\theta, 0)\}$. They are obtained from diagonalisation of the Hamiltonian

$$H = \left(\frac{\hbar^2}{2MR^2}\right)(j_2^2 - K^2) + Bj_2^2 + V_1(\theta;\rho_i) \qquad (5)$$

where

$$V_1(\theta;\rho_i) = \langle \psi_1(\delta;\rho_i) | V(\rho_i,\delta,\theta) | \psi_1(\delta;\rho_i) \rangle \qquad (6)$$

Scattering boundary conditions are applied to the solution of the coupled-channel equations at a large value of ρ to give the S-matrix elements $S^{J,K}_{vj,mn}(E)$ for the state-selected transition in the reaction OH(j, K) + HX(v) $\rightarrow$ H_2O(m,n,K) + X.

The state-selected reaction probabilities are $P^{J,K}_{vj,mn}(E) = \left|S^{J,K}_{vj,mn}(E)\right|^2$.

From these, the state-selected integral reaction cross sections are obtained as

$$\sigma(v,j \rightarrow m,n) = \frac{\pi}{(2j+1)k_{vj}^2} \sum_{K=-j}^{j} \sum_{J>K} (2J+1) P^{J,K}_{vj,mn}(E) \qquad (7)$$

The rotational motion of the HX molecule is not treated explicitly in the scattering calculations. This can be compensated for by applying an "adiabatic bend" procedure that follows the approach of Bowman and co-workers [18]. The potential energy surface is minimised, for fixed values of ρ, δ and θ, in the angles γ and ϕ to give $V_m(\rho,\delta,\theta)$. Then the harmonic zero-point energies ε_{in} and ε_{out} for the in-plane (γ) and out-of plane (ϕ) bending vibrations of the OHHX complex that correlate with HX rotational motion are added to $V_m(\rho,\delta,\theta)$ to give the potential $V(\rho,\delta,\theta)$ that is used in the Hamiltonian of Eq. (1).

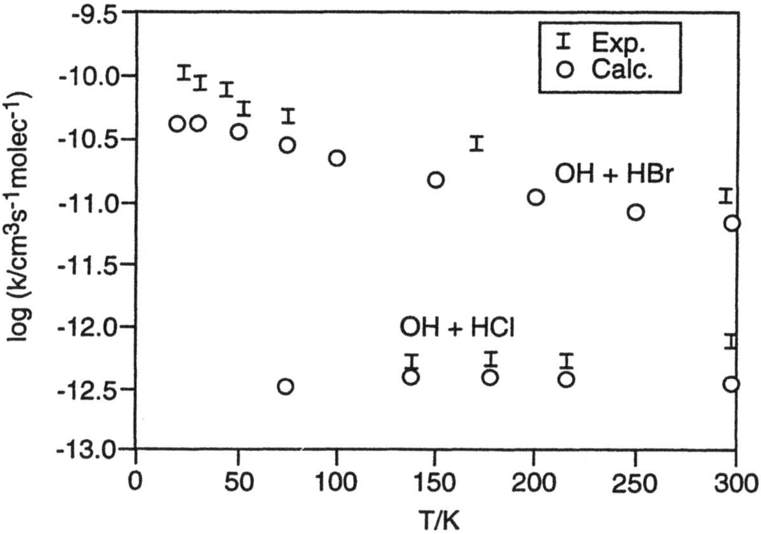

Figure 1: RBA rate constants for the OH+HCl and OH+HBr reactions

Rate constants are calculated by using the following procedure. First, the rate constant *k'(T)* is calculated by Maxwell-Boltzmann averaging the product of the initial velocity and the cross sections $\sigma(v,j)$ of Eq. (7) summed over all product states, with $v = 0$, $j = 0$ and $K = 0$. The rate constant for reaction out of HCl + OH(j = 0, K = 0) is then calculated as

$$k_0(T) = \frac{k'(T)}{[1-\exp(-\frac{2\varepsilon_{in}}{k_\beta T})][(1-\exp(-\frac{2\varepsilon_{out}}{k_\beta T})]\{1+\exp(-\frac{\Delta E}{k_\beta T})\}} \quad (8)$$

where k_β is the Boltzmann constant and the term in curly brackets arises as the two electronic states of OH, separated in energy by ΔE, can both be populated.

3 Reactions of OH with HCl and HBr

The simple potential energy surface used for these reactions [19] contained a London-Eyring-Polanyi-Sato (LEPS) function to describe the bonds breaking and forming in these reactions and also included an accurate potential for H_2O. The potential function was parameterised to give quite good agreement with the transition state geometry and frequencies for the OH + HCl reaction calculated *ab initio*. The OH + HBr potential was based on the OH + HCl one and will be much more approximate.

Figure 1 compares the RBA [19] and experimental [8] rate constants for the OH + HCl and OH + HBr reactions. The agreement for OH + HCl is quite good and the flat temperature dependence of the rate constants is obtained. In the case of the faster OH + HBr reaction, the calculated rate constants are slightly below the

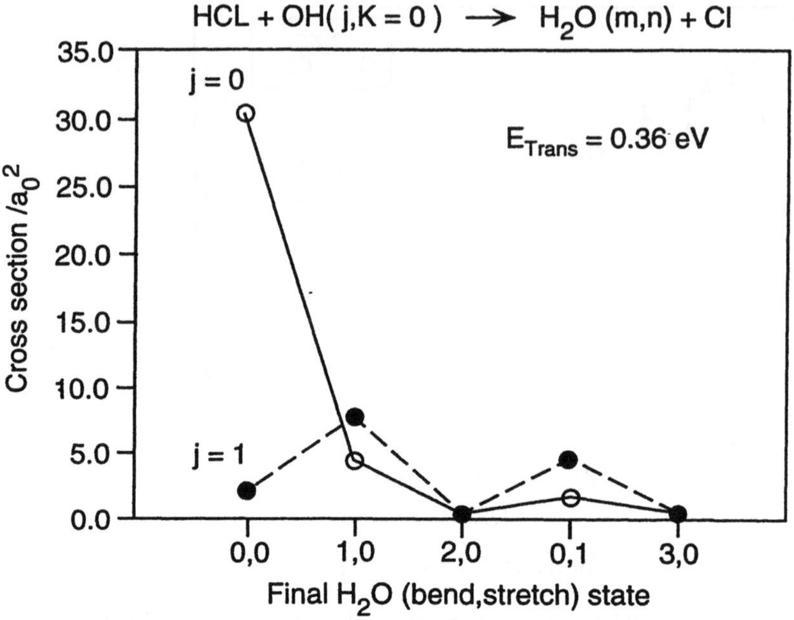

Figure 2: RBA cross sections for the HCl(v = 0)+OH(j, K = 0)→Cl+H_2O(m,n) reaction.

experimental values [9], but the negative temperature dependence of the rate constants is obtained in the RBA calculations.

Figure 2 shows RBA cross sections for the OH(j, K = 0) + HCl → H_2O(m,n) + Cl reaction in which m is a bending mode quantum number for H_2O and n is a local OH stretching mode quantum number. It can be seen that the total reactive cross sections for j=0 is much larger than that for j = 1 over the whole thermal energy range. It is also seen that the ground vibrational state (0,0) of H_2O is the only state with a large cross section out of OH(j = 0) while the state (1,0) with one quantum of energy in the H_2O bending mode is the most likely product state to be formed from OH(j = 1, K = 0). Thus there is a strong correlation between the initial rotational state of OH and the final bending-mode quantum number of H_2O. This can be explained by an examination of avoided crossings between the hyperspherical adiabats $\{E_K(\rho)\}$ correlating with particular reactant or product states [19].

The calculations also show[19] that the cross sections at small translational energies for both OH + HCl and OH + HBr with $|K| = j$ are close in magnitude to those for (j = 0, K = 0) and the cross sections for $j>|K|$ are much smaller than those for $j = |K|$. This suggests that, at lower temperatures, the degeneracy averaged rate constant k_j is obtained as

$$k_j(T) = \frac{2k_0(T)}{(2j+1)}, \quad (j \neq 0) \tag{9}$$

Maxwell-Boltzmann averaging $k_j(T)$ over all j then gives the final rate constant as

$$k(T) \approx \frac{k_0(T)\ [1+2\sum_{j=1} \exp(-\frac{E_j}{k_\beta T})]}{\sum_{j=0}(2j+1)\exp(-\frac{E_j}{k_\beta T})} \approx k_0(T)\sqrt{\frac{B_{HX}\ \pi}{k_\beta T}} \quad (10)$$

where B_{HX} is the rotor constant of the HX molecule.

In OH + HCl, $k_0(T)$ has a slightly positive temperature dependence which is matched by the $T^{-1/2}$ term to produce an overall $k(T)$ that is almost independent of temperature. Furthermore, the ratio for the rate constants for the OH+HCl and OD + HCl reactions is predicted by this formula to be approximately $\sqrt{2}$ and this is in good agreement with experiment [8]. For the OH + HBr reaction, the main dependence of $k_0(T)$ on temperature is due to the electronic partition function of OH and Eq. (10) then suggests the simple form for the rate constant of

$$k(T) = \frac{A}{T^{1/2}[1+\exp(-\Delta E / T)]} \quad (11)$$

where A is a constant. This provides rather a good explanation of the strong negative temperature dependence observed in the experimental rate constants.

This brief summary of the RBA calculations on the reactions of OH with HCl and HBr has aimed to show that quantum treatment of the state-to-state aspects of a four atom reaction can yield new understanding of chemical reaction dynamics and can also provide an insight into useful quantities such as the temperature dependence of rate constants. As the RBA has been applied to reactions with more than four atoms such as $OH + CH_4 \rightarrow H_2O + CH_3$ [16] and $Cl^- + CH_3Br \rightarrow CH_3Cl + Br^-$ [17] this is clearly going to be a useful theoretical tool for the general study of chemical reactions of polyatomic molecules.

4 Acknowledgements

The author is very grateful to Gunnar Nyman who collaborated with the calculations described here.

5 References

[1] Crim, F. F.; Hsiao, M. C.; Scott, J. L.; Sinha, A.; Vander Wal, R. L., Philos. Trans. R. Soc. London Ser. 1990, **A 332**, 259.
[2] Bronikowski, M. J.; Simpson, W. R.; Girard, B.; Zare, R. N., J. Chem. Phys. 1991, **95**, 8647.
[3] Schatz, G. C., J. Chem. Phys. 1979, **71**, 542.
[4] Clary, D. C.; Echave, J. in ***Advances in Molecular Vibrations and Collision Dynamics,*** Vol. 2A; (Bowman, J. M., Ed.) JAI, Greenwich, 1994; p. 203.
[5] Wang, D.; Bowman, J. M. in ***Advances in Molecular Vibrations and Collision Dynamics,*** Vol. 2B (Bowman, J. M., Ed.) JAI, Greenwich, 1994, to be published.
[6] Casavecchia, P.; Balucani, N.; Volpi, G.G. in ***Research in Chemical Kinetics,*** (Compton, R. G.; Hancock, G. Eds.) Elsevier: Amsterdam, 1994; Vol. 1, p. 1.
[7] Clary, D. C., Molec. Phys. 1984, **53**, 3.
[8] Sharkey, P.; Smith, I. W. M. J. Chem. Soc. Faraday Trans. 1993, **89**, 631.

[9] Sims, I. R.; Smith, I.W.M.; Clary, D. C.; Bocherel, P.; Rowe, B. R., J. Chem Phys. 1994, **101**, 1748.
[10] Brooks, A. N.; Clary, D. C., J. Chem. Phys. 1990, **92**, 4178; Brooks, A. N, Ph.D. Thesis, University of Cambridge, 1989.
[11] Sun, Q.; Bowman, J. M., J. Chem. Phys. 1990, **92**, 5201.
[12] Clary, D. C., J. Chem. Phys. 1991, **95**, 7298 .
[13] Clary, D. C., J. Phys. Chem., 1994, **98**, 10678.
[14] Clary, D. C.; Schatz, G. C., J. Chem. Phys. 1993, **99**, 4578.
[15] Clary, D. C., J. Phys. Chem., 1995, in press.
[16] Nyman, G.; Clary, D. C., J. Chem Phys. 1994, **101**, 5756.
[17] Clary, D. C., to be published.
[18] Bowman , J. M., J. Phys. Chem. 1991, **95**, 4960.
[19] Clary, D. C.; Nyman, G.; Hernandez, R., J. Chem. Phys. 1994, **101**, 3704.

Effects of OH Rotation on the $CH_4 + OH \rightarrow CH_3 + H_2O$ Reaction

G. Nyman
Department of Physical Chemistry
Göteborg University, S-41296 Göteborg, Sweden

Abstract
Quantum scattering calculations on the $CH_4 + OH \rightarrow CH_3 + H_2O$ reaction can now be performed. This article discusses such calculations using the Rotating Bond Approximation, treating CH_3 as a pseudoatom. The OH rotation and a reactive C–H stretch of CH_4 are treated explicitly as well as the bending motion and one OH local stretch vibration of H_2O. An adiabatic approach is used to account for all degrees of freedom not explicitly treated in the scattering calculations. Exciting the reactant OH rotation decreases the reaction cross sections.

1 Introduction

Methane is a greenhouse gas which plays a significant role in the heat balance of the atmosphere [1]. It is thus important to assess the rate of removal of methane from the atmosphere, for which the reaction of methane with hydroxyl radicals

$$CH_4 + OH \rightarrow CH_3 + H_2O \qquad (R1\ ),$$

is recognized as the main process [1]. This reaction has accordingly received much interest both experimentally [1–10] and theoretically [11–19]. The results discussed here concern the dynamics of this reaction studied with approximate quantum scattering theory.

By 1990, the quantum mechanical treatment of reactive scattering had progressed from the first accurate converged cross sections for $H + H_2 \rightarrow H_2 + H$ in 1976 [20] to $F + H_2 \rightarrow HF + H$ [21]. The same year the first quantum scattering calculations for realistic four-atom reactions in linear geometries were published [22, 23]. The first quantum scattering calculation for a four-atom reaction with non-linear geometries were carried out by Clary for the $OH + H_2 \rightleftharpoons H + H_2O$ reaction in 1991 [24]. The method used in that work has since been called the Rotating Bond Approximation (RBA) [25, 26].

In this article it is described how the RBA has been extended to treat reactions involving more than four atoms [17]. In the scattering calculations, CH_3 is treated as a pseudoatom Q, whereby reaction (R1) can be written

$$QH(v) + OH(j) \rightarrow Q + H_2O(m, n)\ .$$

The quantum numbers v, j, m and n represent the QH vibration, the OH rotation, the H_2O bend and a local OH stretch of H_2O respectively. These degrees of freedom, and the relative translation, are explicitly treated in the scattering calculations by using the RBA. All other degrees of freedom are treated by an adiabatic approach, to be described later.

Springer Series in Chemical Physics, Volume 61
Gas Phase Chemical Reaction Systems
Eds.: J. Wolfrum, H.-R. Volpp, R. Rannacher, and J. Warnatz

The quantum scattering treatment described here is expected to be reasonable for CH_4 as *ab initio* calculations show that at the transition state the methyl group has a very small barrier to internal rotation [14] meaning that the main interaction with OH is through the reactive hydrogen in CH_4. Perhaps the main effect of the methyl group comes from its contribution to the change in zero point energy as the reaction proceeds. This is accounted for by an adiabatic approach, which in our implementation means that the interaction potential is designed such that it contains the zero point energy of all modes not explicitly treated in the RBA.

The reduced dimensionality potential energy surface RDP1 [17] was used in the work discussed here. We apply the term reduced dimensionality as this surface is an explicit function of a reduced number of coordinates. Some details on RDP1 are reviewed in Section 3.

2 Theory

The RBA method as used to treat four-atom reactions is described in the article by D. C. Clary in this issue. This is not repeated here, but some extensions and the adiabatic approach which is used to account for the degrees of freedom not explicitly treated by the RBA are described.

2.1 The RBA Calculations

In the work discussed here the RBA is applied to the seven-atom reaction $CH_4 + OH \rightarrow CH_3 + H_2O$ [17]. From the point of view of the quantum scattering calculations, the CH_4 moiety is treated as a quasi-diatom QH, where Q is a quasi-atom with the mass of CH_3. In this way the OH + CH_4 reaction is effectively reduced to a four-atom problem. The effect of the other atoms enters the scattering calculations only through adiabatic zero-point energy in the potential employed, which is described in Sec. 3. Therefore, the theoretical development for the scattering calculations is essentially the same as that given by D. C. Clary in a separate article in this issue.

We now discuss some aspects of the approximations made in regards to the treatment of CH_4. For all geometries relevant to this discussion, one of the three principal moments of inertia is clearly smaller than the other two. We therefore refer to the corresponding principal axis as the unique axis, which in the RBA is denoted the R_1 axis.

A particular feature of the RBA is that the rotation of OH is explicitly treated. This allows the study of mode-selective effects with respect to this rotation. Treating this rotation explicitly without allowing it to couple to other angular degrees of freedom, however, requires justification. Firstly, the high symmetry of CH_4 results in a quite unhindered OH rotational motion about the unique axis at the transition state and thus, coupling of this motion to CH_4 is weak. This agrees with the Melissas and Truhlar calculation of the barrier height to internal rotation, which was found to be only 0.02 kcal/mol [14]. Secondly, while the OH rotational motion in a plane containing the unique axis is strongly coupled to CH_4 at the transition state, this is approximately treated in the RBA through the interaction with the reactive hydrogen atom in CH_4.

Another aspect concerns the OH rotational energy associated with its rotation about the unique axis. This energy is conserved if the OH rotation about this axis is uncoupled from CH_4 and the average of the inverse of the OH moment of inertia about it does not change as the reaction proceeds. We expect this to be a reasonable approximation which justifies that in the RBA the initial OH rotational energy about $\mathbf{R_1}$ is conserved, whereby it does not contribute energy to, or withdraw energy from, the reaction coordinate.

2.2 The Adiabatic Approach

Most degrees of freedom in the CH_4 + OH reaction are not explicitly treated in the RBA. Instead they are treated by a modified version of the adiabatic bend theory due to Bowman [27,28]. Bowman's approach is to treat vibrational stretch motions explicitly while bends are treated adiabatically. In brief, the Schrödinger equation is solved approximately in two steps. First it is solved for the Hamiltonian associated with the bending degrees of freedom for a potential that refers to fixed values of the radial coordinates. The eigenvalues obtained are then added to the Hamiltonian for the radial coordinates, for which the close coupling equations are solved. This treatment implicitly assumes that the bending degrees of freedom instantaneously adjust to changes in the radial coordinates, in the same way as the electrons are taken to instantaneously adjust to nuclear motions in the Born-Oppenheimer approximation.

The adiabatic bend treatment of Bowman is accurate if the degrees of freedom not explicitly treated only influence the reaction by supplying or withdrawing energy due to a change in their frequency along the reaction path. The main difference here is that the degrees of freedom we treat explicitly are two stretches and a bend. Both types of adiabatic approaches have been shown to work well for the H_2 + OH reaction [29,30].

Collectively denoting the adiabatically treated degrees of freedom a and the others e, we may write

$$\hat{H}_a = \hat{T}_a + V(a;e), \tag{1}$$

where $\hat{H}_a$ is the Hamiltonian for the a modes, $\hat{T}_a$ is the kinetic energy operator of these modes and $V(a;e)$ is the potential energy surface. $V(a;e)$ is a function of fifteen coordinates even though the ones associated with the e modes are held fixed. To solve the Schrödinger equation for $\hat{H}_a$ repeatedly for different values of the fixed coordinates requires that $V(a;e)$ is globally known.

The *ab initio* potential energy information that is available for reaction R1 is the barrier height, exothermicity and harmonic vibrational frequencies of reactants, products and transition state [13,14]. With only this information, the eigenvalues $E_{a,\mathbf{n}}$, where $\mathbf{n}$ collectively labels the quantum states of the a modes, of $\hat{H}_a$ are simply taken to be the sum of the harmonic energies associated with each relevant vibrational frequency, *i.e*

$$E_{a,\mathbf{n}} = \sum h\nu_i(n_i + \frac{1}{2}). \tag{2}$$

Here ν_i is a harmonic vibrational frequency and the sum is over the adiabatically treated modes.

It is assumed that the reaction is dominated by the shape of the adiabatic potential associated with the ground states of these degrees of freedom.

Therefore the known ground state eigenvalues $E_{a,n=0}$ are added to the *ab initio* energy values before these are used to fit RDP1. In this way, $\hat{H}_a$ need not be diagonalized, a global potential energy surface is not needed and yet, an adiabatic treatment of all modes not explicitly treated in the RBA is obtained at no additional computational expense.

3 Potential Energy Surface

The potential energy surface used for the $CH_4 + OH \rightarrow CH_3 + H_2O$ reaction combines an accurate potential function for H_2O [31] with a London-Eyring-Polanyi-Sato (LEPS) function to describe the C–H and OH reactive bonds. The potential has accurate reactant and product ro-vibrational energy levels, correct bond dissociation energies and transition state geometries in reasonable accord with *ab initio* data [13,14]. It also incorporates the zero point energies of all modes not explicitly treated in the RBA calculations.

The potential energy surface is expanded in the form

$$V = V_{LEPS}(r_{OH}, r_{HQ}, r_{OQ}) + V_{H_2O}(r_{OH}, r_{OH'}, r_{HH'}) - V_{OH}(r_{OH}) \quad (3)$$

where r_{AB} is the distance between atoms A and B. V_{LEPS} is the LEPS potential function [17,32]. V_{H_2O} is the potential energy surface due to Murrell and Carter [31] (the prime on the unreactive hydrogen in H_2O is only used when needed for clarity). V_{OH} is the potential function in V_{H_2O} describing the two-body O—H interaction. It is subtracted out as such a term already exists in V_{LEPS}. In the potential for H_2O in [31] there is a term in $r^2_{OH}r^2_{HH'}$ and in $r^2_{OH'}r^2_{HH'}$ which were removed to avoid an unphysical interaction in the CH_4OH potential when $r_{HH'}$ and r_{OH} or $r_{HH'}$ and $r_{OH'}$ are quite large. This has no significant effect on the vibrational energy levels of H_2O.

There is a single parameter to vary in the potential, viz. the Sato parameter in the LEPS function. The usual procedure is to vary this parameter to obtain a desired value for the classical barrier height. Here it is instead varied to obtain a reduced dimensionality barrier (RDB) height in that it is adapted to a quantum calculation treating explicitly only a reduced number of degrees of freedom. The RDB height is obtained by adding the zero point energy of all modes not explicitly treated in the quantum scattering calculations, to the classical potential at the transition state and to the reactants at equilibrium. The RDB height is such that RDP1 has a vibrationally adiabatic groundstate (VAG) barrier height of 5.9 kcal/mol which is in accord with the *ab initio* calculations of Melissas and Truhlar [14].

4 Results and Discussion

In Fig. 1 reaction cross sections out of the $CH_4(v = 0) + OH(j)$ states for $j \leq 7$ are shown as a function of translational energy. These cross sections are summed

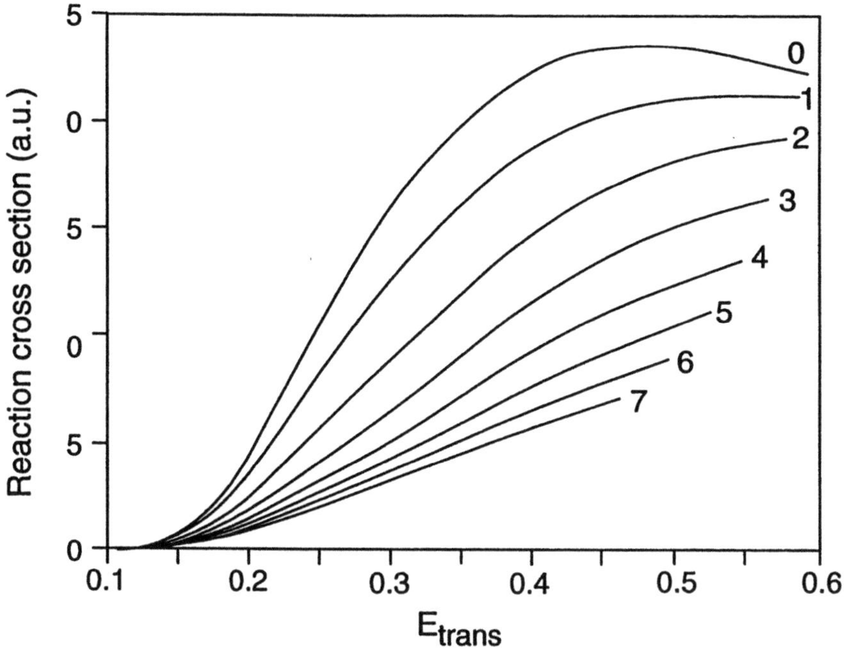

Figure 1: Reaction cross sections for $CH_4(v = 0) + OH(j) \rightarrow CH_3 + H_2O$, summed over all product states and averaged over Ω, are shown as a function of translational energy. The value of j is indicated for each curve.

over all product $CH_3 + H_2O$ states and averaged over Ω, the projection quantum number of j on $\mathbf{R_1}$. The cross sections decrease with j, which results from averaging them over Ω as will be illustrated next.

The cross sections are large for reaction out of $OH(j = |\Omega|)$ but smaller when $j > |\Omega|$. At the energies most important for the thermal rate constant at room temperature the differences are quite large, which is illustrated in Fig. 2 for $j = 1$ and the same trend is seen for other values of j [17]. This is also in agreement with our observations for the HCl + OH and HBr + OH reactions [33]. These observations can be explained by noting that for $j = |\Omega|$, the OH rotation is perpendicular to the QH bond. Therefore, the unreactive H in OH does not get close to the H in QH, *i.e.* it does not interfere with the formation of a new OH bond. When $j > |\Omega|$ on the other hand, the OH rotation is such that the H in OH can come close to the H in QH, thereby obstructing the formation of a new OH bond.

For the OH + CH_4 reaction there is at low energies a quite strong increase in the $j = |\Omega|$ cross sections with j. This is probably due to the large activation energy present in the CH_4 + OH reaction whereby the increase in total energy with increasing j enhances reaction. The combination of the two effects now discussed produces the j dependence seen in Fig. 1.

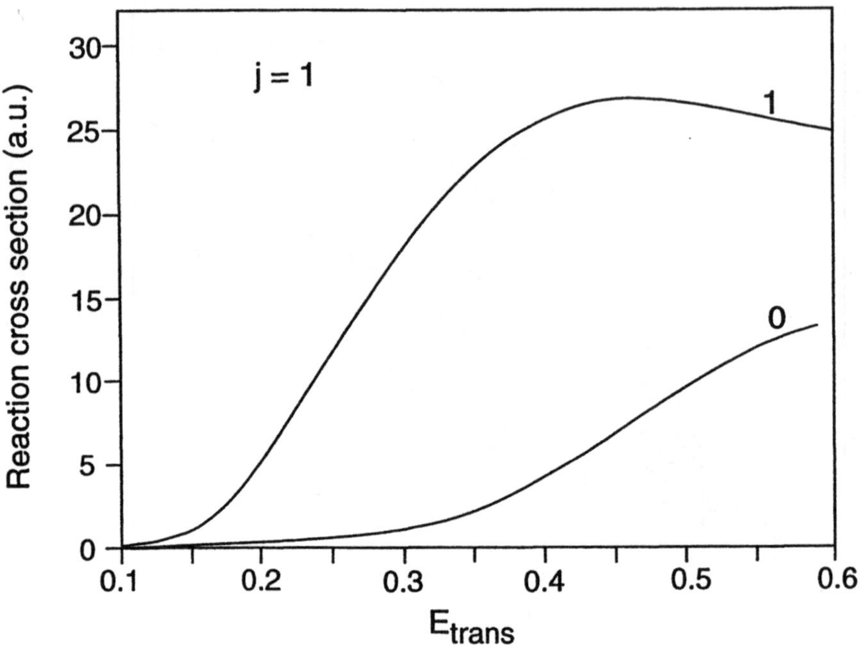

Figure 2: Reaction cross sections for $CH_4(v = 0) + OH(j = 1, \Omega) \rightarrow CH_3 + H_2O$ summed over all product states are shown as a function of translational energy. The value of Ω is indicated to the right.

5 Conclusions

Rotating bond approximation calculations as applied to the $CH_4 + OH \rightarrow CH_3 + H_2O$ reaction have been discussed. The scattering calculations are performed within a model which treats CH_3 as a pseudo-atom and CH_4 as a pseudo-diatom. The rotation of OH, a reactive C–H stretch of CH_4, two vibrations of H_2O and relative translation are treated explicitly in the scattering calculations. An adiabatic approach is used to account for all other degrees of freedom. The reduced dimensionality potential energy surface RDP1 [17] was employed in the calculations.

The reaction cross sections for $OH(j = |\Omega|)$ are substantially larger than those for $OH(j > |\Omega|)$ which has also been seen for HCl + OH and HBr + OH [33]. The $OH(j = |\Omega|)$ cross sections increase with j. The combination of these two effects results in OH rotational excitation, averaged over Ω, lowering the reaction cross sections. The increase in the $j = |\Omega|$ cross sections is probably due to the fairly large activation energy for the $OH + CH_4$ reaction whereby the increase in total energy with increased j enhances reaction.

Many studies have now been performed using the RBA which is proving to be a practical and realistic theoretical tool for treating the reaction dynamics of polyatomic molecules.

6 Acknowledgment

The author is grateful to David Clary for collaboration on the work described here.

7 References

[1] G. L. Vaghjiani and A. R. Ravishankara, *Nature* **350**, 406, 1991.

[2] D. D. Davis, S. Fischer, and R. Schiff, *J. Chem. Phys.* **61**, 2213, 1974.

[3] J. J. Margitan, F. Kaufman, and J. G. Anderson, *Geophys. Res. Lett.* **1**, 80, 1974.

[4] S. Gordon and W. A. Mulac, *Int. J. Chem. Kinet.* **1**, 289, 1975.

[5] K. M. Jeong and F. Kaufman, *J. Phys. Chem.* **86**, 1808, 1982.

[6] D. L. Baulch, M. Bowers, D. G. Malcolm, and R. T. Tuckerman, *J. Phys. Chem. Ref. Data* **15**, 465, 1986.

[7] A. R. Ravishankara, *Ann. Rev. Phys. Chem.* **39**, 367, 1988.

[8] P. Sharkey and I. W. M. Smith, *J. Chem. Soc. Faraday Trans.* **89**, 631, 1993.

[9] W. B. DeMore, *J. Phys. Chem.* **97**, 8564, 1993.

[10] J. R. Dunlop and F. P. Tully, *J. Phys. Chem.* **97**, 11148, 1993.

[11] C. Gonzalez, J. J. W. McDouall, and H. B. Schlegel, *J. Phys. Chem.* **94**, 7467, 1990.

[12] T. N. Truong and D. G. Truhlar, *J. Chem. Phys.* **93**, 1761, 1990.

[13] K. D. Dobbs, D. A. Dixon, and A. Komornicki, *J. Chem. Phys.* **98**, 8852, 1993.

[14] V. S. Melissas and D. G. Truhlar, *J. Chem. Phys.* **99**, 1013, 1993.

[15] V. S. Melissas and D. G. Truhlar. *J. Chem. Phys.* **99**, 3542, 1993.

[16] W.-P. Hu, Y.-P. Liu, and D. G. Truhlar, *J. Chem. Soc. Faraday Trans.* **90**, 1715, 1994.

[17] G. Nyman and D. C. Clary, *J. Chem. Phys.* **101**, 5756, 1994.

[18] G. Nyman, D. C. Clary, and R. D. Levine, *Chem. Phys.* **191**, 223, 1995.

[19] G. Nyman, *Chem. Phys. Lett.* **240**, 571, 1995.

[20] G. C. Schatz and A. Kuppermann, *J. Chem. Phys.* **65**, 4668, 1976.

[21] J. M. Launay and M. Le Dourneuf, *Chem. Phys. Lett.* **169**, 473, 1990.

[22] A. N. Brooks and D. C. Clary, *J. Chem. Phys.* **92**, 4178, 1990.

[23] Q. Sun and J. M. Bowman, *J. Chem. Phys.* **92**, 5201, 1990.

[24] David C Clary, *J. Chem. Phys.* **95**, 7298, 1991.

[25] G. Nyman and D. C. Clary, *J. Chem. Phys.* **99**, 7774, 1993.

[26] D. C. Clary and G.C. Schatz, *J. Chem. Phys.* **99**, 4578, 1993.

[27] J. M. Bowman and D. Wang, *J. Chem. Phys.* **96**, 7852, 1992.

[28] J. M. Bowman and A. F. Wagner, in **The Theory of Chemical Reaction Dynamics**, p. 47 (Ed. D. C. Clary; Reidel, Dordrecht 1986).

[29] D. C. Clary, *J. Phys. Chem.* **98**, 10678, 1994.

[30] U. Manthe, T. Seideman, and W. H. Miller, *J. Chem. Phys.* **101**, 4759, 1994.

[31] J. N. Murrell and S. Carter, *J. Phys. Chem.* **88**, 4887, 1984.

[32] I. W. M. Smith, **Kinetics and Dynamics of Elementary Gas Reactions** (Butterworths, London, 1980).

[33] D. C. Clary, G. Nyman, and R. Hernandez, *J. Chem. Phys.* **101**, 3704, 1994.

Pathways of Vibrational Relaxation of Diatoms in Collisions with Atoms: Manifestation of the Ehrenfest Adiabatic Principle

E.E. Nikitin
Department of Chemistry, Technion – Israel Institute of Technology, Haifa 32000, Israel

Abstract

Different pathways of vibrational relaxation of diatoms in thermal collisions with atoms are discussed in the framework of the Ehrenfest adiabatic principle and generalized Landau-Teller model. Since the efficiency of different energy-transfer channels depend very strongly on the value of the Ehrenfest exponent, it is possible to assign, for given collision partners and the heat-bath temperature, the vibrational energy transfer events to VT, VRT or VR processes.

1 Introduction

The importance of the energy exchange between molecules in a chemically reacting gas was recognized at early stage of the development of chemical kinetics. Max Bodenstein in his pioneering paper of 1913 [1], after having discussed the kinetics of chain reactions, asked the question: *"Für den Stoff stimmt die Rechnung vollkommen, aber stimmt sie auch für die Energie?"*

In 1934, N. Semenov, in his book on chain reactions [2] strongly emphasized the role of the collisional energy transfer in gas-phase chemical kinetics, particularly paying attention to different kind of molecular energy, electronic, vibrational, rotational and translational. However, it was not until the work by Landau and Teller in 1936 [3] when it was realized that the collisional energy transfer should be described in terms of kinetics of populations of individual energy levels. Later on, the discussion of the energy transfer become indispensable sections of comprehensive texts on chemical kinetics as exemplified by the Kondratiev book [4].

The process of vibrational excitation and deexcitation of a diatom in a collision with an atom represents a simplest example from the host of processes which are relevant to gas-phase chemical kinetics. Experimental techniques available now allow one to measure directly state-to-state energy transfer rate coefficients. Theoretically, it is possible to accomplish completely *ab initio* calculation of these coefficients. One can therefore, regard the existing models of the vibrational relaxation from a new standpoint as a means for helping to understand more clearly the dynamics of the energy transfer provided that all the models are related to a single fundamental principle. This is the Ehrenfest adiabatic principle as formulated by Landau and Teller in the application to the collisional vibrational transitions of diatomic molecules.

Landau and Teller demonstrated that the transfer of the vibrational energy of a diatomic molecule into translational energy of colliding partners (say, a diatomic molecule and an noble gas atom) is a very inefficient process, and this property of the collision is related to the Ehrenfest adiabatic principle [5] of mechanics. In its simplest version, the adiabatic principle asserts that the change δI, of the action variable I, of a classical system under the influence of a slow external perturbation of

Springer Series in Chemical Physics, Volume 61
Gas Phase Chemical Reaction Systems
Eds.: J. Wolfrum, H.-R. Volpp, R. Rannacher, and J. Warnatz

duration τ is proportional to the exponential function of $\omega\tau$ provided $\omega\tau$ is large, $\omega\tau \gg 1$, *viz.*

$$\delta I = I^* \exp(-\omega\tau). \tag{1}$$

Here I^* is a quantity that is related to the strength of a perturbation and ω is the frequency of the unperturbed system associated with the action variable. Mathematically, the r.h.s of Eq. (1) is related to the Fourier component of frequency ω of a time-dependent perturbation acting on a system. In the limit $\omega\tau \to \infty$, δI vanishes implying that I is an adiabatic invariant. By the correspondence principle, Eq. (1) translates into the quantum mechanical language as the exponentially small probability of transition between quantum states associated with classical variable I. Of course, the implicit assumption in applicability of Eq. (1) is that the perturbation itself does not bring about a major change in the vibrational frequency of an unperturbed system. This is the case for low-lying vibrational states of a diatomic molecule in a nondegenerate electronic state colliding with chemically inert atom in a closed electronic state.

It seems appropriate, in connection with the general idea of the Symposium, "Experiments and Models", to use the adiabatic principle for a qualitative discussion of energy transfer mechanisms in atom-diatom collisions in terms of the existing models. In this paper, we will illustrate the significance of the Ehrenfest adiabatic principle in the energy transfer processes by way of example of vibrational deactivation of a diatomic molecules out of its first vibrational state (n = 1 → n = 0) induced by collisions with an atom. We assume also that the translational-rotational reservoir is in thermal equilibrium at a given temperature T. Thus, the quantity under discussion will be the average transition probability $p_{10}(T)$ per collision or the state-to-state rate coefficient $k_{10}(T)$. Some representative systems will be taken to illustrate specific pathway of the vibrational energy transfer. We will be using the conventional nomenclature calling the transfer of vibrational to translational energy (with insignificant participation of rotation) VT process, to translational and rotational energy VRT process, and to rotational energy (with insignificant participation of translation) VR processes.

2 Recent Development of the Landau-Teller Model

The Landau-Teller model considers a linear collision of a structureless particle A with a harmonic oscillator BC within an approach which by now is known as a semiclassical method: the relative particle-oscillator motion (coordinate R) is described classically and the vibrational motion of the oscillator (coordinate x) by quantum mechanics; the interaction between incoming particle A and the nearest end B of the oscillator BC is taken to be exponential, $U(R_{AB}) \propto \exp(-\alpha R_{AB})$. The expression for the transition probability in the near-adiabatic limit was found [4] to have the following generic form:

$$P_{10} = C \exp(-2\pi\omega / \alpha v) \tag{2}$$

where ω is the vibrational frequency of the oscillator and v is the relative collision velocity of A and BC. In the near-adiabatic limit, when the condition $2\pi\omega / \alpha v \gg 1$ is satisfied, the preexponential factor C depends on the parameters that enter into the exponent much weaker than the exponential. Eq. (2) expresses the Ehrenfest adiabatic principle (*cf.* Eq. (1)) and identifies the "collision time" with a quantity proportional to $1/\alpha v$.

Rewritten through the energy change in an one-quantum transition, $\Delta E = \hbar\omega$, Eq.(2) assumes the form

$$P_{10} = C\exp(-2\pi\Delta E/\hbar\alpha v) \quad (3)$$

which is also known as the energy-gap law.

Note that the velocity v that enters into Eqs.(2) and (3) corresponds to the asymptotic motion of particles in an *elastic* collision. It provides a certain approximation to the initial velocity $v' = \sqrt{2E'/\mu}$ and the final velocity $v'' = \sqrt{2E''/\mu}$ which are related by the energy conservation equation $E'' - E' = \Delta E$ (here μ is the reduced mass of A and BC). The above approximation is acceptable when the energy transfer is small, $\Delta E \ll E'', E'$.

After averaging over the Maxwell-Boltzmann velocity distribution, the equation for the average transition probability assumes the form:

$$p_{10}(T) \propto \exp[-3(\theta^{LT}/T)^{1/3}] \quad (4)$$

where the Landau-Teller temperature reads:

$$\theta^{LT} = \pi^2\omega^2\mu/2k\alpha^2. \quad (5)$$

The major deficiency of Eq. (4) and of its counterpart for the reverse process, $n = 0 \rightarrow n = 1$, is that they predict equal probabilities for down and up transitions, $p_{10} = p_{01}$. This can be traced back to inability of the semiclasiscal approach to account for the change of the relative collision energy E in the course of inelastic collision. A general method was formulated recently to resolve this shortcoming of the semiclassical approximation [6]. Let the semiclasical (SC) transition probability $P_{10}^{SC}(E)$ for an arbitrary interaction potential be represented in the form:

$$P_{10}^{SC}(E) = C^{SC}(\omega, E)\exp[-\omega\tau(E)] \quad (6)$$

where $\tau(E)$ is a collision time for the elastic scattering. A much better approximation, called improved semiclassical (ISC) approximation [7,8] can be recovered from Eq. (6):

$$P_{10}^{ISC} = C^{SC}[\Delta E/\hbar, (E'-E'')/2]\exp[-\int_{E'}^{E''}\tau(E)dE/\hbar,] \quad (7)$$

where $E'' = E' + \Delta E$. Eq. (7) relates trajectories of the elastic scattering for different energies E to an inelastic event, when initial, E', and final, E'', energies differ by ΔE. In particular, the ISC counterpart of Eqs. (4) and (5) are:

$$P_{10}^{ISC}(T) \propto \exp[-\int_{E^*}^{E^*+\Delta E}\tau(E)dE/\hbar - E^*/kT] \quad (8)$$

where E^* has to be found from the equation:

$$\tau(E^*) - \tau(E^* + \Delta E) = \hbar / kT. \qquad (9)$$

The average transition probabilities calculated from Eq. (8) and its reverse satisfy correct detailed balance relation, $p_{10} = p_{01}\exp(\Delta E / kT)$.

Now, Eq. (9) defines an optimal energy E^*, and therefore a single trajectory, which consists of two branches. A rather unusual property of this trajectory is most simply interpreted when one adopts the Landau idea of analytical continuation of the interaction potential into the classically forbidden region [9]. Then it becomes clear that it is the repulsive part of the interaction potential which is important for the vibrational transition on the near adiabatic regime.

The above short presentation brings to the fore the following important properties of the generalized Landau-Teller exponent:

i. The value of the exponent is determined by a single, optimal, trajectory,

ii. The value of the exponent depends on the properties of the interaction potential close to the turning points of two branches of the optimal trajectory.

Recent studies on the improved semiclassical approximation showed that the expression (8) is valid far beyond the limitations imposed by the simple perturbative treatment of the energy exchange [10], and that the major modifications of the transition probability which are due to the deviation of the interaction potential from a simple exponential function are reproduced by changes in the Ehrenfest exponent [8].

These two properties of the generalized Landau-Teller exponent suggest that it can be used for qualitative discussions of the mechanisms of the vibrational energy transfer in atom-diatom collisions provided the value of α is available. If the whole potential surface is known, α can be identified with the logarithmic derivative of the interaction at the middle turning points of two branches of the optimal trajectory. Unfortunately, the present information on the potential energy surfaces derived from *ab initio* calculations are in most cases not sufficiently reliable. One then can try another approach based on the so-called asymptotic method of calculation of the exchange interaction [11,12]. For the interaction between two atoms, this method expresses the exchange repulsion in terms of the asymptotic parameters of the wave functions of the valence electrons of these atoms. For instance, for the interaction of an atom B from the second row of the Periodic Table with a noble gas atom A possessing high ionization potential (He and Ne), α can be expressed *via* the ionization potential I_B of B and the distance of the closest approach R^* between A and B as:

$$\alpha = 2\sqrt{I_B / I_H} / a_0 - 2(\sqrt{I_H / I_B} - 1) / R^* - \delta\alpha \qquad (10)$$

where a_0 is the Bohr-radius, and I_H is the ionization potential of the ground-state hydrogen atom. Here, the first two terms come from the repulsive exchange interaction, and $\delta\alpha > 0$ is a small (relative to α) correction which is due to the long-range attraction. If the condition $I_A \gg I_B$ is relaxed, and I_A decreases, the steepness of the interaction is also decreases. When I_A becomes equal to I_B, Eq. (10) transforms into:

$$\alpha = 2\sqrt{I_B / I_H} / a_0 - [(7/2)\sqrt{I_H / I_B} - 1) / R^* - \delta\alpha \qquad (11)$$

It is seen that the long-range attraction makes the interaction softer; the most hard interaction corresponds to A = He. For this case, the second and third terms in Eq. (10) provide a very small corrections to the leading first term. Besides, if I_B is close to I_H, Eq.(10) yields $\alpha \approx 2/a_0 \approx 4Å^{-1}$. For the case of Ar interacting with second-row homonuclear diatoms or the second-row hydrides, Eq. (11) is more appropriate. Then the neglect of $\delta\alpha$ will provide an upper estimate of α; assuming $I_B \approx I_H$ and $Ra_0 \approx 5$ we get $\alpha < 3.5Å^{-1}$. We therefore can infer an important conclusion on reasonable values of α or their upper limits for different collision partners.

Turning back to practical application of the generalized Landau-Teller model, one can assert that with the help of the improved semiclassical approximation for the transition probability and the asymptotic method for calculation of the exchange interaction it is possible to get a reliable estimate of the adiabatic Ehrenfest exponent for a collinear atom-diatom collision. This implies of course, a non-empirical prediction, within the exponential accuracy, of the temperature dependence of the average transition probability. It remains to be seen how well these ideas can be extended for a three-dimensional collision.

3 VT Energy Transfer. Breathing Sphere Model and Infinite Order Sudden Approximation.

A simple well-known extension of a collinear Landau-Teller model is provided by the breathing sphere (BSP) model of Schwartz, Slawsky and Herzfeld [13] which fully retains the Landau-Teller exponential factor. More consistent treatment, which approximately takes into account the anisotropic character of the atom-molecule interaction, is based on the so-called infinite order sudden approximation (IOSA) [14] with respect to rotational transitions that accompany the vibrational transition. Within this approximation the rotation of the relaxing molecule plays the role of a spectator, which insignificantly modifies the exponent in Eq. (8) through quite unimportant redefinition of α. If, in addition, the quasiclassical correction to the semiclassical Landau-Teller exponent is small and the effect of the attractive part of the potential is weak, one can write the following simple expression for the deactivation rate constant within BSP or IOSA approximation:

$$k_{10}^{\mathrm{BSP}}(T) \cong k_{10}^{\mathrm{IOSA}}(T) \propto \exp[-3(\theta^{\mathrm{LT}}/T)^{1/3} + \Delta E/2kT] \qquad (12)$$

where θ^{LT} is given by Eq. (5) with α defined as the logarithmic derivative of the spherical part of the potential. Here the second term in the exponential represents the first quasiclassical correction to the semiclassical exponent as derived from the general expressions (8) and (9). However, Eq. (12) contains no indication as to what extent the above models (BSP and IOSA) provide a reasonable approximation to a realistic energy-transfer event beside the general expectation that these models can be applied to the vibrational relaxation of homonuclear diatoms in collisions with noble gas atoms (a weak interaction anisotropy) provided that the collision time τ is considerably shorter than the rotational period of the molecule (rotation as a spectator).

According to Landau and Teller, the temperature dependence of the preexponential factor of the transition probability or the rate coefficient should be considerably weaker than that coming from the exponential. Indeed, the BSP model predicts that the preexponential factor is proportional to $T^{1/3}$ [15]. Eq. (11), in which only the first, leading, term in the exponent is retained provides the basis of the celebrated Landau-Teller plot: a linear dependence of the logarithm of the rate coefficient $k_{10}(T)$ on $T^{-1/3}$. The approximate validity of this relation is well-documented in standard texts [15-17]. The first significant modification to the Landau-Teller "$\ln k_{10}(T)$ vs. $T^{-1/3}$" linear dependence comes from the quasiclassical correction to the semiclassical Ehrenfest exponent which begins to show up at lower temperatures.

An example of such a process is provided by the nitrogen-helium collisions:

$$N_2(n=1)+He \rightarrow N_2(n=0)+He. \tag{13}$$

Figure 1 shows a theoretical curve joining together the high-temperature and low-temperature experimental data. Both the linear part of the plot and its deviation from the linear extrapolation to room temperature is clearly seen. The theoretical curve corresponds to the *ab initio* value $\alpha = 3.8 Å^{-1}$ [18] which is close to the asymptotic estimate of α. A The overall agreement between the BSP model and experimental data is evident. A similar conclusion holds for other collisional processes with the participation of homonuclear diatomics and noble gas atoms (see, *e.g.* review [19]).

With increasing anisotropy, IOSA approximation breaks very early which indicates, of course, the importance of the dynamical contribution of rotation to the vibrational transition. One should expect that this contribution will show up in a substantial modification of the Landau-Teller exponent. Once it is realized that the Ehrenfest-Landau-Teller semiclassical exponential factor is proportional to the square of the Fourier component of the external time-dependent perturbation, and that the respective generalized Landau-Teller exponent can be recovered from the classical exponent, the strategy of finding the most efficient energy-transfer pathway becomes clear: one should simply look for a mode which will provide the largest high-frequency Fourier components of the time-dependent perturbation that simulates a collision.

4 VRT Energy Transfer. Breathing Shell Model and Effective Mass Method

Moore [23] pointed out that the fast rotation of a hydride molecule should be very effective in causing its vibrational relaxation. This follows from the observation that if the reduced mass μ in the expression (5) for θ^{LT} is replaced by the mass of a hydrogen atom (which means, of course, replacement of the translational mode by the rotational mode), the modified Landau-Teller temperature θ^{MLT} will be lower, and the respective $p_{10}^{MLT}(T)$ will be larger. Later on, this idea was developed further with the aim to account explicitly for the details of the potential energy surface. It was assumed that the vibrational transition in a diatom BC is mainly induced by a component of the relative velocity of A and BC which is directed along the gradient of the potential energy surface (the so-called breathing shell model, BSH, or the effective mass method, EMM). Since the motion in this direction is a combination of the relative motion of the colliding partners and the rotational motion of the diatom, the relevant, configuration-dependent, Landau-Teller parameter contains, instead of the reduced mass μ, and the steepness parameter α, the effective quantities μ_{eff}, and

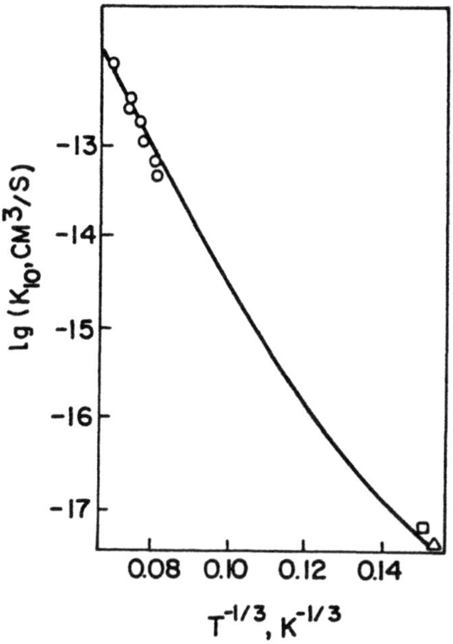

Figure 1: Rate constant for vibrational relaxation of N_2 in He. Experimental data: open circles [20], square [21], triangle [22]. Full curve corresponds to the BSP model [19].

α_{eff}. The former quantity depends on μ, the moment of inertia of a diatom I, and the shape of the equipotential line at the optimal energy E^* for an arbitrary collision configuration $\mathfrak{R}$, and the latter is close to the logarithmic derivative of the potential at the same energy and configuration. It turns out that for hydride-noble gas collisions, the transition probability is very sensitive to the collisional configuration, and there exists an optimal configuration for which the transition probability attains its maximum. Within the exponential accuracy, the deactivation rate constant can be represented as

$$k_{10}^{\mathrm{BSH}}(T) \propto \exp[-3(\theta^{\mathrm{BSH}} / T)^{1/3} + \Delta E / 2kT]. \tag{14}$$

Here,

$$\theta^{\mathrm{BSH}} = \min\{\pi^2\omega^2\mu_{\mathrm{eff}} / 2k\alpha_{\mathrm{eff}}^2\} \equiv \pi^2\omega^2\mu^{\neq} / 2k(\alpha^{\neq})^2 \tag{15}$$

where the minimum should be found with respect to the collision configuration $\mathfrak{R}$. It is seen that BSH calculation are much more demanding compared to BSP calculations since for its implementation more information on the potential surface is needed.

The BSH model can answer the question: "what are the conditions of applicability of BSH and IOSA approaches?" For a given potential surface, one simply calculates the correction to the Landau-Teller exponential which is due to the difference $\delta\theta^{BSH} = \theta^{BSH} - \theta^{LT}$. The correction is small provided $\delta\theta^{BSH}/T \ll 1$. This condition is indeed fulfilled for nitrogen-helium collisions [22] at high temperatures, and therefore the IOSA description of the energy transfer collisions (10) should be appropriate.

Within the plausible assumption that $\alpha^{\neq} \approx \alpha$, one has the following expression for the ratio of θ^{BSH} to θ^{LT}:

$$\theta^{BSH} / \theta^{LT} = \mu^{\neq} / \mu . \qquad (16)$$

The most impressive example when the condition $\delta\theta^{BSH}/T \ll 1$ is violated corresponds to the vibrational energy transfer from diatomic hydrides colliding with heavy noble gas atoms, as *e.g.* in a collision:

$$\mathrm{HCl}(n=1) + \mathrm{Ar} \rightarrow \mathrm{HCl}(n=0) + \mathrm{Ar} . \qquad (17)$$

For this collision, the optimal effective mass $\mu^{\neq}$ varies from 7 to 10 atomic mass units depending on the interaction potentials adopted [27], while μ is 19. This means that for this system θ^{BSH} is about three times less (provided $\alpha^{\neq} \approx \alpha$) than θ^{LT}. It implies in turn, that at room temperature $k^{BSH}(T)$ is about two orders of magnitudes larger than $k^{BSP}(T)$. A careful theoretical analysis of HCl + Ar collisions shows [27] that the experimental results on $1 \rightarrow 0$ transition within the temperature range 296–1953K can be reproduced by the BSH model provided α is chosen to be equal to 4Å^{-1}. This value of α seems to be too large, in view both of the asymptotic estimate and some *ab initio* calculations. Accidentally, the Moore model reproduces experimental data assuming $\alpha = 2.86$Å^{-1}.

However, one can not simply accept the Moore model in favor of BSH model since the latter is formulated on a more sound dynamical basis. The only weak point of the BSH model is the assumption that the three-dimensional collision can be reduced to a one-dimensional scattering problem in the direction of the gradient of the potential energy surface. This assumption can be probed by studying an alternative model that chooses a different mode as responsible for the vibrational deactivation.

5 VR Energy Transfer. Free Rotor and Perturbed Rotor Models

Another way to generalize the Moore idea is based on the assumption that a mode which induces a vibrational transition in a diatom corresponds to the pure rotation of the latter. This approach was adopted in a free-rotor (FR) model by Billing [28]. It was demonstrated later on that the higher-order anisotropy corrections to the interaction potential contribute much to the transition probability [29]; however, the actual significance of these corrections shows itself not in the small modification of the Landau-Teller exponential but rather in a complete redefinition of this quantity [30]. The essential difference in the collision dynamics between BSH and FR models is that in the former case the collision is accompanied by a change in the sense of the rotation of a diatom, whereas in the latter case the sense of rotation is not changed. These two dynamically different pictures were unified within the perturbed-rotor (PR) model [31-33].

The PR model considers, as a zero approximation, a motion of a diatom perturbed by the interaction with an atom placed at a fixed distance R from the center of mass of a diatom. For a particular case of a planar collision, this motion corresponds either to vibration or to hindered rotation of the diatom. Adiabatically-perturbed rotational-vibrational states of a collision complex are coupled by the relative radial motion. This coupling induces transition between different adiabatic channel states; the nature of these transitions is such that near-resonant channels are strongly favored [31-33]. The latter dynamical property allows one to regard the R-mode as a spectator mode and to consider the energy transfer as a pure VR event.

A simple PR model adopted in [32] simulates the repulsive interaction of the H atom on its "rotational" approach to the A atom by an angle-dependent potential $U(\vartheta) \propto 1/\cosh^2(\vartheta/2\gamma)$ where ϑ is the angle between the molecular axis and the collision axis. For this PR model, the following formula for the vibrational deactivation rate coefficient is valid:

$$k_{10}^{\mathrm{PR}}(T) \propto \exp[-3(\theta^{\mathrm{PR}}/T)^{1/3} + \Delta E/2kT]. \tag{18}$$

Here

$$\theta^{\mathrm{PR}} = \pi^2\omega^2 I\gamma^2/2k \tag{19}$$

with I being the moment of inertia of a diatom. Interestingly, the Ehrenfest adiabatic parameter remains the same for either type of motion, vibration or hindered rotation. In other words, the branching between two types of motion affects only the preexponantial factor but not the adiabatic exponential. Since in one limit, the PR model describes the collisions in which the direction of rotation is changed, it is meaningful to compare this model with the BSH model, just by refering to their Ehrenfest exponents. In order to do so one has to relate the key parameters of the interaction potential which enter into the Ehrenfest exponential, α for BSH, and γ for PR. If one accepts a pair-wise exponential potential that describes repulsion between the atom A and atoms B and C of the diatom BC, one has [32]:

$$\gamma = 1/\sqrt{2r_e\alpha} \tag{20}$$

where r_e is the equilibrium internuclear distance in the BC molecule. With this relation between α and γ, it is possible to write a simple expression for the ratio of Ehrenfest exponents for PR and BSH models. If, in addition, one assumes $m_B \ll m_A, m_C$, and $I = m_B r_e^2$ (which is valid for hydrides), one has:

$$\theta^{\mathrm{PR}}/\theta^{\mathrm{BSH}} = (m_H/\mu^{\neq})(r_e\alpha/2) \tag{21}$$

Typical values of the product $r_e\alpha$ for hydrogen-halide molecules are between 4 and 6. Putting $\theta^{PR} / \theta^{BSH} \cong 2.5(m_B / \mu^{\neq})$ we can easily find out the value of this ratio for HCl-Ar collision (see Eq. (17)). With the effective mass $\mu^{\neq}$ between 7 and 10 (*vide supra*), and $m_B = 1$, the ratio $\theta^{PR} / \theta^{BSH}$ is about three times less than unity. Taking into account that the transition probability depends very strongly on the value of the Ehrenfest exponent, the value of this ratio unambiguously indicates that in this case the rotational mode is more efficient in bringing about the vibrational transition than the "parallel-to-gradient" mode, and therefore the PR model applies. For DCl+Ar collisions, we have again $\theta^{PR} / \theta^{BSH} < 1$ that is the PR model provides again the major pathway for the vibrational relaxation. Finally, for TCl+Ar collisions, $\theta^{PR} / \theta^{BSH}$ is about unity, and one can not decide solely from the Ehrenfest exponents, which pathway is more efficient.

To illustrate the above we collected in Fig. 2 the results of different approximations for calculating the vibrational relaxation time τ_v of HCl in Ar using a simple repulsive pair-wise potential with $1/\alpha = 0.35$Å. This value of α is in agreement both with ab-initio calculations and asymptotic estimates. With this value of α, the BSP model (IOS approximation) yields values of τ_v which are represented by the curve 1, and BSH model (EM method) by curve 2. The PR model yields curve 3, (where $\gamma = 0.37$) which should be compared with the experimental data represented by curve 4. There is no doubt, that a very slight decrease in γ will bring the curve 3 in a very close agreement with the curve 4. A substantial change in $1/\alpha$, from 0.35Å to 0.25Å brings curve 2 (BSH and EMM) in agreement with curve 4. It is the latter value which was used in [27] for the interpretation of the experimental results within the EM method. However there are two objections to this value of $1/\alpha$: first, with this value of $1/\alpha$, the PR model will yield much higher relaxation rate which are above experimental values; second, this value of $1/\alpha$ is not compatible with asymptotic estimation of the exchange repulsion. Yet another support for the PR model comes from the study of the isotope effect: the parameter θ^{PR} remains unchanged upon isotopic substitution of hydrogen since the product $I\omega^2$ is invariant under the variation of the mass of a light atom. Indeed, the available experimental data indicate that the vibrational relaxation time of DCl in Ar is about the same as that of HCl (see Fig. 2, curves 4 and 5).

We end this section by noting that many experimental results on the vibrational relaxation of halogen-halide molecules in collisions with noble gas atoms were successfully interpreted within the PR model (see, *e.g.* a review [36]).

6 Conclusion

By way of comparing the values of the Ehrenfest adiabatic exponents for different collisonal models, it has been shown, how the vibrational energy transfer mechanism in atom-diatom collisions can be ascribed to VT, VRT or VR pathways. For heavy-heavy + light mass combination, VT mechanism is the most effective; for heavy-light + heavy mass combination, VR mechanism strongly dominate. In these two limiting cases, the improved semiclassical approximation and the asymptotic method of calculation of the exchange interaction allows one to predict approximately, without further knowledge of the potential surface, the temperature dependence of the vibrational relaxation rate constant under near adiabatic conditions. For arbitrary mass combina-

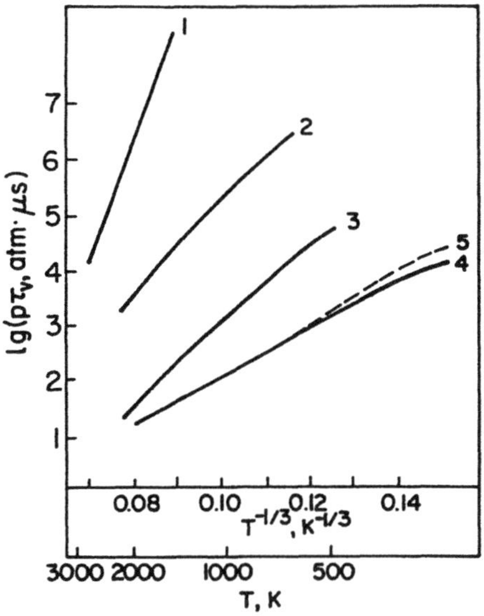

Figure 2: Vibrational relaxation time of HCl in Ar. Theoretical curves calculated with the potential from [28]: curve 1 represents BSP model, curve 2 BSH model, curve 3 PR model. Experimental data: curve 4 from [34]; dashed curve 5 represents experimental data for DCl relaxation [35].

tion, the vibrational energy of a diatom appears, as a result of a deactivating collision, in the form of the relative translational energy of colliding particles and the rotational energy of a diatom. The Ehrenfest adiabatic exponent depends in this case on the shape of the equipotential lines of the potential energy surface that describes the interaction of an atom with a nonvibrating diatom.

Though the Ehrenfest exponent alone can not be used for a reliable estimate of the energy transfer rate coefficient, its value provides an important indication at the most efficient pathway of the vibrational relaxation through the adiabaticity constraint on the energy exchange. To what extent a similar approach can be used for more complicated sysems is still an open question though it seems to be operative also in the vibrational energy transfer in collisions of triatomic [37] and polyatomic molecules with atoms [38].

7 Acknowledgment

This work was supported by the Fund for Promotion of Research at the Technion.

8 References

[1] M. Bodenstein, Z. phys. Chem. **85**, 329 (1913).

[2] N. N. Semenov, ***Tsepnye Reaktsii*** (Goskhimtekhizdat, Leningrad, 1934); English translation: ***Chemical Kinetics and Chain Reactions*** (Clarendon Press, Oxford, 1935).

[3] L.D. Landau, and E. Teller, Phys. Z. Sow. **10**, 34 (1936)
[4] V. N. Kondratiev, ***Kinetika Khimicheskich Gazovykch Reaktsii*** (Nauka, Moscow, 1958); English translation: ***Chemical Kinetics of Gas Reactions*** (Clarendon, Oxford, 1964).
[5] P. Ehrenfest, Versl. Akad. Kon. Amsterdam, **25**, 412 (1916).
[6] E. E. Nikitin, and L. Pitaevskii, Phys. Rev. A **49**, 695 (1994).
[7] E. E. Nikitin, C. Noda, and R. N. Zare, J. Chem. Phys. **98**, 47 (1993).
[8] Y. Karni, and E. E. Nikitin, J. Chem. Phys. **100**, 194 (1994).
[9] L. D. Landau, and E. M. Lifshitz, ***Quantum Mechanics*** (Pergamon Press, Oxford, 1977).
[10] Y. Karni, and E. E. Nikitin, Chem. Phys. **191**, 235 (1995).
[11] B. M. Smirnov, ***Asimptoticheskie Metody v Teorii Atomnykh Stolknovenii*** (Atomizdat, Moscow, 1973).
[12] E. E. Nikitin, and S.Ya. Umanskii, ***Theory of Slow Atomic Collisions*** (Springer, Berlin, 1984).
[13] R. N. Schwartz, Z. J. Slawsky, and K. F. Herzfeld, J. Chem. Phys. **20**, 1591 (1952).
[14] F. Gianturco, ***The Transfer of Molecular Energies by Collision,*** **Lecture Notes in Chemistry, 11** (Springer, Berlin, 1979).
[15] E. E. Nikitin, ***Theory of Elementary Atomic and Molecular Processes in Gases*** (Clarendon Press, Oxford, 1974).
[16] J. D. Lambert, ***Vibrational and Rotational Relaxation in Gases*** (Clarendon Press, Oxford, 1977).
[17] J. T. Jardley, ***Introduction to Molecular Energy Transfer*** (Academic Press, New York, 1980).
[18] A. Banks, D. C. Clary, and H.-J. Werner, J. Phys. Chem. **84**, 3788 (1986).
[19] E. E. Nikitin, A. I. Osipov, and S. Ya. Umanskii, in ***Khimiya Plazmy*, 15** (Atomizdat, Moscow, 1989), p.3.
[20] D. R. White, J. Chem. Phys. **48**, 525 (1968).
[21] R. Frey, J. Lukasik, and J. Ducuing, Chem. Phys. Lett. **14**, 514 (1972).
[22] M. M. Maricq, E. A. Gregory, C. T. Wickham-Jones, D. J. Cartwright, and C. J. S. M. Simpson, Chem. Phys. **75**, 347 (1983).
[23] C. B. Moore, J. Chem. Phys. **43**, 1395 (1965).
[24] G. A. Kapralova, E. E. Nikitin, and A. M. Chaikin, Chem. Phys. Lett. **2**, 581 (1968).
[25] A. Miklavc, and S. F. Fisher, J. Chem. Phys. **69**, 281 (1978).
[26] A. Miklavc, and I. W. M. Smith, J. Chem. Soc. Faraday Trans. 2 **84**, 227 (1988).
[27] T. D. Sewell, S. Nordholm, and A. Miklavc, J. Chem. Phys. **99**, 2767 (1993).
[28] G. D. Billing, J. Chem. Phys. **57**, 5241 (1972).
[29] I. V. Lebed', and S. Ya. Umanskii, Teor. Eksp. Khim. **14**, 102 (1978).
[30] M. Ya. Ovchinnikova, and D. V. Shalashilin, Khim. Phys. **3**, 340 (1984).
[31] M. Ya. Ovchinnikova, Chem. Phys. **93**, 101 (1985).
[32] E. E. Nikitin, and M. Ya. Ovchinnikova, Khim. Fiz. **5**, 291 (1986).
[33] E. E. Nikitin, and D. V. Shalashilin, Khim. Fiz. **11**, 1471 (1992).
[34] R. V. Steele, and C. B. Moore, J. Chem. Phys. **60**, 2794, (1974).
[35] D. J. Seery, J. Chem. Phys. **58**, 1796 (1973).
[36] E. E. Nikitin, S. Ya. Umanskii, and D.V. Shalashilin, Khim. Fiz. **8**, 1011 (1989).
[37] I. Koifman, E. I. Dashevskaya, E. E. Nikitin, and J. Troe, J. Phys. Chem. (to be published).
[38] E. I. Dashevskaya, E. E. Nikitin, and I. Oref, J. Phys. Chem. **98**, 9397 (1993).

Part V

Heterogeneous Reactions

Reactions at Surfaces: Bodenstein's Impact and Some Current Aspects

G. Ertl
Fritz-Haber-Institut der Max-Planck-Gesellschaft,
Faradayweg 4-6, D-14195 Berlin (Dahlem), Germany

Abstract

More than 100 years ago Bodenstein investigated the influence of solid surfaces on the kinetics of gas reactions, leading to a series of papers "Heterogene katalytische Reaktionen". His ideas about the underlying mechanisms had later largely to be revised, however. This contribution concentrates mainly on one of the reactions studied by Bodenstein, namely oxidation of carbon monoxide on Pt(110) single crystal surface as example. There the microscopic mechanism leads to a rich variety of phenomena of nonlinear dynamics, including oscillatory and chaotic kinetics, as well as spatio-temporal pattern formation on mesoscopic length scales.

1 Introduction

In the course of his studies on the kinetics of homogeneous gas reactions Bodenstein found that for example the rate of hydrogen oxidation was affected by the geometry and material of the reaction vessel [1]. These observations prompted him later to perform a series of investigations on the kinetics of reactions catalyzed by solid surfaces whose results and conclusions he published between 1903 and 1907 in a sequence of papers entitled "Heterogene katalytische Reaktionen" [2-6]. He found quite frequently rate laws in which the reaction rate r was inversely proportional to the concentration (partial pressure) of one of the species involved in the reaction, such as $r \sim [SO_2]/[SO_3]$ for the oxidation of SO_2 [5], or $r \sim [O_2]/[CO]$ for the oxidation of CO [4]. His main conclusion was that the catalytic reaction at the surface of the solid was actually very fast, but that the molecules involved in the reaction were forming a condensed film on the surface with gradual decreasing density which consisted predominantly of one of the species and whose thickness increased with the concentration of this species. Diffusion of the reactants through this film was considered to be slow and determining the overall rate, thus accounting for its dependence on the inverse of the concentration (Fig. 1). In his initial work on hydrogen oxidation he had even concluded that a wind of water vapour continuously blows off the surface against the molecules arriving there from the gas phase and thereby retards the reaction.

As we now know, these ideas had later to be revised: heterogeneously catalyzed reactions are confined to the monolayer formed by chemisorbed particles, and diffusion through this adsorbed phase does not play any role. (Diffusion processes may, however, be of decisive importance for the transport in highly porous materials with high specific surface area as applied in industrial catalysis). The decisive insights were achieved by I. Langmuir who, for example, wrote on April 21, 1915 into his laboratory notebook:

Springer Series in Chemical Physics, Volume 61
Gas Phase Chemical Reaction Systems
Eds.: J. Wolfrum, H.-R. Volpp, R. Rannacher, and J. Warnatz

M. Bodenstein

Heterogene katalytische Reaktionen *I - V*

Z. physikal. Chemie (1903 - 1907)

For example: $CO + {}^1/_2\, O_2 \rightarrow CO_2 \quad r \sim [O_2] / [CO]$

$SO_2 + {}^1/_2\, O_2 \rightarrow SO_3 \quad r \sim [SO_2] / [SO_3]$

Idea:

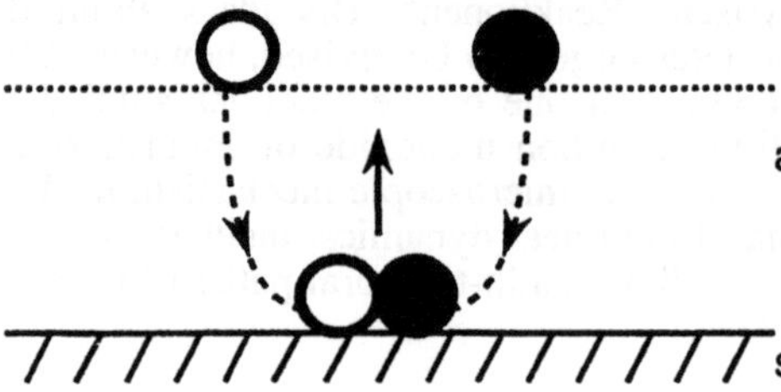

- Reaction at surface very fast
- Rate of overall reaction determined by diffusion through adsorbed layer

Figure 1: Bodenstein's view about the principal mechanism of reactions catalyzed by solid surfaces.

"1. The surface of a metal contains atoms spaced according to a surface lattice.

2. Adsorption films consist of atoms or molecules held to the atoms forming the surface lattice by chemical forces".

2 The Surface Science Approach: Catalytic Oxidation of Carbon Monoxide as an Example

In his studies on the kinetics of catalytic CO oxidation on platinum Langmuir found again (like Bodenstein) a rate law of the type $r \sim pO_2/pCO$ [8]. In his view (Fig. 2) the reaction proceeds through dissociative chemisorption of oxygen, $O_2 + * \rightarrow 2O_{ad}$ ($*$ denotes a free adsorption site on the surface), followed by formation of CO_2 by recombination of a chemisorbed O atom with a CO molecule impinging from the gas phase, $O_{ad} + CO \rightarrow CO_2 + *$. Chemisorbed CO, on the other hand, was considered to block adsorption sites and hence to inhibit the reac-

I. Langmuir

Trans. Faraday Soc. *17* (1922), 607

$$CO + {}^{1}/_{2}\,O_2 \rightarrow CO_2/Pt$$

$$r \sim P_{O_2}/P_{CO}$$

Idea:

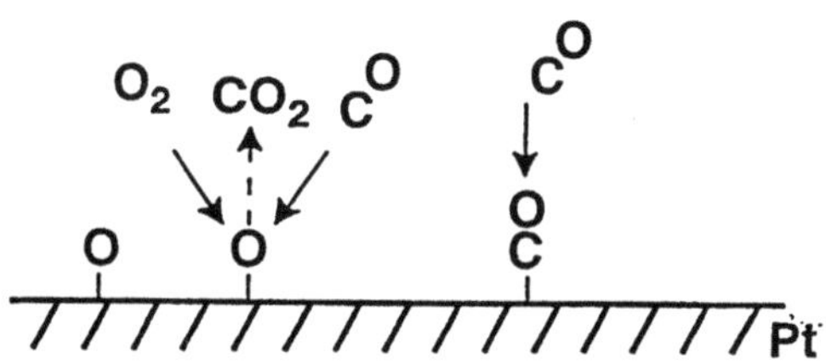

Figure 2: I. Langmuir's view about the mechanism of the catalytic oxidation of carbon monoxide on platinum, taking place within the monolayer of chemisorbed species. The observed rate law led him to the (erroneous) conclusion of the operation of the Eley-Rideal mechanism.

tion, thus accounting for the reciprocal dependence of the rate on the CO partial pressure. This type of reaction is nowadays denoted as Eley-Rideal mechanism. While it may operate in certain (highly energetic) surface reactions, it is not governing CO oxidation on platinum which in fact proceeds via the so-called Langmuir-Hinshelwood mechanism in which *both* reactants interact from the chemisorbed state [9] (Fig. 3).

The essential point is that "*" has a different meaning for oxygen and carbon monoxide: while CO tends to from densely packed adlayers which inhibit dissociative oxygen adsorption beyond a critical coverage, chemisorbed O atoms, on the other hand, form rather open adlayer phases, into which CO may still adsorb. Hence the rate does not decrease with increasing pO_2, even if the surface is saturated with O_{ad}. The schematic potential diagram in Fig. 3 sketches the progress of the reaction on platinum surfaces: most of its exothermicity is released to the heat bath of the solid during chemisorption of the reactants, while recombination of $CO_{ad} + O_{ad}$ is associated with a fairly small activation barrier (< 100 kJ/mole) and the CO_2 formed is instantaneously released into the gas phase.

Information about the atomic structure, electronic properties, energetics and dynamics of solid surfaces and adsorbed phases are nowadays obtained by apply-

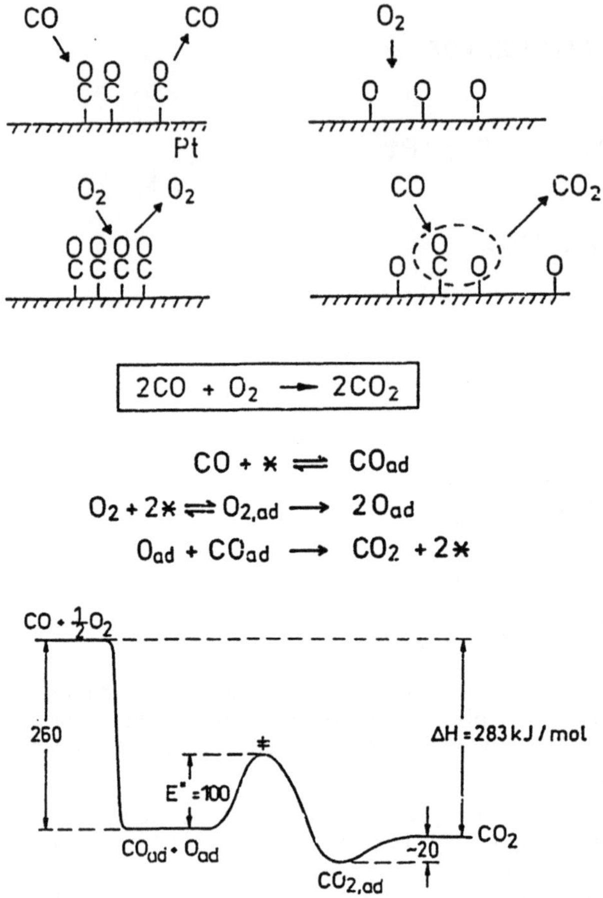

Figure 3: The actual Langmuir-Hinshelwood mechanism underlying CO oxidation on platinum.

ing the large arsenal of surface physical methods to well-defined single crystal surfaces [10]. For simple systems, such as dissociative chemisorption of H_2, theory is now able to provide reliable potential energy surfaces [11], and theoretical treatment of the dynamics [12] reproduces the results from state-selected experiments fairly well [13]. Combination with modern laser techniques, on the other hand, now opens up the field of 'femtochemistry' also to the study of photoexcited processes at surfaces [14].

3 Nonlinear Dynamics: Oscillatory Kinetics and Spatio-Temporal Pattern Formation

Any textbook on chemical kinetics makes use of the Bodenstein stationary principle: if in a system of consecutive reaction steps $A \rightarrow B \rightarrow C$ the concentration of the intermediate B is small, [B] << [A], [C], then d[B]/dt ≈ 0 and [B] remains approximately constant. If the reactions run in a flow reactor where [A] and [C] are kept constant, then of course also [B] = constant.

This almost trivial conclusion may, however, become invalid if the kinetics of a more complex reaction are no longer governed by a set of *linear* ordinary differential equations. Such a case is, for example, given by the CO oxidation reaction at a Pt(110) single crystal surface where for certain sets of control parameters (pO_2, pCO, T) and by operation in a flow system the kinetics may become oscillatory or even chaotic. This is illustrated by Fig. 4 which shows the variation of the work function (which is a measure for the O-coverage as well as for the reaction rate) as a function of time for three slightly differing sets of control parameters [15]. While this quantity varies periodically with time in a), it is chaotic in b) and even more in c). The latter data reflect in fact a case of 'hyper-chaos', in which Lyapounov exponents are positive.

Phenomena of this type had been known for long times of homogeneous reactions in solution (e.g. the famous Belousov-Zhabotinski reaction [16], for electrode processes [17], as well as in heterogeneous catalysis [18]).

All these effects can be traced back to the common theoretical framework of nonlinear dynamics. We are dealing with open systems far from equilibrium for which the kinetics can be described by a set of coupled *nonlinear* ordinary differential equations of the form

$$\frac{dx_i}{dt} = F_i\left(x_j, p_k\right)$$

where p_k are the external control parameters (in our case the temperature and the partial pressures of the reactants O_2 and CO), and x_i are the state variables (i.e. the surface concentrations of the intermediate adsorbed species CO_{ad} and O_{ad}). Even if the p_k are kept constant, the x_i might not only vary with time, giving rise to oscillatory of chaotic overall kinetics, but also on the spatial coordinates. Concentration differences within the system then give rise to diffusion fluxes and the complete description has to be formulated in terms of coupled partial differential equations

$$\frac{\partial x_i}{\partial t} = D_i \Delta x_i + F_i\left(x_j, p_k\right)$$

where D_i is the diffusion coefficient of species i and Δ the Laplace operator.

The fundamental consequences for chemical systems were explored in detail by Prigogine [19] who introduced the concept of "dissipative structures", that means spatio-temporal self-organisation of the species involved in the reaction. This leads to the formation of propagating concentration waves whose length scales are determined by the product of a diffusion coefficient and an effective rate constant which was, remarkably, recognized already in 1906 by Bodenstein's colleague R. Luther in Leipzig [20].

With the CO oxidation reaction on Pt(110) a rich variety of concentration patterns was observed by means of photoemission electron microscopy (PEEM) with typical dimensions in the μm-range [21]. This technique is based on the different dipole moments of adsorbate complexes (O_{ad} and CO_{ad}) giving rise to variations of the local work function. This in turn affects the yield of photoemitted electrons which is imaged, spatially resolved, on a fluorescent screen.

As an example, Fig. 5 shows a sequence of images recorded from a Pt(110)

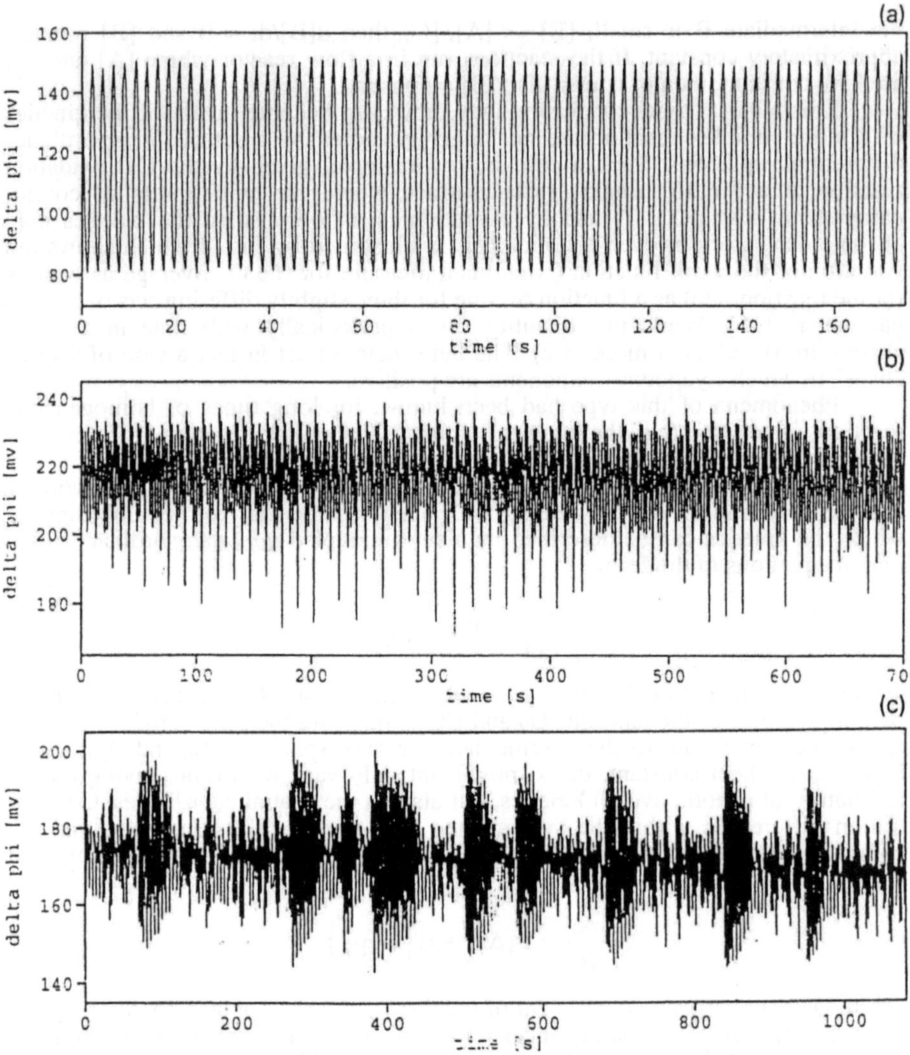

Figure 4: Variation of the work function of a Pt(110) surface (reflecting the surface concentration of adsorbed O atoms as well as the reaction rate) during CO oxidation for three different sets of fixed control parameters (T = 540 K), (a) $pO_2 = 9.8 \times 10^{-5}$ mbar, $pCO = 4.9 \times 10^{-5}$ mbar (periodic oscillations), (b) $pO_2 = 9.8 \times 10^{-5}$ mbar, $pCO = 4.5 \times 10^{-5}$ mbar (chaos), (c) $pO_2 = 8.0 \times 10^{-5}$ mbar, $pCO = 4.0 \times 10^{-5}$ mbar (hyperchaos).

surface in intervals of 15 s during catalytic CO oxidation, whereby dark areas are predominantly covered by O atoms and bright areas by CO molecules [22]. With stationary control parameters as indicated, continuously expanding spirals are formed (first image) which, however, change their character markedly under the influence of an external perturbation such as a periodic modulation of the temperature by as little as ± 0.5 K which was switched on after the first image.

Effects of this type are by no means restricted to uniform and well-defined single crystal planes, but they are also observed with polycrystalline foils as well as with the fine tips used for field ion microscopy (FIM) [23]. There the individual crystal planes have diameters of only a few tens of nm, so that these systems may be regarded as good models for the small particles applied in 'real' catalysis.

4 Concluding Remarks

In this summary of course only a few aspects of current research on reactions at solid surfaces could be roughly sketched. In particular, no account of the intense theoretical activities accompanying the experimental studies was given. These theoretical studies still rely heavily on the concepts established by Bodenstein and his contemporaries about hundred years ago, namely on the description of the temporal evolution of a reacting system in terms of concentrations and (concentration independent) rate constants. While such a continuum-type approach may be well

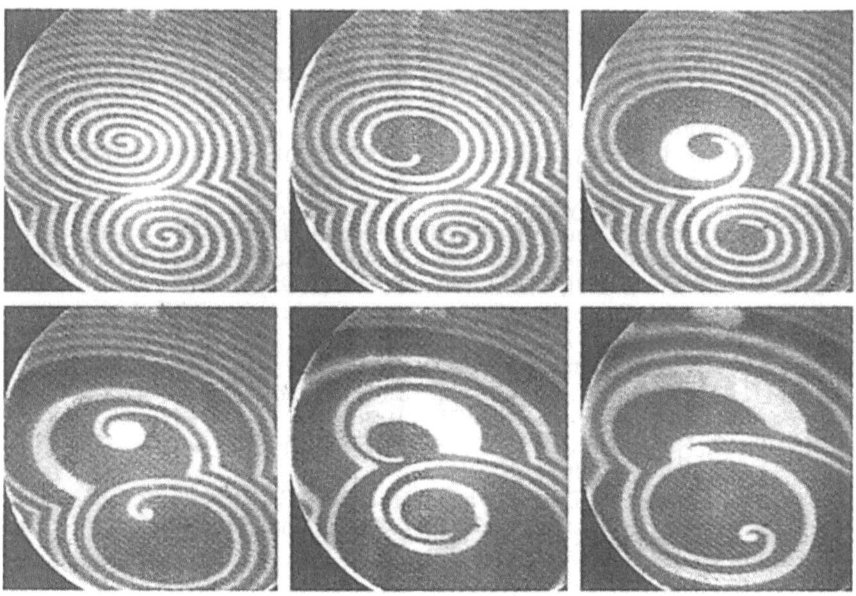

Figure 5: Sequences of PEEM images (0.17×0.17 mm^2) from a Pt(110) surface during CO oxidation at $pO_2 = 4.0 \times 10^{-4}$ mbar, $pCO = 5.6 \times 10^{-5}$ mbar, taken in intervals of 15 s. After the first image the temperature T = 463 K was modulated by $\pm$ 0.5 K with a periodicity of 10 s.

justified for non-interacting and randomly distributed particles in homogeneous systems, with progressing insight into the mechanisms of atomic interactions involved in surface reactions also the need for extensions of these concepts becomes evident. Some consequences of such a more microscopic approach will be demonstrated in another contribution to this symposium.

5 References

[1] M. Bodenstein, Z. phys. Chem. **29**, 665 (1889).
[2] M. Bodenstein, Z. phys. Chem. **46**, 725 (1903).
[3] M. Bodenstein, Z. phys. Chem. **49**, 41 (1904).
[4] M. Bodenstein and F. Ohlmer, Z. phys. Chem. **53**, 166 (1905).
[5] M. Bodenstein and C. G. Fink, Z. phys. Chem. **60**, 1 (1907).
[6] M. Bodenstein and C. G. Fink, Z. phys. Chem. **60**, 46 (1907).
[7] See also I. Langmuir, J. Am. Chem. Soc. **38**, 2221 (1916).
[8] I. Langmuir, Trans. Faraday Soc. **17**, 607 (1922).
[9] T. Engel and G. Ertl, Adv. Catalysis **28**, 1 (1979).
[10] See e.g. Surface Science: ***"The first thirty years"*** (C.B. Duke, Ed.), North-Holland, Amsterdam 1994.
[11] B. Hammer, M. Scheffler, K.W. Jacobsen and J.K. Nørskov, Phys. Rev. Lett. **73**, 1400 (1994).
[12] A. Gross, B. Hammer, M. Scheffler and W. Brenig, Phys. Rev. Lett. **73**, 3121 (1994).
[13] H.A. Michelsen, C.T. Rettner, D.J. Auerbach and R.N. Zare, J. Chem. Phys. **98**, 8294 (1993).
[14] E. Knoesel, T. Hertel, M. Wolf and G. Ertl, Chem. Phys. Lett. **240**, 409 (1995).
[15] M. Eiswirth, Th. Kruel, G. Ertl and F.W. Schneider, Chem. Phys. Lett. **165**, 305 (1992).
[16] A. M. Zhabotinsky, Chaos **1**, 379 (1991).
[17] a) W. Ostwald, Z. phys. Chem. **35**, 3 (1900); b) K.F. Bonhoeffer, Z. für Elektrochem. **51**, 24 (1947).
[18] a) G. Ertl, Adv. Catalysis **37**, 213 (1990); b) F. Schüth, B.E. Henry and L. D. Schmidt, Adv. Catalysis **39**, 51 (1993); c) M. M. Slinko and N. Jaeger, ***Oscillatory heterogenous catalytic systems***, Elsevier Amsterdam (1994); d) R. Imbihl and G. Ertl, Chem. Rev. **95**, 697 (1995).
[19] G. Nicolis and I. Prigogine, ***Self-organisation in nonequilibrium systems,*** Wiley, New York (1977).
[20] R. Luther, Z. für Elektrochem. **12**, 596 (1906).
[21] a) H.H. Rotermund, W. Engel, M.E. Kordesch and G. Ertl, Nature **343**, 399 (1990); b) G. Ertl, Science **254**, 1756 (1991).
[22] M. Bär, S. Nettesheim, H.H. Rotermund, M. Eiswirth and G. Ertl, Phys. Rev. Lett. **74**, 1246 (1995).
[23] a) V. Gorodetskii, W. Drachsel and J.H. Block, Catal. Lett. **29**, 223 (1993); b) V. Gorodetskii, J. Lauterbach, H H. Rotermund, J.H. Block and G. Ertl, Nature **370**, 277 (1994).

Heterogeneous Chemistry in the Atmosphere

D. M. Golden, C. A. Rogaski[1] and L. R. Williams
Molecular Physics Laboratory, SRI International
Menlo Park, CA 94025

Abstract

Chemical kineticists have long been involved in the study of the chemistry of the atmosphere. This involvement is both natural and important and has led to quantitative mechanistic understanding of many atmospheric phenomena. Historically the atmosphere had been regarded as essentially a gas phase reactor, with surface phenomena a laboratory hindrance to obtaining measurements that were applicable to the real situation. Relatively recent understanding has changed this view. In this paper, emphasis will be given to results from our laboratory using Knudsen cell techniques to study interactions of gaseous species with atmospherically relevant liquid sulfuric acid-water solutions and soot particles representative of those found (or postulated) in some specific atmospheric venues.

We have studied the interactions of HNO_3, HCl and HBr with liquid sulfuric acid surfaces using both time dependent uptake and equilibrium vapor pressure methods in a Knudsen cell reactor equipped with mass spectrometric detection. Measured solubilities will be presented as Henry's law coefficients along with thermochemical parameters.

Soot particles emitted by the current and projected fleet of subsonic aircraft may impact both the chemistry and radiative properties of the upper troposphere and lower stratosphere by providing nucleation centers for water-based aerosols. We have studied the uptake of water on soot samples before and after exposure to exhaust gas species such as sulfur dioxide and nitrogen dioxide as well as to ozone. In experiments performed to date, we see no change in the interaction of soot with water after such exposure. However, when the soot is exposed to HNO_3 or H_2SO_4, water uptake is greatly enhanced. Additionally, we find some interesting chemical transformations when HNO_3 interacts with soot.

1 Introduction

The "atmosphere" is a large subject and for purposes of this discussion we will refer to the portions from ground level to about 50 km. Even within this restriction the chemical nature varies significantly with altitude. In the usual way [2] we refer to the altitudes up to about 8-17 km as the troposphere and the region above as the stratosphere. These regions are often further subdivided and in the troposphere the differences between urban and remote regions are quite significant. For purposes of this paper, discussion will be limited to the stratosphere.

Springer Series in Chemical Physics, Volume 61
Gas Phase Chemical Reaction Systems
Eds.: J. Wolfrum, H.-R. Volpp, R. Rannacher, and J. Warnatz

1.1 Stratospheric Ozone Chemistry

At first glance the elementary chemical processes thought to be important in both the troposphere and the stratosphere would seem to be purely gas-phase reactions. Thus the original suggestion of Chapman [3] that the dynamic balance of ozone production and loss could be explained by the sequence:

$$O_2 + h\nu \rightarrow O + O \quad (1)$$
$$O + O_2 + M \rightarrow O_3 + M \quad (2)$$
$$O_3 + h\nu \rightarrow O + O_2 \quad (3)$$
$$O + O_3 \rightarrow O_2 + O_2 \quad (4)$$

Subsequently it was realized [4] that this sequence would be enhanced by the homogeneous catalytic destruction of O_3 (step 4) via the sequence:

$$X + O_3 \rightarrow XO + O_2 \quad (5)$$
$$O + XO \rightarrow O_2 + X \quad (6)$$

where X has been suggested to be HO, NO, Cl, Br and possibly I.

There are many more reactions that have to be, and have been, considered [5]. When all these are combined with appropriate atmospheric dynamics and solar fluxes the resulting model falls short of quantitative confirmation of the ever-increasing atmospheric data base and would lend very little confidence with respect to predictions. Fortunately, it has been well established that heterogeneous reactions catalyzed on and in atmospheric particulates can be invoked to reconcile models and measurements.

1.2 Antarctic Ozone Hole

The importance of heterogeneous processes in the chemical balance of the stratosphere has been dramatically illustrated by the annual appearance of the ozone hole during the Antarctic spring [6]. Heterogeneous reactions on particle surfaces in the polar stratospheric clouds (PSCs) convert chlorine reservoir molecules into species that photolytically yield active chlorine molecules that cause ozone destruction. In addition, heterogeneous reactions remove odd nitrogen that would normally sequester active chlorine in stable reservoir molecules. Similar heterogeneous chemistry on the background stratospheric sulfate aerosol (the "Junge Layer") affects global ozone levels [7].

The following five heterogeneous reactions are considered to be key contributors to the Antarctic ozone hole:

$$ClONO_2 + H_2O \rightarrow HOCl + HNO_3 \quad (7)$$
$$N_2O_5 + H_2O \rightarrow 2\ HNO_3 \quad (8)$$
$$ClONO_2 + HCl \rightarrow Cl_2 + HNO_3 \quad (9)$$
$$N_2O_5 + HCl \rightarrow ClNO_2 + HNO_3 \quad (10)$$
$$HOCl + HCl \rightarrow Cl_2 + H_2O \quad (11)$$

Work in the Atmospheric Chemistry Group at SRI and elsewhere has shown that these reactions occur efficiently on water ice and nitric acid trihydrate, the materials believed to make up polar stratospheric clouds [8–12]. Reactions (7), (9), (10) and (11) convert the stable chlorine reservoir species, $ClONO_2$ and HCl, into the more easily photolyzed species, HOCl, Cl_2 and $ClNO_2$. When the sunlight returns during the Antarctic spring, photolysis releases active chlorine radicals that destroy ozone via catalytic cycles. In addition, reactions (7)-(10) generate nitric acid, which is highly soluble in water ice. This deposition of odd nitrogen on PSCs (denitrification) reduces the formation of the $ClONO_2$ reservoir that sequesters active chlorine by reducing the availability of NO_2 to react with ClO.

1.3 Global Ozone Depletion

Ozone depletion is not limited to the Antarctic polar region. Both ground-based and satellite measurements indicate that global ozone concentrations are decreasing about 5 percent per decade at northern mid-latitudes [7]. Moreover, model calculations including only gas phase chemistry are unable to simulate this decrease [7]. Heterogeneous chemistry occurring on the surfaces, or in the bulk, of the ubiquitous stratospheric sulfate aerosol particles has been shown to contribute to ozone depletion on a global scale [7]. Of the five reactions above, reaction (8) appears to be most important in the global stratosphere; its rate is fairly large and relatively independent of temperature or sulfuric acid concentration [13–16]. Reaction (8) operates by reducing the amount of NO_x. This in turn reduces the amount of chlorine sequestered in the $ClONO_2$ reservoir and increases the impact of the chlorine-based catalytic ozone destruction cycles.

2 Experimental Methods

There is a growing literature [17,18] on theoretical attempts at a basic understanding of the chemistry and kinetics of these atmospherically important heterogeneous processes. For the most part, however, the emphasis is on experimental determination of both mechanism and rate data. There have been several techniques developed in this pursuit, many of which have been outlined previously [19,20]. In this paper, the Knudsen cell technique is described in some detail, since the experiments we wish to describe have been performed using this apparatus.

2.1 Knudsen Cell Reactor

Heterogeneous reaction rates and solubilities of trace species can be measured using a Knudsen cell reactor shown schematically in Figure 1 [21,22]. The experimental apparatus consists of two chambers separated by a valve. The material of interest is placed in the bottom chamber, which can be cooled down to stratospheric temperatures (200 to 240K). The gas-phase species is introduced into the top chamber which has a small escape aperture leading to a differentially pumped mass spectrometer detection system. The concentration in the top chamber is kept low enough that molecular flow applies, so the residence time in the top chamber is determined by the size of the escape aperture. When the valve between the two chambers is opened, loss of the gas-phase species to the surface competes with escape through the aperture and is observed as a decrease in the mass spectrometer signal. Loss of molecules to the surface can be due to uptake by the surface, reaction on the surface, and diffusion into the bulk.

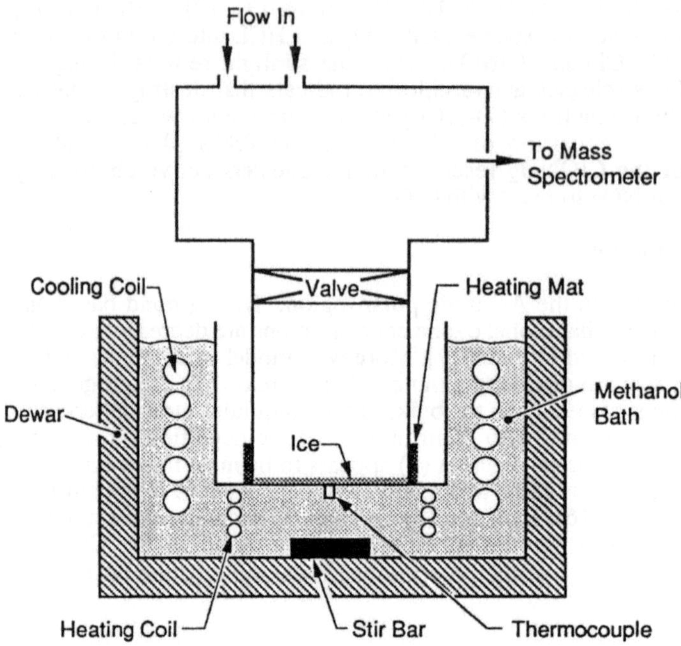

Figure 1: Knudsen cell reactor with low temperature bath.

The quantity measured in the Knudsen cell experiment is the uptake coefficient, γ, defined as:

$$\gamma = \frac{\text{Number of molecules lost to surface (molec/sec)}}{\text{Number of gas - surface collisions (molec/sec)}} \tag{12}$$

The number of molecules lost to the surface is measured by the change in flow through the escape aperture upon exposing the surface, (F^0-F), where F^0 and F are the reactant flows (molec/s) out of the cell in the absence and presence of the surface, respectively.

Since the number of collisions, ω, of a gaseous species of concentration [C] and average velocity $\overline{v}$ with a surface of area A is given by gas kinetic theory as,

$$\omega = \overline{v}A[C]/4 \tag{13}$$

the concentration in the Knudsen cell is related to the reactant flow, F, and the size of the escape aperture, A_h, by:

$$[C] = \frac{4F}{A_h \overline{v}} \tag{14}$$

Combining Eqs. 12, 13, and 14 yields the following expression for the net uptake coefficient in the Knudsen cell experiment.

$$\gamma = \frac{A_h}{A_s} \frac{F_0 - F}{F} \tag{15}$$

The mass spectrometer signal intensity for a given molecule is, in turn, linearly proportional to its flow out of the Knudsen cell. By varying the relative size of A_h and A_s, a range of γ between 1 and 1×10^{-4} can be measured accurately.

Typical data for the measurement of an uptake coefficient using the Knudsen cell reactor is shown in Figure 2 [23]. In this case, the reactant $ClONO_2$ is exposed to a 55 wt% sulfuric acid surface at 223 K. The top panel shows the decrease in mass spectral signal for $ClONO_2$ (*m/e* 46, NO_2^+) when the valve is opened at time = 0 seconds. $ClONO_2$ reacts with H_2O to form HOCl and HNO_3, Reaction (7), and the slow increase with time in gas-phase HOCl (*m/e* 52, $HOCl^+$) product is shown in the lower panel. The signal at *m/e* 46 is used to calculate an uptake (or reaction) coefficient of $\gamma = 0.012$.

The uptake coefficient need not be constant with time. A heterogeneous interaction on a solid surface could result in the adsorbate or its reaction products accumulating on the surface. This could lower the number of available surface sites and thereby decrease the net uptake coefficient. On liquid surfaces, solubility limitations can lead to time-dependent uptake coefficients. As the surface layer of the liquid saturates with gas, re-evaporation starts to compete with adsorption and the net uptake coefficient decreases with time. The time-dependence of the uptake coefficient can be analyzed to yield the effective Henry's law solubility.

The Knudsen cell reactor is a versatile tool in the study of heterogeneous atmospheric chemistry. The rate of interaction with the surface is measured relative to a physical process, escape through an aperture, which is straightforward to calibrate. This puts the gas-surface rate measurement on an absolute basis. In addition, the low operating pressure means that complicated corrections for gas-phase diffusion do not need to be made.

However, keeping the pressure in the Knudsen cell regime means that surfaces with high vapor pressures cannot easily be studied. A second limitation of the technique concerns the detection sensitivity of the mass spectrometer system which may necessitate the use of reactant concentrations that are higher than found in the stratosphere. In some cases, such as HCl uptake on ice surfaces, high concentrations appear to lead to changes in the surface which greatly increase the uptake. Finally, the time scale for experiments in the Knudsen cell reactor is many seconds, although recently, work has been done to modify a Knudsen cell that incorporates a pulsed valve which allowed millisecond time resolution [25]. Even with the increased time resolution the pulsed valve provides, it is virtually impossible to measure uptake on truly clean surfaces (which may not be representative of atmospheric surfaces anyway). On the other hand, the experiment lends itself well to the measurement of slower processes such as the solubility limitation of HCl in sulfuric acid.

The Knudsen cell reactor has been used successfully to measure reaction and uptake rates on solid and liquid surfaces, including ice, nitric acid trihydrate, soot and concentrated sulfuric acid [8,9,16,26,27]. Recent measurements of uptake and reactivity on soot surfaces are particularly intriguing. In these experiments, funded by NASA's Subsonic Assessment Program, we are investigating the impact of solid particles found in the exhaust of aircraft, i.e., soot, on stratospheric chemistry.

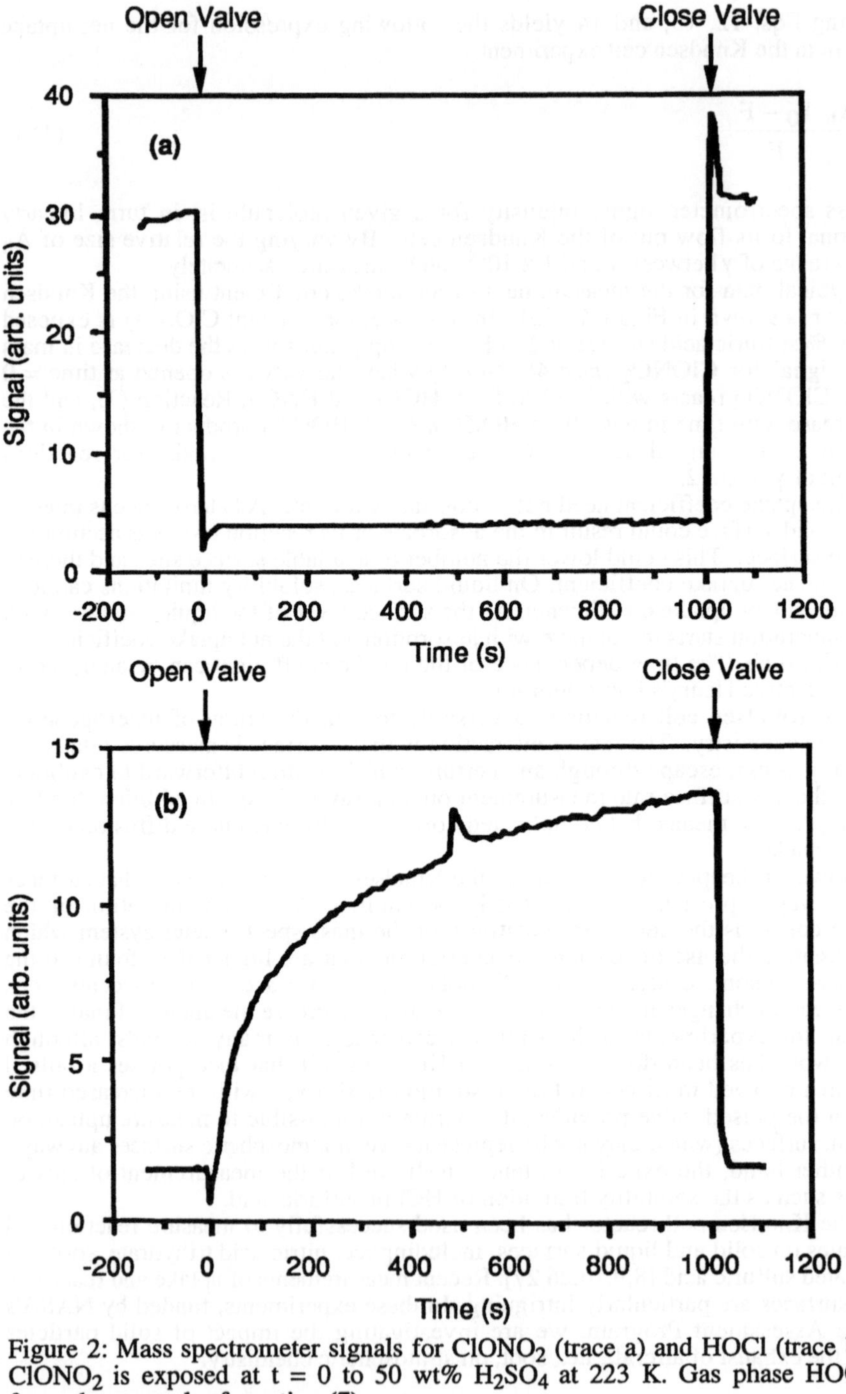

Figure 2: Mass spectrometer signals for $ClONO_2$ (trace a) and HOCl (trace b) as $ClONO_2$ is exposed at t = 0 to 50 wt% H_2SO_4 at 223 K. Gas phase HOCl is formed as a result of reaction (7).

3 Representative Measurements

3.1 Solubility of HCl, HBr, and HNO_3 in Sulfuric Acid

One of the factors limiting the rate of heterogeneous reactions such as (9), (10) and (11) is the availability of the second reactant at the surface. For a trace species like HCl, this may be determined by the solubility of the species in sulfuric acid. Measurements of the effective Henry's law constant for HCl in sulfuric acid solutions between 50 wt.% and 60 wt.% [24] have been made. Because the solubility is low, the range of temperatures and compositions over which the Knudsen cell experiment could be performed is limited. However, extrapolation to room temperature measurements [28] looks reasonable, and the agreement with the results of other groups [29,30] is good.

The solubility increases with decreasing temperature and increases with decreasing sulfuric acid concentration. These trends combine to make HCl most soluble in the coldest regions of the stratosphere where the sulfate aerosol is most dilute. The temperature dependence of the Henry's law constant is described by:

$$\ln H = -\Delta H_{sol}/RT + \Delta S_{sol}/R \qquad (16)$$

where ΔH_{sol} (kcal/mol) is the heat of solvation of HCl, ΔS_{sol} (cal/mol-K) is the change in entropy upon solvation, and R is the gas constant [31]. H in equation (16) is expressed as the mole fraction of the solute in the solvent assuming a standard state of 1 atm for the partial pressure of the gas phase solute molecules. For a log plot of H^* in our units (M/atm), the intercept is equal to ($\Delta S/4.58$ + log [solvent]) where [solvent] is the molarity of the sulfuric acid. The temperature dependence of the effective Henry's law constant measured here yields a ΔH and ΔS which reflect both solvation and dissociation, of the HCl. The temperature range of the data is not large enough to justify a least squares fit to determine both ΔH and ΔS. We use the ΔS for HCl in the gas phase going to H^+ and Cl^- ions in aqueous solution (-30 cal/mol-K) [32] to fix the intercept of the fit. The resultant ΔH's are given in Table 1 and are similar to ΔH's determined from the temperature dependence of H^* measured by other groups [29, 30].

For the 60-80 wt.% H_2SO_4 concentrations representative of stratospheric aerosol particles, the solubility of HCl is quite low – a typical aerosol particle will contain only a few HCl molecules. The extremely small amount of HCl available in solution means that reactions (9), (10) and (11) are likely to be unimportant on the sulfate aerosol at mid-latitudes. This is consistent with model calculations that show that most of the discrepancy between measured and calculated ozone at mid-latitudes can be accounted for by including only heterogeneous reaction (8) in the model calculations [33, 34]. In the laboratory, reactions (9) and (11) have been observed to occur, although under conditions of very high HCl concentration [26, 15]. In extremely cold parts of the stratosphere where the sulfate aerosol contains more water and the HCl solubility is correspondingly higher, these reactions could start to become important.

The solubility of HNO_3 in sulfuric acid will determine whether the products of reactions (7) to (10) remain in solution or enter the gas phase. The effective Henry's law constant, H^*, for HNO_3 in sulfuric acid as a function of temperature for solutions between 58 wt.% and 87 wt.% has been determined [35]. The data in Ref. [35] have been reanalyzed to correct an error of a factor of the square root of π in one of the equations, and with a steeper dependence of the viscosity (from which the diffusion coefficient is calculated) on decreasing temperature [36]. As a result, the H*'s increase more rapidly with decreasing temperature than reported in reference [35]. Fits to the data yield ΔS's for > 66wt.% sulfuric acid that are between -23 and -25 cal/mol-K, in

Table 1: Temperature dependence of effective Henry's Law constants for several species in H_2SO_4. [(a)]A standard state correction was used to convert ΔS = −30 cal/mol-K to an intercept. Assuming a standard state of 1 atm, the intercept is equal to (ΔS/4.58 + log [solvent]), where [solvent] is the molarity of the sulfuric acid. [(b)]A standard state correction was used to convert ΔS = −21 cal/mol-K to an intercept. Same as in (a). [(c)]A standard state correction was used to convert ΔS = −27 cal/mol-K to an intercept. Same as in (a).

HCl(a)		HNO_3(b)		HBr(c)	
H_2SO_4 (wt%)	ΔH (kcal/mol)	H_2SO_4 (wt%)	ΔH (kcal/mol)	H_2SO_4 (wt%)	ΔH (kcal/mol)
50	-11.5	58	-11.3	54	-12.3
55	-10.5	66	-10.1	60	-11.4
60	-10	74	-9.5	66	-10.1
		87	-8.3	72	-8.8

reasonable agreement with what one would estimate using Trouton's rule (≈ -21 cal/mol-K). We expect Trouton's rule to be a reasonable estimate because less than 10% of the HNO_3 dissociates in solution over this concentration range of sulfuric acid. The data is fit with the intercept determined by ΔS = -21 cal/mol-K. The resultant ΔH's are given in Table 1. Reanalysis of reference [35] data brings the H*'s into much better agreement with the results of reference [29].

The solubility of HNO_3 is low enough that reactions (7) to (10) will produce predominantly gas phase HNO_3. These heterogeneous reactions convert odd nitrogen to the more stable species nitric acid, but will not remove nitrogen completely from circulation as is observed during Antarctic ozone hole formation when denitrification by polar stratospheric clouds occurs.

The solubility of HBr in sulfuric acid has been studied as well [37]. Bromine radicals contribute to ozone destruction through a catalytic reaction cycle involving BrO and ClO. Thus heterogeneous chemistry of bromine containing species merits some attention, even though most of the stratospheric bromine is already present in active species. Table 1 shows these results in a manner analogous to the HCl and HNO_3 results.

3.2 Reactions on/with Soot

We are currently studying the heterogeneous interactions of exhaust and atmospheric gas-phase species with the goal of understanding the chemical modifications that transform newly-formed, hydrophobic soot into aged, hydrophilic soot that might act as condensation sites for aqueous aerosols. These experiments demonstrate the applicability of the Knudsen cell technique to work with solid substrates consisting of small particles.

The aim of these experiments is to determine the ability of various engine exhaust gases such as NO_2, SO_2, HNO_3 and H_2SO_4 as well as atmospherically important gases like O_3 to process soot from being hydrophobic to hydrophilic. Initially, soot produced in combustion processes is hydrophobic and, over time, becomes hydrophilic. Very little is known concerning the processes that allow soot to become a possible nucleation center for aerosols in the atmosphere, we hope to shed some light on this topic as well.

Soot samples of specific geometric surface areas were first exposed to H_2O, for which no measurable uptake is observed, then a specific gas and then, again, H_2O to see if there has been any change in the ability of the soot sample to adsorb H_2O. The chemicals tested can be separated into two classes: Acids, HNO_3 and H_2SO_4, and non-acids, NO_2, SO_2, and O_3. None of the non-acids altered the soot's ability to adsorb H_2O, i.e no change in the H_2O uptake measurements was observed after treatment with a non-acid; however, after treatment with an acid, the H_2O uptake on soot changed dramatically. These results are shown in Fig. 3.

The uptake coefficient for HNO_3 on soot appears to be $\gamma \approx 0.15$. Using a stratospheric surface area of ~ 4 x 10^{-10} cm^2/cm^3 (about one and a half orders of magnitude smaller than sulfate aerosol in the stratosphere) leads to a loss rate that competes with photolysis and reaction with OH in the daytime and which should be larger at night.

In performing these experiments, we observed a curious and interesting behavior. When soot, or more specifically, Degussa FW2 channel black, is exposed to HNO_3, after a period of time, a burst of NO_x and H_2O appears and then constant NO_x production is observed. An example of HNO_3 uptake on soot is shown in Figure 4. In this figure, the mass spectrometer signals have been converted into molecular flows

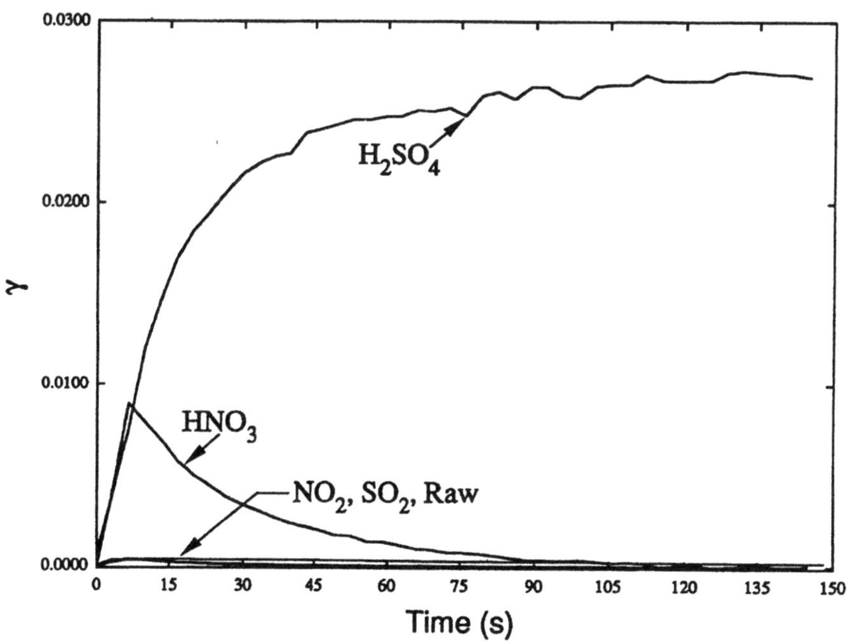

Fgure 3: H_2O uptake on soot exposed to various species

out of the Knudsen cell reactor. Integrating the various traces to determine the amount of HNO_3 adsorbed and NO_x produced reveals the value for this ratio:

$$\frac{HNO_3[\text{adsorbed}]}{NO_x[\text{produced}]} = \frac{3}{2}$$

The NO_x (and H_2O) peak appearance time depends upon the initial HNO_3 flow rate and the sample mass. It was observed that if the sample mass was kept constant and experiments performed at various flow rates, a constant number of molecules per geometric surface area is required before the burst appears. Choosing a 120 mg sample size, $(1.7 \pm 0.3) \times 10^{18}$ molecules/cm^2 are present on the soot sample when the burst appears.

The currently accepted models of the effects of supersonic aircraft on the atmosphere suggest that stratospheric ozone destruction via NO_x reactions, engendered by the extra NO_x from engine emissions, is mitigated by the heterogeneous hydrolysis of N_2O_5 to make HNO_3, thus sequestering the NO_x as stable nitric acid. If some part of the soot processing were to release NO_x from nitric acid, this would obviate the NO_x scavenging. Presently, more work is being done to elucidate what is occurring during the interaction of soot with nitric acid.

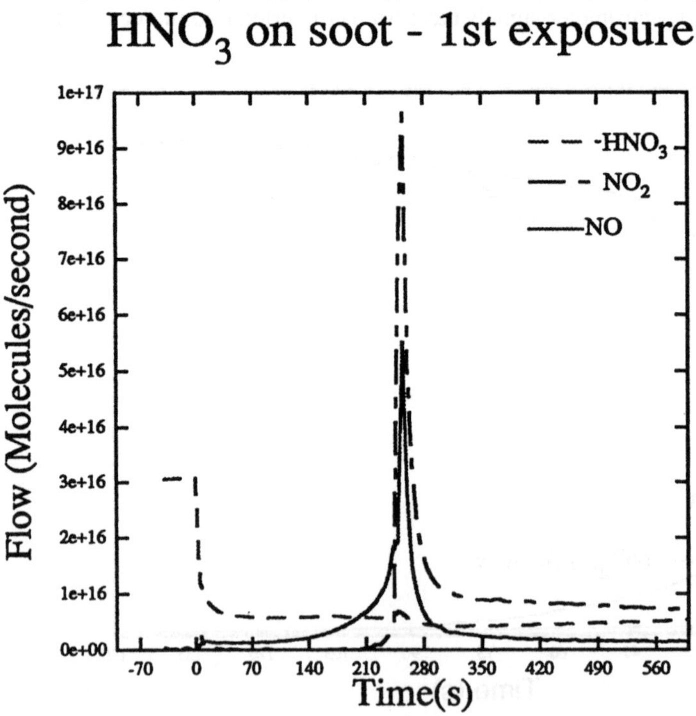

Figure 4: Chemical changes upon exposure of soot to HNO_3.

In addition to measuring reaction rates on soot surfaces in the Knudsen cell, we have also used FTIR spectroscopy to investigate the changes in the surface functional groups after the soot is exposed to different species. The FTIR spectrum of the original soot sample, and the changes that occur when the soot is exposed to sulfuric acid vapors have been observed. Some of the soot peaks disappear and new peaks near the sulfate vibrational frequencies clearly show that the surface has been chemically modified.

4 Conclusions

This paper has explored some Knudsen cell experiments designed to elucidate heterogeneous processes of possible importance to stratospheric chemistry. Detailed mechanistic understanding of the processes involved is in its infancy.

5 Acknowledgements

This research was supported by the NASA Upper Atmosphere Research Program and the NASA Subsonic Assessment Program

6 References and Notes

[1] Department of Energy Global Change Distinguished Postdoctoral Fellow.
[2] R. P. Wayne, ***Chemistry of Atmospheres*** (2nd Edn), Clarendo Press, Oxford 1991.
[3] S. A. Chapman, Mem. Roy. Meteorol. Soc., **3**, 103 (1930).
[4] D. R. Bates and M. Nicolet, J. Geophys. Res., **55**, 301 (1950).
[5] W. B. DeMore, S. P. Sauder, D. M. Golden, R. F. Hampson, M. J. Kurylo, C. J. Howard, A. R. Ravishankara, C. E. Kolb, and M. J. Molina, JPL Publication 94-26, 1994. Jet Propulsion Laboratory, California Institute of Technology, Pasadena, CA.
[6] S. Solomon, Rev. Geophys., **26**, 131 (1988).
[7] WMO, 1995: Scientific Assessment of Ozone Depletion: 1994; Global Ozone Research and Monitoring Project–Report No. 37 (World Meteorological Organization, Washington, DC, 1995).
[8] M. A. Tolbert, M. J. Rossi, R. Malhotra, and D. M. Golden, Science, **238**, 1258 (1987).
[9] M. A. Tolbert, M. J. Rossi, and D. M. Golden, Science, **240**, 1018 (1988).
[10] M. J. Molina, T. Tso, L. T. Molina, and F. C. Y. Wang, Science, **238**, 1253 (1987).
[11] D. R. Hanson and A. R. Ravishankara, J. Phys. Chem., **96**, 2682 (1992).
[12] M. T. Leu, Geophys. Res. Lett., **15**, 12 (1988).
[13] M. Mozurkewich and J. G. Calvert, J. Geophys. Res., **93**, 15889 (1988).
[14] J. M. Van Doren, L. R. Watson, P. Davidovits, D. R. Worsnop, M. S. Zahnisher, and C. E. Kolb, J. Phys. Chem., **95**, 1684 (1991).
[15] D. R. Hanson and A. R. Ravishankara, J. Geophys. Res., **96**, 17307 (1991).
[16] L. R. Williams, J. A. Manion, D. M. Golden, and M. A. Tolbert, J. Appl. Meterology, **33**, 785 (1994).

[17] T. Peter and P. Crutzen, in ***Low Temperature Chemistry of the Atmosphere***, NATA ASI Series, Springer-Verlag, 1994, pg. 499 (G. K. Moortgat, A. J. Barnes, G. LeBras, and J. R. Sodean, Eds.).
[18] B. Luo, K. S. Carslaw, T. Peta, and S. L. Clegg, Geophys. Res. Lett., **22**, 247 (1995).
[19] C. E. Kolb, D. R. Worsnop, M. S. Zahniser, P. Davidovitz, L. F. Keyser, M-T. Leu, M. J. Molina, D. R. Hanson, A. R. Ravishankara, L. R. Williams, and M. A. Tolbert, in ***Current Problems in Atmospheric Chemistry***, JAI Press, Greenwich, Ct, 1995 (J. R. Barker, Ed.), in press.
[20] D. M. Golden and L. R. Williams, in ***Low Temperature Chemistry of the Atmosphere***, NATA ASI Series, Springer-Verlag, 1994, p. 235 (G. K. Moortgat, A. J. Barnes, G. LeBras, and J. R. Sodeau, Eds.).
[21] D. M. Golden, G. N. Spokes, and S. W. Benson, **12**, 534 (1973).
[22] M. A. Quinlan, C. M. Reihs, D. M. Golden, and M. A. Tolbert, J. Phys. Chem., **94**, 3255 (1990).
[23] D. M. Golden, J. A. Manion, C. M. Reihs, and M. A. Tolbert, in ***The Chemistry of the Atmosphere: Its Impact on Global Change***, *Blackwell*, Oxford, 1994, p. 39 (J. G. Calvert, Ed.).
[24] L. R. Williams and D. M. Golden, Geophys. Res. Lett., **20**, 2227 (1993).
[25] K. Tabor, L. Gutzwiller, and M. J. Rossi, J. Phys. Chem., **98**, 6172 (1994).
[26] M. A. Tolbert, M. J. Rossi, and D. M. Golden, Geophys. Res. Lett., **15**, 847 (1988).
[27] C. Rogaski, D. M. Golden, and L. R. Williams (in preparation).
[28] L. R. Watson, J. M. VanDoren, P. Davidovitz, D. R. Worsnop, M. S. Zahniser, and C. E. Kolb, J. Geophys. Res., **95**, 5631 (1990).
[29] R. Zhang, P. J. Woolridge, and M. J. Molina, J. Phys. Chem., **97**, 8541 (1993).
[30] D. R. Hanson and A. R. Ravishankara, J. Phys. Chem., **97**, 12309 (1993).
[31] E. Wilhelm, R. Battino and R. J. Wilcock, Chem. Rev., **77**, 219 (1977).
[32] R. P. Bell, The Proton in Chemistry, Cornell Univ. Press, Ithaca, NY (1973).
[33] F. Arnold, Th. Buhrke, and S. Qiu, Nature, **348**, 49 (1990).
[34] J. M. Rodriguez, M. K. Ko, and N. D. Sze, Nature, **352**, 134 (1991).
[35] C. M. Reihs, D. M. Golden, and M. A. Tolbert, J. Geophys. Res., **95**, 16,545 (1990).
[36] L. R. Williams and F. S. Long, J. Phys. Chem., **99**, 3748 (1995).
[37] L. R. Williams, D. M. Golden, and D. L. Huestis, J. Geophys. Res., **100**, 7329 (1995).

Simulation of Heterogeneous Reaction Systems

F. Behrendt, O. Deutschmann, B. Ruf, R. Schmidt, J. Warnatz
Universität Heidelberg
Interdisziplinäres Zentrum für Wissenschaftliches Rechnen
Im Neuenheimer Feld 368, D-69120 Heidelberg, Germany

Abstract

Bodenstein was the first in the field of catalysis identifying correctly the concentrations of the adsorbed species as controlling element for many heterogeneous reaction systems. While failing to be correct in all details, Bodenstein prepared the ground for many others on their way to a detailed understanding of heterogeneous catalysis kinetics.

In this contribution, the interaction between surface and gas-phase reactions and their coupling by molecular transport is investigated numerically. Two examples are discussed more detailed: heterogeneous ignition and diamond formation by chemical vapour deposition.

1 Introduction

The heterogeneous systems this work focuses on consist of a gas phase flowing over a catalytically active surface. Numerous natural and technical processes can be described by such systems:

- The catalytically active surface opens another, and in many cases faster, reaction path for the gas phase reactants. Therefore, heterogeneous catalysis is used in numerous technical processes, e.g., in the production of basic industrial chemicals as well as the cracking and reforming of crude oil [1–3]. Catalysts are also useful in stabilizing flames and in fuel combustion at low temperatures, a precondition for reducing pollutants (e.g., NO_x) [4–6].

- The deposition of gas phase species on the surface creates thin layers of well defined crystallographic structure. Therefore, the combustion of a CH_4/H_2-mixture or of acetylene can lead to diamond deposition on a catalytically active surface [7–11].

- There are even some processes where the surface of the solid does not only act as a catalyst but takes part in the reaction as a reactant, like the coal combustion or caustic processes. The surface is consumed during this kind of process, so for the continuity of the process permanent solid supply is necessary.

In contrast to homogeneous catalysis, in the heterogeneous case reactants and catalyst do not exist in a single but in two different phases, separated by an interface, which is an additional area of high reactivity. The main principle of a catalyst is to accelerate a chemical reaction by diminishing its activation energy

Springer Series in Chemical Physics, Volume 61
Gas Phase Chemical Reaction Systems
Eds.: J. Wolfrum, H.-R. Volpp, R. Rannacher, and J. Warnatz

without influencing its thermodynamical equilibrium. On surfaces, many chemical reactions exhibit reaction rates orders of magnitudes larger than in the gas phase. Additionally, while the catalyst does not influence the thermodynamical equilibrium, its influence on the product selectivity can be high.

The heterogeneously catalysed reaction takes place on the surface of the solid and can be subclassified into five steps:

- Diffusion of the reactants to the surface of the catalyst
- Adsorption of the reactants on the surface
- Reaction between the adsorbed reactants
- Desorption of the products
- Diffusion of the products into the gas phase

The concentrations of the reactants and the products on the surface are connected with the adsorption and desorption equilibria. These depend on the concentrations of reactants and products in the gas phase, which in turn are controlled by the thermodynamical reaction equilibrium for the gas phase and by transport processes. Therefore, the concentrations depend on external conditions like temperature, pressure, initial concentration, and flow conditions. Hence, different partial processes become rate determining in the overall reaction system.

For a quantitative understanding of heterogeneous reaction systems the coupling of the different partial processes and the description of detailed models are necessary. In technical applications multidimensional spatial models are taken, which are suitable for the description of complicated and sometimes turbulent flows. For this purpose, it is unavoidable to reduce rigorously the complexity of some physical-chemical processes like diffusive transport or reaction kinetics. A second possibility consists of using detailed reaction and transport models by simplifying the flow model.

The mathematical modelling of heterogeneous reaction systems in the present paper is done with one-dimensional instationary models for laminar flows. The chemical reactions in the gas phase and on the surface are described by elementary reactions. The mass transport in the gas phase and between gas phase and surface is described by a molecular multi-species transport model.

The modelling and simulation of reactive flows evolved rapidly over the two decades and led to impressive results. One aspect of this work has been the modelling of flows and the development of simulation programs [12] while at the same time detailed reaction mechanisms for the gas phase [13] as well as numerical techniques for solving the resulting stiff differential equation systems have been developed and improved [14]. In contrast to the gas phase reaction mechanisms only a few complete surface reaction data sets have been derived. In the last years, a number of elementary surface reaction steps have been detected with the help of numerous spectroscopical and microscopical methods based on interactions of molecules with single crystal surfaces. Such kind of research has been carried out with well-defined single-crystal surfaces under extremely low pressure (mostly UHV conditions). The transfer of these reaction data to technical applications where high pressure (pressure gap) and polycrystalline catalyst material (material gap) poses a major challenge.

This transfer problem points to an important chance for modelling to fill this gap. The examination of elementary surface reaction steps under UHV conditions enables the researcher to set up principle surface reaction mechanisms which are used as a first trial input for computer simulations. Simulation of experiments carried out under technical conditions lead to the validation or improvement of the proposed reaction mechanisms by comparing general quantities like ignition or extinguishing temperature with the measured ones. Computer simulations, as mentioned before, represent a good link between the microscopic and the macroscopic processes. This means that by using detailed models one is able to make out complex correlations between the overall processes, as they are observed in the technical applications, and the basic elementary processes.

2 Coupling of Reactive Flow and Surface Reactions

Computational tools analyzing the elementary chemical processes which take place at the gas-surface interface and couple them to the surrounding gas phase have been recently developed. The mathematical models are based on the numerical solution of the Navier-Stokes equations.

Below, a reactive flow system will be discussed which can be described by the one-dimensional governing equations using the geometry of the problem. So, the resulting independent variables are the time and the distance normal to the catalytic surface. Detailed models for the chemical reactions as well as for the molecular transport are used. In order to include surface chemistry, the gas-phase problem is closely coupled with the transport to the gas-surface interface and the reaction thereon. The elementary-reactions concept is extended to heterogeneous reactions. Therefore, the boundary conditions for the governing equations at the catalyst become more complex compared to the pure gas-phase problem.

In a time-dependent formulation, the mass fraction of a gas-phase species i at the surface is determined by the diffusive and convective processes as well as the production or depletion rate $\dot{s}_i$ of that species by surface reactions. This boundary condition can be written in an integral form with respect to a control volume adjacent to the surface as

$$\int \rho \frac{\partial Y_i}{\partial t} \mathrm{d}V = - \int (\vec{j}_i + \rho Y_i \vec{u})\, \vec{n}\, \mathrm{d}A + \int \dot{s}_i M_i \mathrm{d}A \qquad (i = 1, ..., N_\mathrm{g}), \tag{1}$$

where ρ is the density, Y_i the mass fraction of species i in this control volume, $\vec{j}_i$ the diffusive flux (including thermal diffusion), M_i the molecular mass of the ith species, N_g the number of gas-phase species, and $\mathrm{d}A$ the surface area. $\vec{n}$ is the outward-pointing unit vector normal to the surface. In non-reacting continuum fluid mechanics the flow velocity normal to a solid wall is zero. However, with chemical reactions occurring at the wall, the velocity can be non-zero. This so-called Stefan velocity $\vec{u}$ is given by

$$\vec{n}\,\vec{u} = \frac{1}{\rho} \sum_{i=1}^{N_\mathrm{g}} \dot{s}_i M_i \ . \tag{2}$$

The catalytic surface itself is described by its coverage with species adsorbed and its temperature using a mean field approximation.

The temperature of the catalyst is derived from various contributions of an energy balance at the interface. The conductive, convective, and diffusive energy transport from the gas phase adjacent to the surface as well as the chemical heat release at the surface, the thermal radiation and a possible external heating (here resistive heating) of the catalyst are included. This results in

$$\int \left(\rho c_p + \rho_{\mathrm{cat}} c_{\mathrm{cat}}\right) \frac{\partial T}{\partial t} \mathrm{d}V = \int \lambda \frac{\partial T}{\partial \vec{r}} \mathrm{d}A - \sum_{i=1}^{N_{\mathrm{g}}} \int h_i \left(\vec{j}_i + \rho Y_i \vec{u}\right) \vec{n} \mathrm{d}A \tag{3}$$

$$- \sum_{i=N_{\mathrm{g}}+1}^{N_{\mathrm{g}}+N_{\mathrm{s}}} \int \dot{s}_i M_i h_i \,\mathrm{d}A - \int \sigma \epsilon \left(T^4 - T_{\mathrm{ref}}^4\right) \mathrm{d}A + I^2 R.$$

Here, λ is the thermal conductivity and h_i the specific enthalpy of species i either in the gas phase or at the surface. In the radiation term, σ is the Stefan-Boltzmann constant, ϵ is the temperature-dependent surface emissivity, T_{ref} is the reference temperature to which the surface radiates. The term I^2R represents an energy source corresponding to resistive heating of the catalyst, where I is the current and R the electrical resistance depending on temperature. N_{s} is the number of surface species, c_p the specific heat capacity of the gas at the wall, and c_{cat} denotes the specific heat capacity of the catalyst while ρ_{cat} is the density of the catalyst material.

The variation of the surface coverage Θ_i (i.e., the fraction of surface sites covered by species i) is given by

$$\frac{\partial \Theta_i}{\partial t} = \frac{\dot{s}_i}{\Gamma}, \tag{4}$$

where Γ is the surface site density of the catalyst.

The numerical solution is performed by the method of lines. Spatial discretization of the partial differential-equation system using finite differences on statically adapted grids leads to large systems of ordinary differential and algebraic equations. This system of coupled equations is solved by an implicit extrapolation method using the software package LIMEX [14]. The code computes species mass-fraction and temperature profiles in the gas phase, fluxes at the gas-surface interface, and surface temperature and coverage as function of time.

3 Chemical Kinetics

The chemistry is modelled using a set of elementary reactions for the gas phase as well as for the surface. The rate coefficients are given in the Arrhenius form as

$$k_f = A \exp \frac{-E_{\mathrm{a}}}{RT} = A' T^{\beta} \exp \frac{-E_{\mathrm{a}}}{RT} \tag{5}$$

where A is the pre-exponential factor and E_{a} the activation energy. For some reactions, the pre-exponential factor also depends on the temperature, leading to an additional factor in Eq. (5) with β describing this dependence. For reversible reactions, the rate coefficients k_r are related to the forward rate constants through the equilibrium constant K_c as

$$k_r = \frac{k_f}{K_c} \tag{6}$$

K_c is calculated from thermodynamic data.

The reaction schemes for the gas phase can be adopted from modelling work on flame chemistry (see e.g. [13]). Its validity has been established through numerous studies on flames, shock tubes, flow reactors, and well stirred reactors [12].

The formalism for the treatment of gas phase reactions have been extended to treat heterogeneous reactions [15, 16]. Then, the production rate $\dot{s}_i$ for each species due to surface reactions (including adsorption and desorption) can be written as

$$\dot{s}_i = \sum_{k=1}^{K} \nu_{ik} (k_{f_k} \prod_{i=1}^{N_g+N_s} [X_i]^{\nu'_{ik}} - k_{r_k} \prod_{i=1}^{N_g+N_s} [X_i]^{\nu''_{ik}}), \tag{7}$$

with

$$\nu_{ik} = \nu''_{ik} - \nu'_{ik}. \tag{8}$$

Here, K is the number of elementary surface reactions (including adsorption and desorption), ν'_{ik} (ν''_{ik}) are the stoichiometric coefficients of the ith species in the kth reaction on the left (right) side, k_{f_k} (k_{r_k}) the forward (reverse) rate coefficient, and $[X_i]$ the concentration of species i. For the adsorbed species, the concentrations are given in [mol m^{-2} s^{-1}].

For adsorption processes, initial sticking coefficients on a clean surface S_0 are given. They are converted to usual rate coefficients using the relation.

$$k_{\text{ads}} = \frac{S_0}{1 - S_0/2} \frac{1}{\Gamma^\tau} \sqrt{\frac{RT}{2\pi M}}, \tag{9}$$

with τ = sum of surface reactants' stoichiometric coefficients, T = gas temperature, and M = molecular mass.

The Arrhenius expressions for the rate coefficients (Eq. 5) of a few surface reactions are found to be modified by the coverage of some surface species (see, e.g., [17]). The reason is that on a fully covered surface, energetic interactions between adsorbed species change the energetic state of the surface and, hence, can lead to an increase or decrease of rate coefficients compared to the low coverage case. Therefore, the rate coefficients are modelled taking this additional coverage dependence into account:

$$k_{f_k} = A_k \exp \frac{-E_{a_k}}{RT} \prod_{j=N_g+1}^{N_g+N_s} \Theta_j{}^{\mu_{jk}} \exp \frac{\epsilon_{jk} \Theta_j}{RT} \ . \tag{10}$$

The term associated with μ_{j_k} provides the production rate $\dot{s}_i$ in reaction k to be proportional to any arbitrary power of a surface species concentration. So, reaction orders different from the molecularities of the reaction can be treated as well (Eq. 7). ϵ_{j_k} represents the decrease of the activation energy E_{a_k} in the case of a fully covered surface by species j.

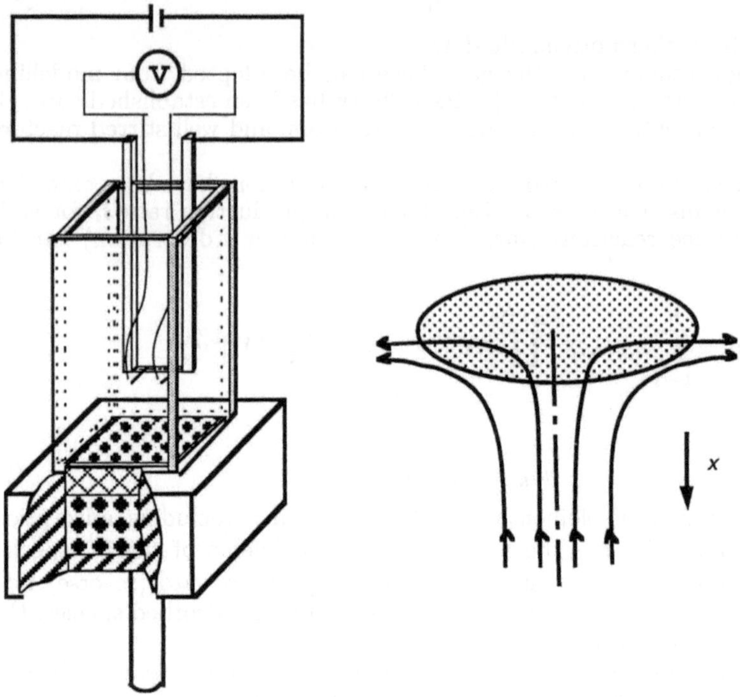

Figure 1: Experimental configuration (left) and flow field model (right) to study catalytic ignition of a stagnation point flow on a catalytic foil.

4 Heterogeneous Ignition

Extensive experimental and theoretical attention has been given to catalytic combustion in the past decade. The potential of catalytic combustion and catalytically supported combustion in reducing emissions of pollutants, improved ignition, and enhanced stability of flames has been recognized. Two simple configurations are often used to investigate catalytic combustion experimentally: the stagnation flow field over a catalytic active foil [18–20, 22, 23] and a chemical reactor with a catalytic active wire inside [24, 25]. Thereby, the temperature of the catalyst can be controlled by resistive heating of the catalytic foil or wire.

Catalytic ignition is a sudden transition from a kinetically controlled system to one controlled by mass transport. Therefore, the complex interactions of the chemical and transport processes in the gas phase and at the surface as well have to be included. Thus, the numerical simulation of catalytic ignition and

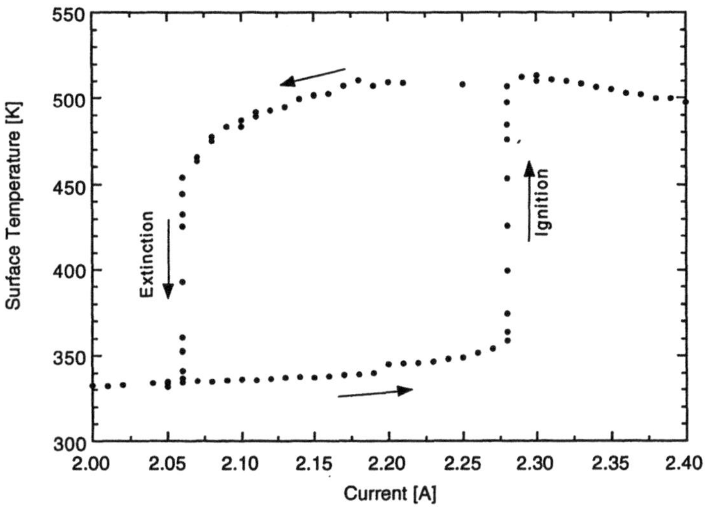

Figure 2: Experimental ignition and extinction curve for hydrogen oxidation on a platinum foil.

the comparison of calculated and experimental results represent a suitable tool to validate the models and the reaction mechanisms proposed.

The procedure determining the catalytic ignition temperature is as follows: The reactive mixture, initially at room temperature, slowly flows towards the catalyst. The temperature of the catalyst is increased by a stepwise increase in the current applied to the catalyst. After each increase, the system is allowed to reach its steady-state temperature. When reaching the ignition temperature, T_i, the temperature of the catalyst rises rapidly due to heat release by the exothermic surface reactions. A few seconds later, a new steady state is established controlled by mass transport of reactants towards and products away from the catalyst. An experimental setup to study catalytic ignition and a model of the flow field are shown in Fig. 1.

After ignition, the current supplied to the foil can be reduced without extinguishing the combustion process until finally extinction occurs for a current much lower than one needed for ignition. Figure 2 shows a typical curve for the catalytic ignition of a hydrogen/oxygen mixture, diluted by nitrogen, on a platinum foil at atmospheric pressure [23]. For other systems (e.g., CH_4/air mixtures on Pt [20, 21, 26]), even an autothermal behaviour was observed for certain conditions, i.e. the catalytic combustion continues after switching off electrical power supply.

Ignition temperatures are a function of fuel/oxgyen molar ratio, here given by $\alpha = p_{\text{fuel}}/(p_{\text{fuel}}+p_{O_2})$. Figure 3 shows the catalytic ignition temperatures for a stagnation point flow of a H_2/O_2 mixture (6 % in N_2 dilution) on a platinum foil. The ignition temperature decreases with increasing H_2/O_2 ratio. A quantitative agreement with experimental results is achieved validating the surface reaction mechanism proposed [23].

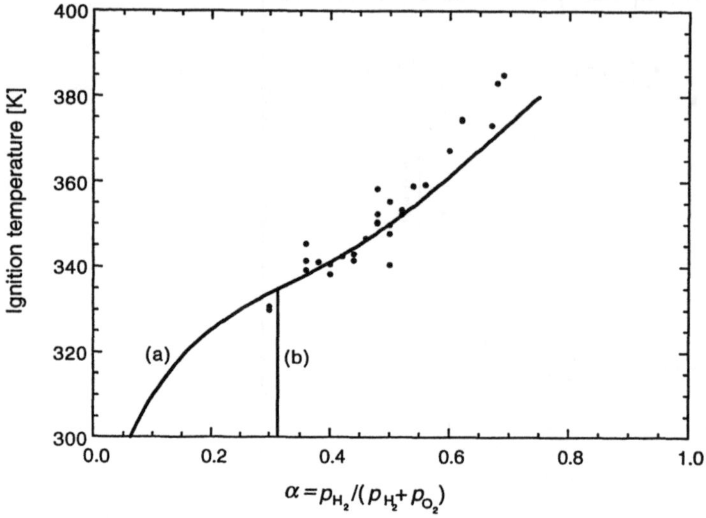

Figure 3: Ignition temperature of $H_2/O_2/N_2$ mixtures over a platinum foil as function of fuel concentration. Comparison of experimental (symbols) and numerical results (line); for a and b see text.

For low hydrogen concentrations ($\alpha = p_{H_2}/(p_{H_2}+p_{O_2}) < 0.07$), the surface is almost completely covered by oxygen, and ignition occurs immediately corresponding to experimental observations. There is a sharp drop in the ignition temperature for this fuel lean conditions due to a kinetic phase transition which has been investigated recently [27]. In our studies, bistable solutions are identified for $0.07 < \alpha < 0.32$, i.e., calculations starting with different initial coverage (hydrogen or oxygen covered surface, or a bare platinum surface, respectively), but otherwise identical conditions end up at different solutions. For calculations starting with the surface covered by hydrogen atoms (curve (a) in Fig. 3), an ignition-like temperature curve is observed and an ignition temperature can be determined. Otherwise, starting with a surface either bare or covered with oxygen (curve (b) in Fig. 3), oxidation of hydrogen begins immediately (i.e., without electrical heating). This bistability reproduces experimental observations for lean mixtures where the measurements have to be started with a hydrogen covered surface, i.e., with a pure hydrogen flow in order to avoid immediate ignition.

For $\alpha > 0.32$, the surface is primarily covered by hydrogen because the H_2 sticking probability is higher than for O_2. Increasing the surface temperature by power supplied to the catalyst leads to a point where the adsorption/desorption equilibrium of hydrogen shifts to desorption, resulting in bare surface sites where oxygen can adsorb. As an illustration Fig. 4 shows the development of surface coverage and temperature during catalytic ignition. The adsorbed O atoms react immediately with the surrounding H atoms to form OH, leading in turn to a relatively fast formation of water which desorbs. So, more and more surface sites are available for O_2 adsorption as well as for further H_2 adsorption causing an increased water production rate. The chemical heat release by this exothermic

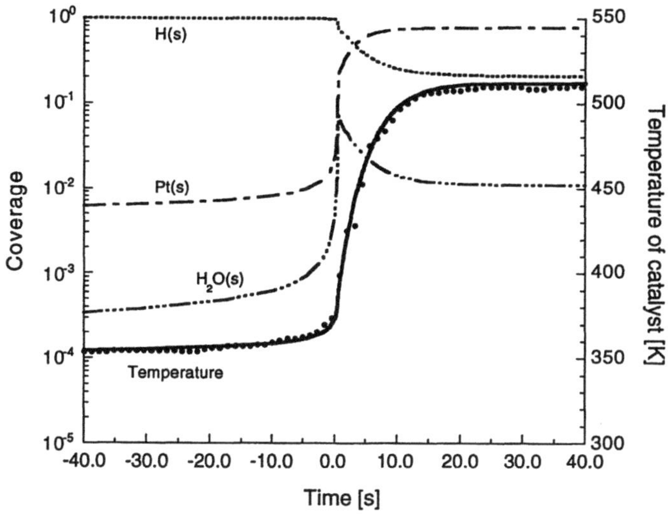

Figure 4: Calculated surface coverage and surface temperature (line = calculation, points = experiment) as a function of time during heterogeneous ignition of the H_2/O_2 on a Pt foil for $\alpha = 0.5$.

reaction causes an increase in surface temperature which further accelerates the reaction, i.e., ignition occurs. A few seconds after ignition a new steady state is established. Now, the global process is controlled by diffusion of reactants to the catalyst and of products away from the catalyst. The surface coverage is lower than before ignition. A sufficient number of bare surface sites (Pt(s) in Fig. 4) is available for adsorption of oxygen and hydrogen. In spite of oxygen excess in the inlet gas, hydrogen coverage is still higher than oxygen coverage due to a higher diffusivity and sticking probability of hydrogen.

The dependence of the ignition temperature on α is hence explained by the adsorption/desorption kinetics. Since the adsorption rate depends on gas phase concentration, higher hydrogen concentrations in the gas phase lead to fewer bare surface sites which is responsible for the rise in ignition temperature with increasing H_2/O_2 ratios.

In contrast to the rise in T_i with increasing fuel/oxygen ratio for the H_2/O_2/Pt system, investigations of the catalytic ignition of CH_4/O_2 show a decrease of T_i with increasing fuel/oxygen ratio. Here, the surface is primarily covered by oxygen before ignition.

5 Diamond Formation by CVD

Chemical vapour deposition (CVD) is a method of paramount importance in various branches of surface related materials sciences, such as microelectronics, optics and tools industries [28]. The interest is in the growth and design of surfaces with outstanding characteristics in surface hardness, chemical and mechanical resistance or electronic properties. For all the mentioned technical applications

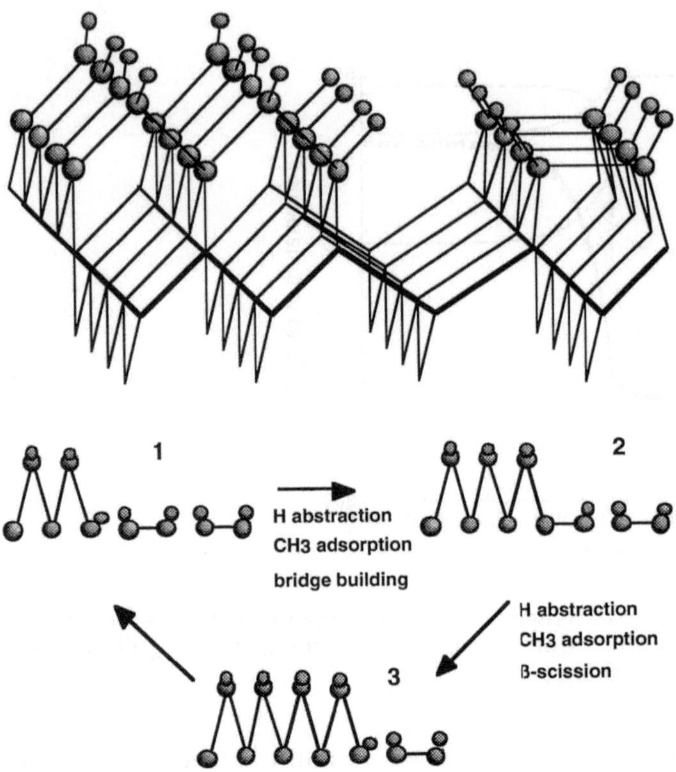

Figure 5: Monoatomic step and essential features of the surface reaction scheme on the reconstructed diamond (100) surface. The surface sites are occupied with H atoms (small balls).

diamond coated surfaces have attracted much attention. Diamond can be grown from the gas phase by plasma discharges [29], from flames stabilized in stagnation flows [30,31] and from hot-filament chemical-vapour deposition (HFCVD) [32,33]. In all variants, reactive species initiating the growth process at the surface sites are produced either by electron collisions, chemical reactions or heterogeneous catalytic processes.

Optimizing growth rates and quality of the diamond films require an understanding of the processes occurring in the gas phase as well as on the surface. In the last years several models explaining diamond growth on different low index planes of diamond, in terms of elementary chemical reaction steps, have been proposed [34–38]. Generally, C_1-hydrocarbons have been assumed to be the important species for diamond growth and isotopic labeling experiments [39] support this point of view.

Based on a model proposed by Harris and Goodwin [36], an elementary surface reaction mechanism for the diamond (100)-surface reconstructed to the (100)-(2x1):H form is developed [11]. The mechanism assumes that diamond growth takes place at surface step sites as shown in Fig. 5. The principle features

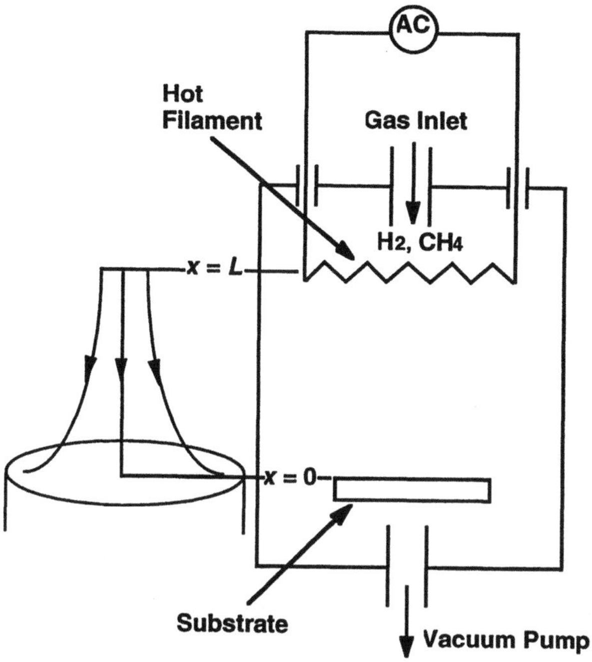

Figure 6: Sketch of a hot-filament CVD system (right) and of modeled stagnation flow field (left).

of the proposed growth mechanism describing step propagation are shown in the lower part of Fig. 5. The growth is initiated by a H-abstraction reaction. This is an Eley-Ridel reaction step in which a H-atom from the gas phase reacts with an adsorbed H-atom on the diamond surface leading to a surface radical and a H_2-molecule in the gas phase. A methyl radical from the gas phase adsorbs on the surface radical site. After two successive H-abstraction reactions a C-C bridge can be build between the two neighbouring surface sites (structure 2 in Fig. 5). In structure 2 a C-C bond on the surface, the so-called dimer bond, has to be broken to continue the growth. This is possible by an adsorbed CH_3-molecule which can loose a H-atom via H-abstraction. The resulting CH_2 radical on the surface can break the surface dimer bond by β-scission [40]. A following C-C bridging between two neighbouring surface sites leads to structure 3 which is equal to structure 1 and the cycle is closed. The whole mechanism consists of 15 elementary reactions and is described in more detail in [11].

Figure 6 shows a hot-filament CVD reactor together with a sketch of a stagnation-point flow the simulation is based on. A mixture of H_2 and CH_4 passes a hot filament placed at a distance L = 1 cm away from the substrate. The pressure is 33.3 mbar, the temperature of the filament 2430 K, the temperature of the substrate is varied between 800 and 1240 K, and the gas composition is 0.4 mole% CH_4 in H_2. The corresponding Navier-Stokes equations [10, 16] of

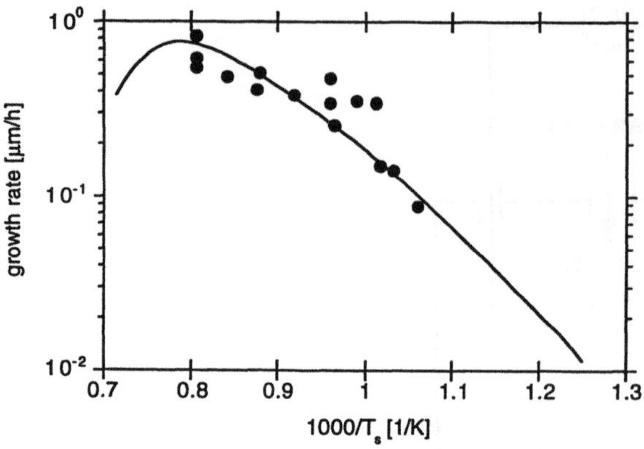

Figure 7: Diamond growth as function of the substrate temperature for 0.4 mole% CH_4. Circles: experiment of Chu (1992); solid line: simulation.

the stagnation point flow are solved. The coupling of the gas phase processes with the surface has a direct influence on the gas phase species profiles as shown in [10]. The results of the simulation using the surface reaction scheme sketched in Fig. 5 and the numerical model mentioned above are shown in Fig. 7. One can see that the simulation describes the experimental results of Chu *et al.* [41] for homoepitaxial growth on the C(100) surface quite well. The decrease of the simulated growth rate at substrate temperatures above 1240 K was also observed for polycristalline diamond by Kondoh *et al.* [42]. Hence, it can be concluded that the proposed surface reaction scheme is appropriate to explain diamond growth on the reconstructed diamond (100) surface.

6 Conclusions

Heterogeneous reaction systems are modelled by detailed chemical kinetics and transport models and numerically simulated for two simple configurations, an catalytically active wire and disk. As special case for the latter configuration, chemical vapour deposition of diamond is simulated. Elementary reaction mechanisms are applied in the gas phase and on the surface, and the coupling of the catalytic surface with the surrounding reactive flow is accounted for by a detailed transport model.

The instationary simulation offers a detailed description of the transition from a kinetically-controlled to a transport-limited regime during catalytic ignition. The onset of ignition is shown to depend on the adsorption/desorption kinetics. Hence, further investigations are recommended to improve the kinetic data for these key reaction steps. Special attention should be given to a possible coverage-dependence of the rate coefficients. For the deposition of diamond growth rates and experimentally observed apparent activation energies can be reproduced and explained in terms of elementary steps.

In the given paper, the surface properties are described by values averaged over the total surface. However, for future work, lateral processes on the surface (e.g., species diffusion on the surface, island formation) will be included in the modelling.

7 Acknowledgment

This work was supported by the Deutsche Forschungsgemeinschaft (DFG) within the Sonderforschungsbereich 359 "Reaktive Strömungen, Diffusion und Transport". Additional support by the DAAD is gratefully acknowledged.

8 References

[1] G. C. Bond, **Heterogeneous Catalysis** (Clarendon Press, Oxford 1987).

[2] E. Fitzer and W. Fritz, **Technische Chemie. Einführung in die Chemische Reaktionstechnik** (Springer Berlin 1989).

[3] M. Baerns, Nachr. Chem. Tech. Lab. **43**, 245 (1995).

[4] L. D. Pfefferle and W. C. Pfefferle, Catal. Rev.-Sci. Eng. **29**, 219 (1987).

[5] J. Warnatz. VDI-Berichte **1205**,1 (1995).

[6] C. M. Friend, Spektrum der Wissenschaften **6**, 72 (1993).

[7] R. C. DeVries, Ann. Rev. Mater. **17**, 161 (1987).

[8] F. G. Celii and J. E. Butler, Annual Rev. Phys. Chem. **42**, 643-684 (1991).

[9] J. Angus, A. Argoitia, R. Gat, Z. Li, M. Sunkara, L. Wang, and Y. Wang, Phil. Trans. R. Soc. Lond. A **342**, 195 (1993).

[10] B. Ruf, F. Behrendt, O. Deutschmann, and J. Warnatz, J. Appl. Phys. **79(8)**, in press (1996).

[11] B. Ruf, F. Behrendt, O. Deutschmann, and J. Warnatz, Surf. Sci., in press (1996).

[12] J. Warnatz, U. Maas, R. W. Dibble. **Combustion** (Springer, New York 1996).

[13] D. L. Baulch, C. J. Cobos, R. A. Cox, C. Esser, P. Frank, Th. Just, J. A. Kerr, M. J. Pilling, J. Troe, R. W. Walker, and J. Warnatz, J. Phys. Chem. Ref. Data **21**, 11 (1992).

[14] P. Deuflhard, E. Hairer, and J. Zugk, Num. Math. **51**, 501 (1987).

[15] M. E. Coltrin, R. J. Kee, and F. M. Rupley, *SURFACE CHEMKIN (Version 4.0): A Fortran Package for Analyzing Heterogeneous Chemical Kinetics at a Solid-Surface - Gas-Phase Interface*, Sandia National Laboratories Report, SAND90-8003B (1990).

[16] J. Warnatz, M. D. Allendorf, R. J. Kee, and M. E. Coltrin. Combust. Flame **96**, 393 (1994).

[17] K. Christmann, **Introduction to Surface Physical Chemistry** (Springer, Berlin 1991).

[18] S. Ljungström, B. Kasemo, A. Rosén, T. Wahnström, and E. Fridell, Surf. Sci. **216**, 63 (1989).

[19] X. Song, W. R. Williams, L. D. Schmidt, and R. Aris, Twenty-Third Symposium (International) on Combustion, p. 1129-1137, The Combustion Institute, Pittsburgh (1990).

[20] W. R. Williams, M. T. Stenzel, X. Song, and L. D. Schmidt, Combust. Flame **84**, 277 (1991).

[21] O. Deutschmann, F. Behrendt, and J. Warnatz, Catalysis Today **21**, 461 (1994).

[22] H. Ikeda, J. Sato, and F. A. Williams, Surf. Sci. **326**, 11 (1995).

[23] O. Deutschmann, R. Schmidt, and F. Behrendt. Proc. 8th International Symposium on Transport Phenomena in Combustion, San Francisco (1995).

[24] P. Cho and C. K. Law, Combust. Flame **66**, 159 (1986).

[25] M. Rinnemo, M. Fassihi, and B. Kasemo. Chem. Phys. Lett. **211**, 60-64 (1993).

[26] F. Behrendt, O. Deutschmann, U. Maas, and J. Warnatz, JVST A **13 (3)**, 1373 (1995).

[27] V. P. Zhdanov and B. Kasemo, Surf. Sci. Rep. **20**, 11 (1994).

[28] P. K. Bachmann, I. M. Buckley-Golder, J. T. Glass, and M. Kamo (Eds.). *Proceedings of the 5th European Conf. on Diam., Diamond-like and Relat. Mater.*, Diamond and Rel. Mat. **4** (1995).

[29] N. Naito, A. Takano, M. Sumia, M. Kawasaki, and H. Koinuma, Appl. Phys. Lett. **66**, 1071 (1995).

[30] K. V. Ravi, Diam. and Rel. Mater. **4**, 243 (1995).

[31] M. Murayana, S. Kojima, and K. Uchida, J. Appl. Phys. **69**, 7924 (1991).

[32] S. J. Harris, A. M. Weiner, and T. Perry, Appl. Phys. Lett. **53**, 1605 (1988).

[33] F. G. Celii, P. E. Pehrsson, H.-T. Wang, and J. E. Butler, Appl. Phys. Lett. **52**, 2043 (1988).

[34] S. J. Harris, Appl. Phys. Lett. **56**, 2298 (1990).

[35] S. J. Harris and D. N. Belton, Thin Solid Films **212**, 193 (1992).

[36] S. J. Harris and D. G. Goodwin, J. Phys. Chem. **97**, 23 (1993).

[37] D. N. Belton and S. J. Harris, J. Chem. Phys. **96**, 2371 (1992).

[38] S. Skokov, B. Weiner, and M. Frenklach, J. Phys. Chem. **98**, 7073 (1994).

[39] P. D'Evelyn, C. J. Chu, R. H. Hauge, and J. C. Margrave, J. Appl. Phys. **71**, 1528 (1992).

[40] B. J. Garrison, E. J. Dawnkaski, D. Srivastava, and D. W. Brenner, Science **255**, 835 (1992).

[41] C. J. Chu, R. H. Hauge, J. L. Margrave, and M. P. D'Evely, Appl. Phys. Lett. **61**, 1393 (1992).

[42] E. Kondoh, T. Ohta, T. Mitomo, and K. Ohtsuka, J. Appl. Phys. **73**, 3041 (1993).

Chemical Kinetic Modelling of Hydrocarbon Ignition

C.K. Westbrook, W.J. Pitz, H.J. Curran, P. Gaffuri,
and N.M. Marinov
Lawrence Livermore National Laboratory
P. O. Box 808, Livermore, CA 94550, U.S.A.

Abstract
Chemical kinetic modelling of hydrocarbon ignition is discussed with reference to a range of experimental configurations, including shock tubes, detonations, pulse combustors, static reactors, stirred reactors and internal combustion engines. Important conditions of temperature, pressure or other factors are examined to determine the main chemical reaction sequences responsible for chain branching and ignition, and kinetic factors which can alter the rate of ignition are identified.

1 Introduction

Ignition of hydrocarbon combustion takes many forms in laboratory-scale and practical systems ranging from shock tubes to internal combustion engines. For example, in spark-ignited automotive engines, two vastly different types of ignition problems constantly compete against each other; the spark plug must ignite the reactive mixture reliably and repeatably in order for the engine to produce power, while ignition of the end-gases (the last portion of reactive mixture to be consumed by the flame in the combustion chamber) can produce knocking behavior with potential to destroy the engine. High efficiency, low polluting pulse combustion systems, used for home space heating, industial processing and propulsion, are controlled almost entirely by the periodic ignition of fresh fuel and oxidizer by mixing with hot product gases. This ignition process involves a crucial interaction between ignition and resonant pressure oscillations in the combustion chamber, and if the ignition and pressure wave are not properly in phase with each other, the pulse combustor will not operate.

Hydrocarbon ignition usually involves complex interactions between physical and chemical factors, and it therefore is a suitable and often productive subject for computer simulations. In most of the studies to be discussed below, the focus of the attention is placed on the chemical features of the system. The other physical parts of each application are generally included in the form of initial or boundary conditions to the chemical kinetic parts of the problem, as appropriate for each type of application being addressed.

2 General Features of Ignition

Ignition in combustion systems consists of enabling the combustible medium to increase or sustain its rate of chemical reaction and heat release. Every different combustion system operates under its own conditions involving energy loss mechanisms, transport mechanisms for energy and mass, and many other properties, and therefore the meaning of ignition will vary widely depending on what system is being considered. In many situations, ignition will result in steady combustion such as ignition of a laminar burner flame, while in other cases it will result in transient and violent behavior such as explosion. The subject of ignition is suitable for entire volumes,

Springer Series in Chemical Physics, Volume 61
Gas Phase Chemical Reaction Systems
Eds.: J. Wolfrum, H.-R. Volpp, R. Rannacher, and J. Warnatz

and the present discussion is sharply focused on purely chemical kinetic elements; the book by Glassman [1] includes a very nice discussion of many of the additional concepts important to ignition for further study. For the purposes of the present discussion, ignition consists of a rapid growth in the radical pool of a chemically reactive mixture which may be entirely gaseous or may include fuel droplets or particles. This growth rate must be sufficient to overcome any appropriate loss mechanisms which may apply to the system in question. While the material below will emphasize the chemical kinetic features of these systems, other processes are occurring which will be noted and must be considered in more complete simulations of such systems.

3 Chemical Kinetic Reaction Mechanisms

Development of detailed chemical kinetic reaction mechanisms is a very large subject area, combining theory, experiment and computer studies. Many of the experimental studies consist of carefully focused work that isolates one reaction at a time, while other experiments provide information on collections of large numbers of reactions and chemical species. The product of these studies is generally an interconnected set of thermochemical data, reaction rate expressions, chemical species data, and often data on transport properties. In many cases the numerical model and solution algorithms are intricately connected to the reaction mechanism as well. These reaction mechanisms are generally developed by beginning with the simplest and most fundamental blocks [2,3], usually the submechanism for H_2 oxidation, followed by the mechanisms for CO, methane, methanol, ethane and continuing as far as the application demands. These mechanisms are then tested extensively through comparisons with experimental data. The most useful such mechanisms are validated by comparisons with data from as wide as possible a range of operating conditions and types of environments. Thus, a mechanism would use data from shock tubes, flow reactors, laminar flames, and many other types of experiments to test its generality and validity. Examples of reaction mechanisms in the recent literature which have been developed and tested in this way can be found in the literature cited [4-8].

The most important elements in such reaction mechanisms are the steps which affect the population (or pool) of reactive radical species. Below, the role which these reactions play in determining practical features of ignition will be outlined.

4 High Temperature Ignition

At temperatures above about 1200 K, hydrocarbon reaction mechanisms are simplified by the fact that alkyl radicals react primarily by means of β-decomposition. Thus the complex sequence (discussed below) of reactions initiated by addition of molecular oxygen to alkyl radicals

$$R + O_2 = RO_2 \tag{1}$$

does not influence the overall process of ignition. Identification of the chain branching sequences is also very much simplified. For example, H atom abstraction from fuel molecules RH by O atoms leads to chain branching

$$RH + O = R + OH \tag{2}$$

because it consumes one O radical and produces two radicals R and OH. In actual practice, chain branching from O atom reactions is not often especially important. However, these reactions can be important in applications where the oxidizer directly produces O atoms, such as when the oxidizer is something like N_2O, NO_2 or ozone [3]. This can also explain the rapid ignition of fuels such as nitrohexane or hexyl nitrate, where decomposition of the fuel produces NO_2 radicals which then decompose further to produce O atoms in large quantities [9]. For example, a compound referred to as "diesel ignition improver" is ethyl-hexyl nitrate, which rapidly releases O atoms into the reactive gas mixture and accelerates ignition by enhancing reaction (2) above.

However, O atoms are not often present in large quantities under most practical ignition conditions, and other reaction sequences are usually more important in determining ignition rates. Under most normal combustion conditions, the most important chain branching reaction is the reaction of H atoms with molecular oxygen [2,3], consuming one H atom radical to produce two radicals

$$H + O_2 = O + OH. \qquad (3)$$

In many practical application environments, the overall rate of combustion or ignition can easily be understood in terms of this reaction. Processes or reactions which increase the H atom concentration will accelerate the rate of combustion and enhance ignition, while processes which consume H atoms or remove H atoms from the radical pool retard ignition. One example of this is the observation that increased pressure retards both ignition and flame propagation [10]. The explanation for this trend is the contribution of a competing reaction

$$H + O_2 + M = HO_2 + M \qquad (4)$$

which competes with Reaction (3) but does not increase the radical pool or lead to chain branching. The rate of reaction (4) increases with pressure relative to the rate of reaction (3), so the rate of chain branching from reaction (3) is effectively reduced by increasing pressure. As discussed below, there are a variety of other chemical effects which provide either competition with reaction (3) to inhibit ignition or provide additional H atoms to promote ignition, and both effects are due to the chain branching nature of reaction (3).

4.1 Hydrocarbon Ignition in Shock Tubes

An example of the influence that H atom production has on shock tube ignition is illustrated in Fig.1, showing ignition delay time for several alkanes as a function of temperature behind reflected shock waves. The experiments were carried out by Burcat *et al.* [11] and the kinetic modelling was done by Westbrook and Pitz [12]. Note that no experiments were carried out for i-C_4H_{10}; the results for isobutane in Fig. 1 are computational results only. Ignition delay times for methane are much longer than for the other fuels and the ignition delay times for all of the larger n-alkanes are approximately equal. Although not shown, ignition delay times for ethane are significantly shorter than for the larger fuels. The model results are in good agreement with the measurements, and the calculations show that the differences between these similar fuels can be best understood in terms of their relative chain branching rates and especially the production of H atoms from the different fuels. Specifically, H atom abstraction from methane produces methyl radicals, which are either oxidized to produce formaldehyde or recombine to produce ethane. Production of ethane results in chain termination due to removal of methyl radicals from the radical pool, not branching. Therefore, the major alkyl radical produced from methane leads not to chain branching but primarily to chain termination. In the case of ethane, the only alkyl radical produced by ethane is the ethyl (C_2H_5) radical; β-decomposition of ethyl radicals produces H atoms *via*

$$C_2H_5\ (+M) = C_2H_4 + H\ (+M). \qquad (5)$$

Therefore, all H atom abstraction reactions from ethane lead to ethyl radicals, and

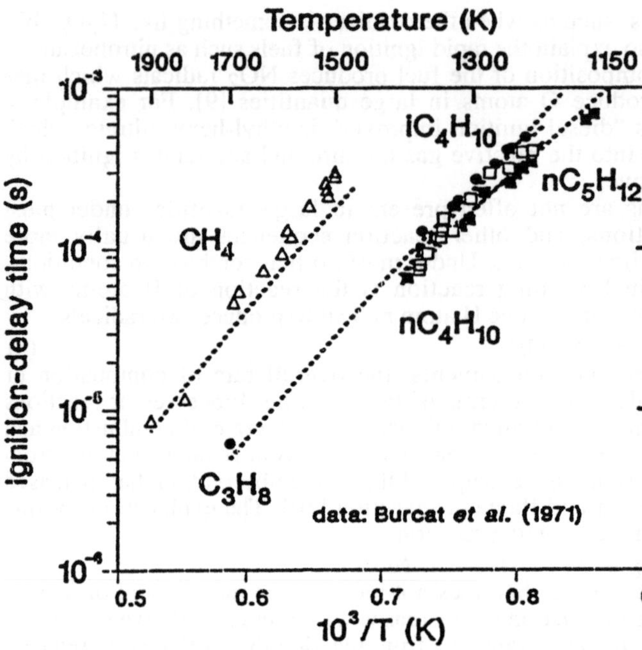

Figure 1: Experimental shock tube ignition delay time measurements (symbols) and model predictions for methane, propane, n-butane and n-pentane. Also shown are computed predictions for iso-butane.

the most important subsequent reaction of ethyl radicals produces H atoms, which under these conditions provide chain branching *via* reaction (3) with O_2. All of the other alkane fuels produce two or more alkyl radicals from H atom abstraction reactions, and some of these alkyl radicals produce H atoms while others produce methyl radicals. For example, propane (C_3H_8) produces i-C_3H_7 and n-C_3H_7 radicals and n-pentane (C_5H_{12}) produces three pentyl radicals, leading to:

$$C_3H_8 \Rightarrow \text{i-}C_3H_7 = C_3H_6 + H \qquad (6)$$

$$\Rightarrow \text{n-}C_3H_7 = C_2H_4 + CH_3 \qquad (7)$$

$$\text{n-}C_5H_{12} \Rightarrow \text{1-}C_5H_{11} = C_2H_4 + C_2H_4 + CH_3 \qquad (8)$$

$$\Rightarrow \text{2-}C_5H_{11} = C_3H_6 + C_2H_4 + H \qquad (9)$$

$$\Rightarrow \text{3-}C_5H_{11} = \text{1-}C_4H_8 + CH_3. \qquad (10)$$

Thus H atom abstraction from alkane fuels larger than ethane, followed by β-decomposition leads to a mixture of chain branching and chain termination reactions. The rate of ignition is then seen to be fastest for ethane, with all of the β-decomposition reactions producing chain branching, followed by the larger alkanes with mixtures of chain branching and termination, followed by methane, the slowest to ignite since only methyl radicals are produced. It must not be forgotten that these alkyl radicals can participate in additional reactions not shown here, so the discussion above represents only the major reaction paths. However, under shock tube conditions, these represent the most significant reaction paths and the overall analysis of the relative rates of ignition in these terms is generally accurate. Most numerical models of shock tube ignition consider the ignition to occur at either constant volume in the case of reflected shock waves or at constant pressure for incident shock waves. Of course, both of these are idealizations and, like corrections for boundary layers and other nonideal conditions, their effects and appropriateness must be considered in the modelling analysis.

4.2 Hydrocarbon Ignition in Detonations

Chemical kinetics of high temperature ignition play a key role in the propagation of detonations. Ignition behind the shock wave in a detonation must be rapid enough for the reactive heat release of the combustible gases to reinforce the shock wave. Kinetic modelling has been valuable in the analysis of detonation phenomena, ranging from full CFD simulations with chemical kinetics [13] to studies using the ZND model [14], which uses kinetic models to calculate characteristic ignition delay times and relate these time scales to properties such as the detonation cell size, critical detonation initiation energy, and critical tube diameter. Most kinetic models of ignition in detonation waves are considered to be adiabatic and constant volume calculations, similar to conditions in reflected shock wave simulations.

The same type of chain branching/chain termination analysis used above for the shock tube applies equally to kinetic analyses of detonations. In particular, modifications in operating conditions that reduce the rate of chain branching will result in longer ignition delay times, which lead to larger detonation cell sizes and greater critical energy for detonation initiation. Eventually, if the ignition delay time becomes sufficiently long, a detonation may eventually no longer be possible for a given fuel/oxidizer mixture. An excellent example of this feature is provided by the effect of adding diluents or chemical inhibitors to otherwise detonable mixtures. If halogenated species such as HBr, CH_3Br, CF_3Br, or others are added to detonable hydrocarbon/air or hydrocarbon/oxygen mixtures, they are found [15] to increase detonation cell sizes, and if enough inhibitor is added, the detonation eventually is extinguished. Kinetic modelling [15] shows that the halogenated inhibitors lead to reaction sequences which catalytically remove radical species from the radical pool. For example, if HBr is the additive, the following reactions are important:

$$HBr + H = H_2 + Br \qquad (11)$$

$$H + Br_2 = HBr + Br \qquad (12)$$

$$Br + Br + M = Br_2 + M. \qquad (13)$$

The overall sum of these reactions is the net reaction $H + H = H_2$. It is interesting to note that this is exactly the same reaction set first analyzed 100 years ago by Max Bodenstein [16], whose work is being recognized in this Symposium. Removal of H atoms is especially important, since as noted above, these H atoms cannot then participate in chain branching through reaction (3). The same processes have been shown [17] to inhibit and extinguish flame propagation through their catalytic activity in providing chain termination and H atom removal from the radical pool.

4.3 Hydrocarbon Ignition in Pulse Combustors

A third example of the role of kinetic modelling in the analysis of practical combustion systems at high temperatures is in simulations of pulse combustion. Pulse combustors have been used for many years in propulsion, industrial processes such as drying, and in home furnaces, where they are valued for their thermal efficiency and low NO_x production rates. However, until kinetic modelling was used to analyze the role of thermal ignition in these systems, it had been impossible to understand the principles of pulse combustion and the real reasons for their good performance.

These studies by Keller and Barr *et al.* [18-20] showed that, in contrast with ignition in shock tubes and detonations, faster ignition did not necessarily lead to enhanced system performance. Instead, the key to pulse combustion is that the periodic ignition of the fuel/oxidizer mixture should occur in phase with the resonant pressure oscillations in the combustion chamber, following Rayleigh's criterion. Enhanced system performance may sometimes require that ignition be accelerated, but in other cases it may be necessary to retard hydrocarbon fuel ignition to improve the system performance. These observations were made in systems fueled by natural gas and methane [18], in which the fuel/air ignition occurred slightly earlier than the peak in the system pressure. Kinetic modelling indicated that system performance could be improved if ignition could be retarded, and this could be accomplished by adding an inert diluent such as CO_2 or N_2, or by decreasing the percentage of ethane in the natural gas, predictions that were then verified by direct experimental measurements.

The kinetic simulations of the pulse combustor ignition can be carried out under conditions which closely approximate those in a continuously stirred tank reactor (cstr). In those calculations, hot product gases are steadily mixed with cold, unburned reactants until the mixtures ignite. The reaction mechanisms used are valid for high temperatures, and the most important, sensitive reaction is reaction (3), and the combined influences of chemical kinetics, acoustics, and fluid dynamics can all be incorporated into a coherent practical design model [20].

Practical ignition systems can also operate at temperatures much higher than those discussed here. Spark plug ignition in automobile engines or plasma jet or torch ignition are examples of systems where, at least for a brief period of time, temperatures are sufficiently high to ionize the reactive gases. In many situations, the period of time when plasma effects prevail is very short, and the gas temperatures rapidly fall into the regime where the above comments on high temperature ignition apply. The general subject of plasma/kinetics interactions is outside the scope of the present paper, but the principle of identifying the major chain branching sequences that lead to ignition is still certainly valid.

5 Intermediate Temperature Ignition

At intermediate system temperatures, different reaction sequences lead to chain branching. The activation energy of reaction (3) is quite high (~17 kcal/mol), so when the temperature falls below about 1000 K, its rate becomes quite slow. Instead, reaction (4) is most important, since it has almost no temperature dependence, producing hydroperoxy radicals HO_2. The dominant chain branching sequence in this temperature range is the series of reactions

$$RH + HO_2 = R + H_2O_2 \tag{14}$$

$$H_2O_2 + M = OH + OH + M \tag{15}$$

which consumes one radical and produces two OH radicals. Production of OH radicals is especially important since OH reacts rapidly with hydrocarbon fuel molecules. The significant parameter in this system is the activation energy for decomposition of hydrogen peroxide (~45.5 kcal/mol), or equivalently the strength of the O-O bond, which translates into an effective decomposition temperature. Under conditions related to internal combustion engine pressures, this decomposition generally occurs at temperatures of about 900 K. Other reactions that take place at about the same temperature range and involve breaking the same type of O-O bond are the decompositions of alkylhydroperoxides, such as

$$CH_3OOH = CH_3O + OH \tag{16}$$

and $$n\text{-}C_6H_{13}OOH = C_6H_{13}O + OH. \tag{17}$$

Numerical modelling studies of engine knock [21,22] in internal combustion engines indicated that some antiknock additives, including tetraethyl lead, act by suppressing this intermediate temperature chain branching reaction sequence. Although tetraethyl lead is no longer used as an antiknock additive in many parts of the world due to its adverse environmental impacts, its mode of effectiveness may provide important insights into the development of future antiknock compounds. The modelling analysis indicated that when tetraethyl lead is added to gasoline, the lead evolves into lead oxide in solid particulate form. These particulates then provide a distributed amount of heterogeneous surface area in the engine end gas where HO_2 and other radical species are adsorbed, thereby removing them from the reactive radical pool. An alternative theory proposed by Benson [23] suggests that the role of tetraethyl lead is to provide gas phase, not solid particulate, oxides of lead, which then similarly eliminate HO_2 from the reactive radical pool. Each HO_2 radical thus removed under these conditions is therefore unable to participate in the chain branching reaction sequence of reactions (13) and (14) and the resulting lower rate of reaction is sufficient to suppress the autoignition of the engine end gases and eliminate knocking. Similarly, additives such as alkylhydroperoxides which decompose at the same conditions where HO_2 chain branching is important, accelerate the overall rate of reaction and promote engine knock.

6 Low Temperature Ignition

At sufficiently low temperatures, alkyl radicals produced by H atom abstraction from the fuel react primarily by addition to molecular oxygen via reaction (1) to produce alkylperoxy radicals. The subsequent sequence of reactions has a great wealth of kinetic and thermochemical detail, some of which will be discussed below. However, the overall process includes reactions which lead to chain branching and other sequences which lead only to chain propagation. A careful and detailed understanding and kinetic description of the rates of these different reaction paths is required to be able to interpret and predict the response of laboratory and applied combustion chemistry problems.

Much of the practical interest in hydrocarbon oxidation under these conditions is motivated by the observation that these conditions are especially important in producing engine knock and ignition in diesel engines. Extensive kinetic analyses of hydrocarbon oxidation under these lower temperature conditions [24] have identified the most important features of these problems. Most models of hydrocarbon autoignition have been motivated by the assumption that the chemical details of hydrocarbon oxidation are so complex that significant simplifications are essential in order to be able to simulate the process. Perhaps most prominent of these simplified model treatments of hydrocarbon ignition are the "Shell Model" [25] and related developments by Keck [26] and Cox and Cole [27]. A recent survey and critical analysis of these simplified approaches by Griffiths [28] has summarized strengths and limitations of these models. However, recent studies in detailed kinetic modelling of hydrocarbon oxidation [22] have made it possible to address a wide variety of issues related to ignition, most of them leading to improved descriptions of ignition in internal combustion engines and engine knock.

6.1 Negative Temperature Coefficient and Low Temperature Reaction Mechanisms

At temperatures lower than approximately 900 K, high activation energies for alkyl radical decomposition make these processes relatively slow. Under such conditions, the most important reactions for alkyl radicals R consist of addition of

$$R + O_2 = RO_2 \tag{1}$$

molecular oxygen. In many ways this reaction is the most important step for low temperature oxidation, although it does not immediately determine the overall rate of chain branching. The activation energy for reaction (1) is approximately zero in the forward direction and quite large (~30 kcal/mol) in the dissociation direction. Therefore the equilibrium constant for this reaction is very strongly temperature dependent. At very low temperatures, reaction (1) proceeds rapidly to produce RO_2 very efficiently; at high temperatures RO_2 dissociates rapidly and the concentration of RO_2 is very small. Benson [29] has defined the concept of a "ceiling temperature" in terms of the equilibrium constant for reaction (1), where

$$[R]\,[O_2]\ /\ [RO_2] = 1 \tag{18}$$

and it can often be convenient to think, rather simplistically, that at temperatures below the ceiling temperature, oxidation of alkyl radicals takes place through addition of molecular oxygen, and at temperatures above the ceiling temperature, alkyl radicals are consumed by β-decomposition. As described below, there is a general sequence of reactions which follow the production of RO_2 that leads to chain branching which, because of the temperature-dependent equilibrium constant of reaction (1), shuts off above the ceiling temperature. The chain branching reactions that operate above the ceiling temperature do not become fast until the temperature exceeds the ceiling temperature by a considerable amount. As a result, there is a temperature range characterized by a so-called "negative temperature coefficient" (NTC) of reaction, over which the overall rate of fuel oxidation actually decreases as temperature is increased. Eventually, higher temperature chain branching reactions become more important and the rate of reaction again increases with increasing temperature. Below, examples of NTC behavior are described.

The RO_2 radicals formed by reaction (1), unless they immediately decompose back to $R + O_2$, then react primarily through transfer of H atoms within the radical species. This isomerization process is rather complex and proceeds through a transition state consisting of a ring-like structure which includes all of the atoms between the H atom being transferred and the O-O radical site. The rate of this process depends on the number of atoms in this transition state ring structure and the type of C-H bond (i.e., primary, secondary, or tertiary) being broken. This process has been described by Pollard [24] and then used [22,30,31] to simulate this process. Overall, this reaction path is denoted by

$$RO_2 = QOOH \tag{19}$$

where Q represents an olefin structure.

The next reaction of these QOOH radicals really determines the overall chain branching rate of combustion at lower temperatures. There are three classes of reactions which provide chain propagation. The QOOH radicals

$$QOOH = \text{Heterocycle} + OH \tag{20}$$

$$= \text{Conjugate olefin} + HO_2 \tag{21}$$

$$= \beta\text{-decomposition} \tag{22}$$

can cyclize to produce a range of epoxide or heterocyclic species products. The rate of this process depends on the ring size of these heterocyclic compounds [22,24].

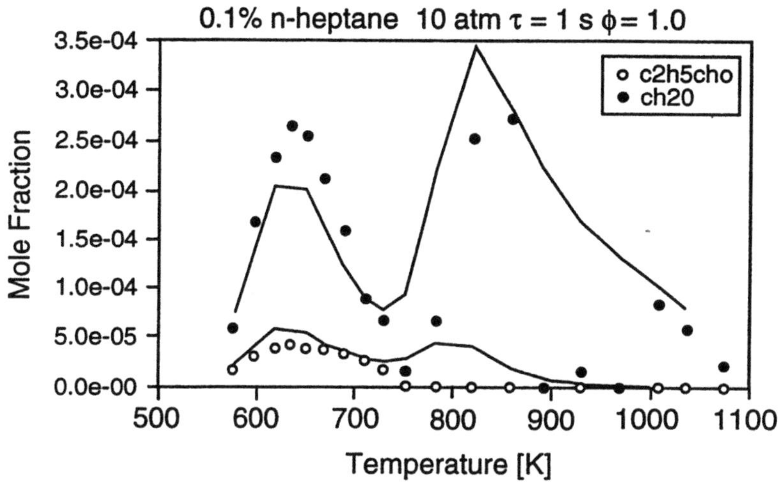

Figure 2: Experimental (circles) and computed concentrations of formaldehyde and propionaldehyde in stirred reactor for n-heptane at 10 atm, residence time of 1 sec.

For certain limited types of QOOH radicals, unimolecular decomposition can lead to olefins or smaller species and oxygenated compounds. All of these product species include one radical, so these reaction sequences which began with one R alkyl radical are effectively chain propagation paths.

A fourth reaction sequence for QOOH is the addition of another O_2 species to produce a O_2QOOH species. This product can then isomerize further to produce a ketohydroperoxide species which is reasonably stable, in addition to OH radicals. When the ketohydroperoxide eventually decomposes, it leads to additional radicals and smaller hydrocarbon species. The overall reaction sequence therefore produces two or more radicals along with smaller stable species, providing overall chain branching. The relative rates of each of these elementary processes will establish the chain branching rate at any given temperature and the overall rate of ignition.

6.2 Stirred Reactor Simulations

An important class of applications is the stirred reactor. This consists of a reaction volume in which incoming reactants are rapidly and thoroughly mixed with residual gases, and the outflow from the reactor consists of the average contents of the reactor. Computationally, the inlet boundary condition generally consists of a fixed mass flux and composition, while the outflow condition consists of a fixed pressure and zero gradients in the species mass fractions and temperature. In many cases, the reaction takes place at nearly constant temperature. Most model simulations are carried out for steady-state conditions, but time-dependent phenomena can be simulated equally well, which is required for phenomena such as those reported by Lignola *et al.* [32]. The time-dependent formulation can lead to complex behavior including periodic ignitions, cool flames and other features.

We have used experimental data from a stirred reactor obtained by Dagaut *et al.* [33] to study the oxidation of n-heptane over a temperature range from 550-1150 K. Steady-state production of formaldehyde and propionaldehyde are shown as functions of temperature in Fig. 2. The presence of a negative temperature coefficient region from about 600-750 K is evident in both the modelling and experimental results.

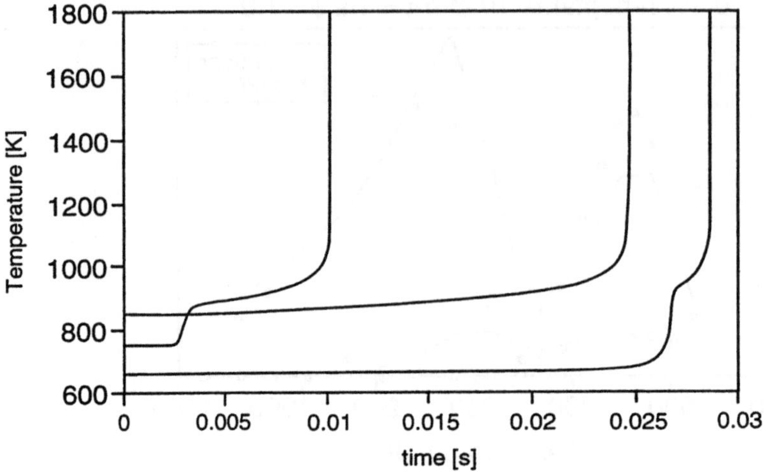

Figure 3: Computed temperature histories in rapid compression machine simulations, stoichiometric n-heptane/oxygen in nitrogen and argon.

6.3 Rapid Compression Machine

Many experimental studies have used a system in which a piston is used to compress a reactive mixture to reaction temperatures and pressures. This system resembles a piston engine which operates with only one compression cycle. The boundary conditions are constant mass and volume, with heat transfer at a rate determined by comparison with measured pressure-time data. In many ways, the rapid compression machine (RCM) is like the shock tube in computational modelling terms, although heat losses in the shock tube are rarely important but can be very significant in RCM simulations. We have examined data from Minetti *et al.* [34] over temperatures from 650-900 K for stoichiometric mixtures of n-heptane and oxygen, diluted by nitrogen and argon. Computed temperature histories [35] from simulations of ignition of n-heptane are shown in Fig. 3. The results again show a significant NTC behavior. A two-stage ignition is observed for an initial temperature of 650K, a more rapid two-stage ignition is seen at an initial temperature of 750K, but then a longer ignition delay time is shown for an initial temperature of 850K, where the first stage, associated with the low temperature sequence of reactions, has disappeared.

6.4 Engine Knock

Perhaps the most dramatic practical example of the importance of low temperature ignition kinetics is the problem of engine knock in spark-ignition engines. Past experimental studies [36] and modelling work [22] have shown how such factors as fuel molecule size and structure influence the onset of engine knock. In a recent study [37] of autoignition kinetics and engine knock, octane ratings and ignition of the five distinct isomers of hexane were studied using motored engine experiments and kinetic modelling. This work showed how RO_2 isomerization, decomposition of QOOH species, and production and decomposition of ketohydroperoxides all combine to determine autoignition in internal combustion engines. Modelling engine knock problems requires following the pressure and temperature experienced by engine end gases, conditions which result from piston motion, turbulent flame propagation, and heat transfer with other processes of lesser importance. In real engines, residual product gases from previous engine cycles also contribute to the details of ignition. When all of these factors are included in model calculations, the details of the reaction mechanisms are sufficient to reproduce the influences of fuel molecule structure and size on ignition rates for a very wide range of alkane, olefin and oxygenated fuels. In addition, these models show how fuel blending agents such as methyl-tert butyl ether (MTBE) are effective in suppressing engine knock.

7 Summary

Current reaction mechanisms are capable of simulating a wide range of ignition phenomena in both laboratory and practical environments. Each class of problems requires a thorough description of the relevant initial and boundary conditions in order to reproduce the physical problem. Perhaps the most important requirement of each kinetic model is the ability to accurately quantify the reaction sequences leading to chain branching, since ignition is most generally defined as a process in which explosive radical growth leads to rapid fuel consumption and heat release.

8 Acknowledgements

The authors thank Prof. Eliseo Ranzi of Politecnico di Milano for support of P. Gaffuri during this work. This work was supported by the DOE Office of Basic Energy Sciences, Chemical Sciences, and was carried out under the auspices of the U.S. Department of Energy by the Lawrence Livermore National Laboratory under contract W-7405-ENG-48.

9 References

[1] Glassman, I., ***Combustion,*** Academic Press, Inc., Orlando, Fla., 1987.
[2] Westbrook, C. K., and Dryer, F. L., *Eighteenth Symposium (International) on Combustion,* p. 749, The Combustion Institute, Pittsburgh, 1981.
[3] Westbrook, C. K., and Dryer, F. L., Prog. Energy Combust. Sci. **10**, 1 (1984).
[4] Baulch, D. L., Cobos, C. J., Cox, R. A., Esser, C., Frank, P., Just, Th., Kerr, J. A., Pilling, M. J., Troe, J., Walker, R. W., and Warnatz, J., J. Phys. Chem. Ref. Data **21**, 411 (1992).
[5] Warnatz, J., in ***Combustion Chemistry***, (W. C. Gardiner, Jr., Ed.), chap. 5, "Rate Coefficients in the C/H/O System", Springer-Verlag, New York, 1984.
[6] Miller, J. A., and Bowman, C. T., Prog. Energy Combust. Sci. **15**, 287 (1989).
[7] Glarborg, P., Miller, J. A., and Kee, R. J., Combust. Flame **65**, 177 (1986).
[8] Dagaut, P., Reuillon, M., Boettner, J.-C., and Cathonnet, M., *Twenty-Fifth Symposium (International) on Combustion*, p. 919, The Combustion Institute, Pittsburgh, 1995.
[9] Tieszen, S. R., Stamps, D. W., Westbrook, C. K., and Pitz, W. J., Combust. Flame **84**, 376 (1991).
[10. Westbrook, C. K., and Dryer, F. L., Combust. Flame **37**, 171 (1980).
[11] Burcat, A., Scheller, K., and Scheller, K., Combust. Flame **16**, 29 (1971).
[12]. Westbrook, C. K., and Pitz, W. J., in ***Shock Waves and Shock Tubes,*** (Bershader , K., and Hanson, R.K., Eds.) Stanford Univ. Press, 1986.

[13] Oran, E. S., Boris, J. P., and Kailasanath, K., ***Numerical Approaches to Combustion Modeling***, p, 421, AIAA, Washington, 1991.
[14] Westbrook, C. K., Combust. Flame **46**, 191 (1982).
[15] Westbrook, C. K., *Nineteenth Symposium (International) on Combustion*, p. 127, The Combustion Institute, Pittsburgh, 1982.
[16] Bodenstein, M., Lind, S. C., Z. phys. Chem. **57**, 168 (1906).
[17] Westbrook, C. K., Combust. Sci. Technol. **23**, 191 (1980).
[18] Keller, J. O., and Westbrook, C. K., *Twenty-First Symposium (International) on Combustion*, p. 547, The Combustion Institute, Pittsburgh, 1986.
[19] Keller, J. O., Bramlette, T. T., Dec, J. E., and Westbrook, C. K., Combust. Flame **79**, 151 (1990).
[20] Barr, P. K., Keller, J. O., Bramlette, T. T., Westbrook, C. K., and Dec, J. E., Combust. Flame **82**, 252 (1990).
[21] Pitz, W. J., and Westbrook, C. K., Combust. Flame **63**, 113 (1986).
[22] Westbrook, C. K., Pitz, W. J., and Leppard, W. R., Society of Automotive Engineers, SAE-912314 (1991).
[23] Benson, S. W., J. Phys. Chem. **92**, 1531-1533 (1988).
[24] Pollard, R. T., Hydrocarbons, ch. 2, ***Comprehensive Chemical Kinetics, Vol. 17, Gas-Phase Combustion,*** (C. H. Bamford and C. F. H. Tipper, Eds.), Elsevier, New York, 1977.
[25] Halstead, M. P., Kirsch, L. J., and Quinn, C. P., Combust. Sci. Technol. **30**, 45 (1977).
[26] Hu, H., and Keck, J. C., Society of Automotive Engineers Trans. **96**, 1987.
[27] Cox, R. A., and Cole, J. A., Combust. Flame **60**, 109 (1985).
[28] Griffiths, J. F., Prog. Energy Combust. Sci. **21**, 25 (1995).
[29] Benson, S. W., Prog. Energy Combust. Sci. **7**, 125 (1981).
[30] Westbrook, C. K., Warnatz, J., and Pitz, W. J., *Twenty-Second Symposium (International) on Combustion,* p. 893, The Combustion Institute, Pittsburgh, 1988.
[31] Chevalier, C., Pitz, W. J., Warnatz, J., Westbrook, C. K., and Melenk, H., *Twenty-Fourth Symposium (International) on Combustion,* p. 93, The Combustion Institute, 1992.
[32] Lignola, P.-G., Reverchon, E., Autuori, R., Insola, A., and Silvestre, A. M., Combust. Sci. Technol. **44**, 1 (1985).
[33] Dagaut, P., Reuillon, M., and Cathonnet, M., Combust. Sci. Technol. **95**, 233 (1994).
[34] Minetti, R., Carlier, M., Rivaucour, M., Therssen, E., and Sochet, L.-R., Combust. Flame **102**, 298 (1995).
[35] Gaffuri, P., Curran, H. J., Pitz, W. J., and Westbrook, C. K., Combustion Institute Central and Western States Section Proceedings, 263 (1995).
[36] Lovell, W. G., Ind. Eng. Chem. **40**, 2388 (1948).
[37] Curran, H. J., Gaffuri, P., Pitz, W. J., Westbrook, C. K., and Leppard, W. R., Society of Automotive Engineers, to be published (1995).

Part VI

Modelling of Flow, Turbulence and Complex Chemical Reactions

Using Direct Numerical Simulation with Detailed Chemistry to Study Turbulent Combustion

M. Baum,[1] M. Hilka,[2] and T.J. Poinsot[3]

[1] CERFACS, 42 av. Gustave Coriolis, 31057 Toulouse Cedex, France
[2] Laboratoire EM2C, Ecole Centrale de Paris, Grande Voie des Vignes, 92295 Châtenay-Malabry, France
[3] Institut de Mécanique des Fluides de Toulouse, av. Camille Soula, 31400 Toulouse, France

Abstract

Direct Numerical Simulation (DNS) has proved to be a valuable tool in addressing fundamental physical questions and deriving models for turbulent premixed combustion. Considering the role of DNS for the modelling of turbulent combustion the field of application of DNS methods is exposed. This paper focuses on DNS with detailed chemical reaction kinetics. Recent work on the structure of hydrogen-air and methane-air flames in turbulent flows, as well as a study of quenching of methane-air flames near a wall are presented to illustrate the applications of DNS in combustion research.

1 Introduction

The basic understanding of turbulent combustion, its prediction in practical applications and its control are fundamental problems in combustion research and technology. Any progress in this area has an immediate impact on many practical devices (e.g. piston engines and jet engines). However, studying turbulent combustion is complicated by the coupling of various complex phenomena, such as turbulence, molecular transport and chemical reactions. Therefore, using only analytical and asymptotic theories turns out to be insufficient to address this kind of problems. Even precise experiments do not always provide the needed information (e.g. flame–wall interactions) and are often quite expensive. In this framework *Direct Numerical Simulation* (DNS) became an attractive tool in the past several years that has proved to be valuable in addressing fundamental physical questions and in the construction of models for turbulent combustion. DNS in the following, designates numerical simulations of the Navier-Stokes equations without any turbulence model. That means that all scales of turbulence and chemical reactions are fully resolved. DNS therefore provides data with the same precision as experiments of the same configuration.

Both experimental and numerical investigations of the role of chemical reaction kinetics in turbulent combustion are complicated by the strong coupling of hydrodynamics with thermochemistry and by resolution requirements: hydrodynamic and thermodynamical spatial and temporal scales span many orders of magnitude in flames with high Reynolds and Damköhler numbers (the latter being the ratio of characteristic flow time scales to chemical time scales). Thus DNS of practical tur-

Springer Series in Chemical Physics, Volume 61
Gas Phase Chemical Reaction Systems
Eds.: J. Wolfrum, H.-R. Volpp, R. Rannacher, and J. Warnatz

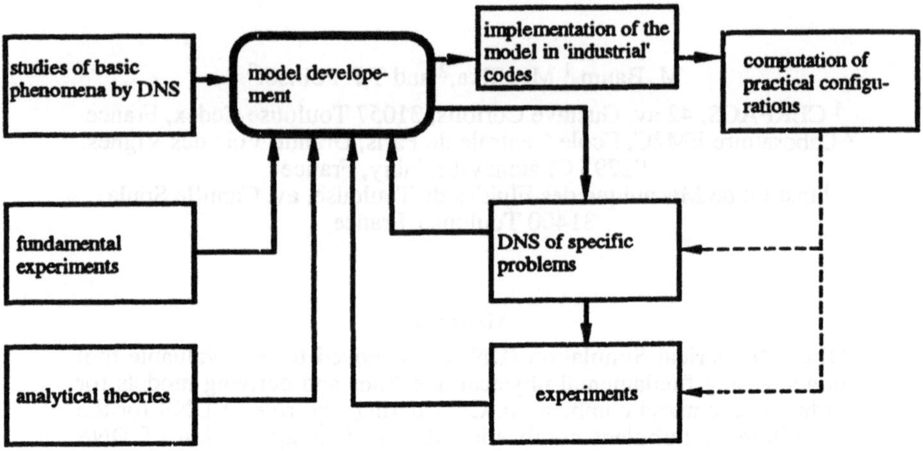

Figure 1: Role of DNS in modelling.

bulent flames is impracticable at present and for the foreseeable future. However, carefully designed model situations are amenable to DNS and therefore DNS appears to be a powerful tool, complementary to experiments and analytical approaches.

One goal of combustion research is to develop a numerical tool that predicts combustion and its characteristics for practical applications. As this cannot be done by solving directly the Navier-Stokes equations for these configurations, simplifications, based on physical models, have to be introduced to represent fluid turbulence and turbulent combustion. The development and validation of these models is still a crucial topic in combustion research. DNS may be employed at two levels in the process of developing and testing models (Fig. 1): (1) Associated to fundamental experiments and analytical theories DNS provides insight into basic physics of combustion, suggesting models or their improvement. (2) DNS permits to validate these models. The validation of models is complementary to the "classical" approach: implementation of the model in an industrial code and computation of a "practical" application. A major problem in this procedure is to determine the origins of differences between computations and reference experiments. They may come from the model itself, from uncertainties in the measurements, or from the difficulty to match initial and boundary conditions in both situations (e.g. initial turbulence). DNS overcomes this difficulty and allows a more direct and quicker testing of models prior to their integration in industrial codes, thus accelerating the whole model development process. Studies using DNS may take advantage of the fact that all variables in the whole domain are available, while experiments can keep track of few data in a delimited domain only. For these reasons DNS plays an important role in combustion modelling, together with analytical theories and experiments.

Available DNS approaches are summarized in Fig. 2. The important question here is to choose the appropriate method for a given problem. The creation of a material surface for example may be studied in cold flows using three dimensional constant density DNS. Even if the flow field is quite different from "real flames" useful information can be obtained for a reasonable price (typically 1hour Cray2 per run). On the other hand three dimensional variable density DNS with simple chemistry and transport mechanisms provide more realistic flames for low Reynolds

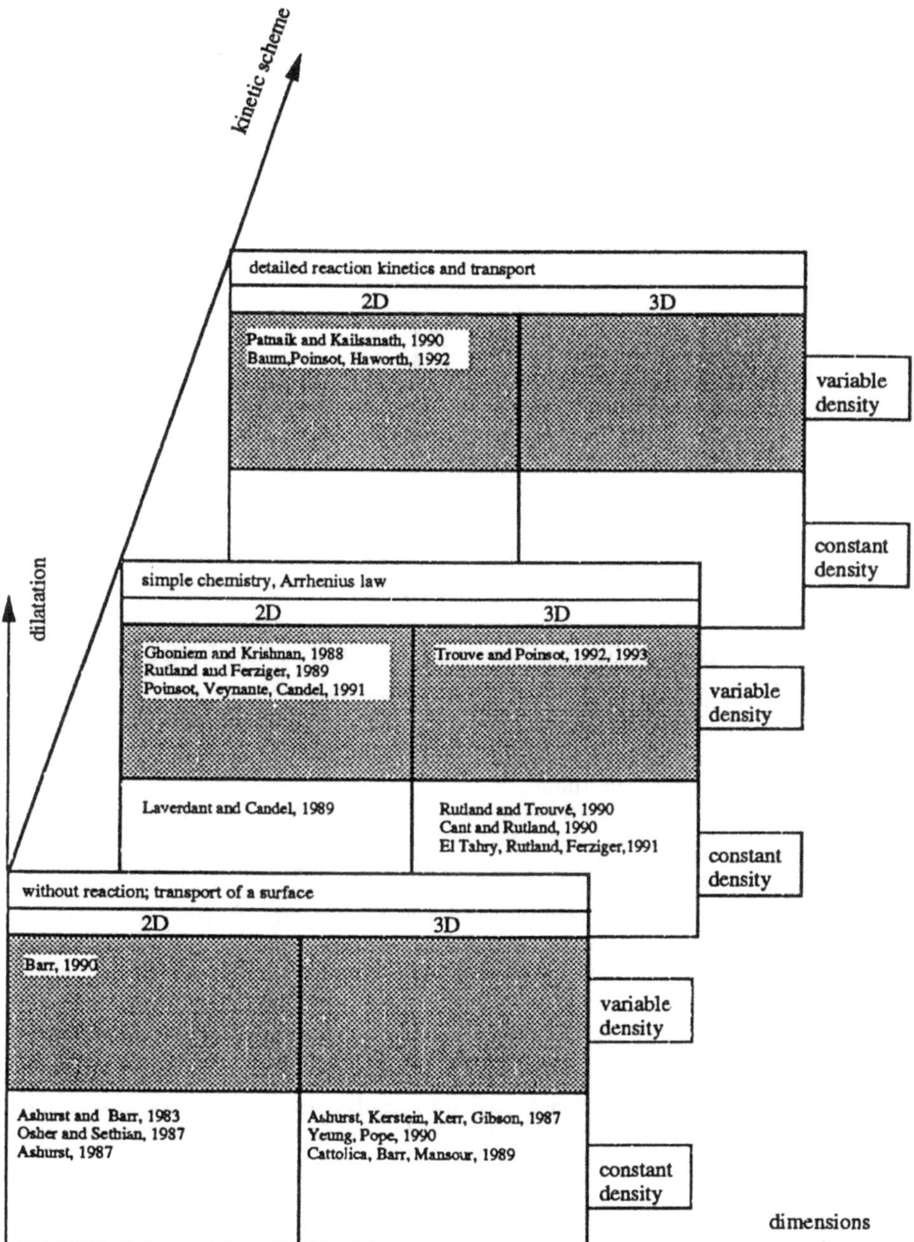

Figure 2: DNS methods for premixed flames (the author's list is not complete).

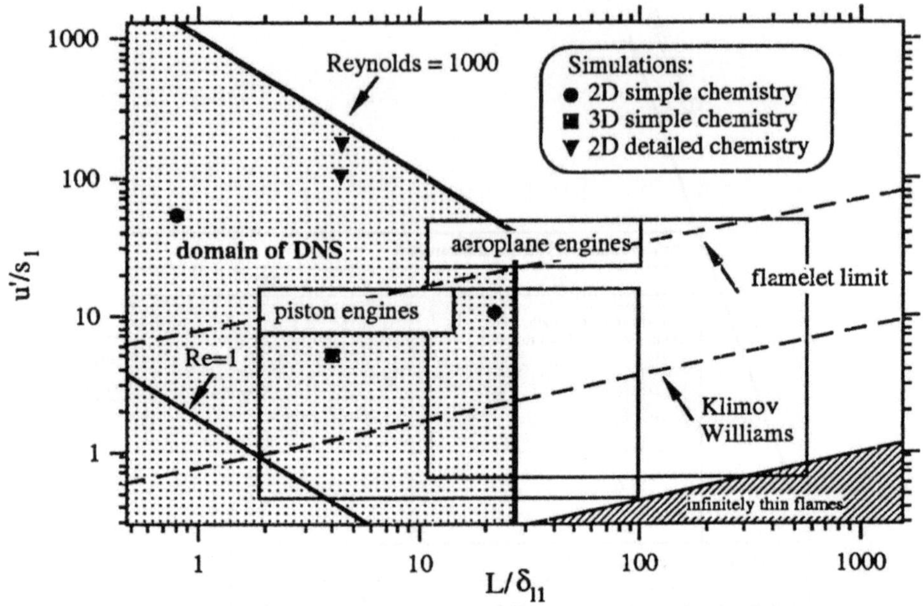

Figure 3: Combustion diagram indicating the domain where DNS applies.

numbers, but at higher computational cost (typically 100 hours Cray2 per run). Variable density DNS with detailed chemistry finally are available for two dimensional configurations only at high cost (typically 100 hours Cray2 per run), but permit to study the local structure of the flame.

The main problem of DNS is the use of computational domains that are large enough to represent realistic flows. Their size is limited by the given computational resources and the characteristic length scales of turbulence and chemical reaction kinetics. The minimum size of the domain is given by the integral length scale L while the minimum spatial resolution is determined by the Kolmogorov scale η and the flame thickness δ_l. For a given number of grid points N in each spatial direction an integral length scale-based Reynolds number may be specified, which cannot be exceeded:

$$\eta > L/N, \quad \eta = (\nu^3 L/u'^3)^{1/4}, \quad Re_L = u'L/\nu \Longrightarrow Re_L < N^{4/3} \tag{1}$$

Supposing that at least 10 points are needed to resolve the flame front (flame thickness δ_l, flame speed s_l, $\delta_l = \nu/s_l$) a chemical Reynolds number Re_c may be introduced which cannot be exceeded:

$$L > N/10 \times \delta_l, \quad Re_c = s_l L/\nu \Longrightarrow Re_c < N/10 \tag{2}$$

The domain where DNS may be applied is shown in a combustion diagram (Fig. 3), where it is superimposed on the operation ranges of piston and jet engines. Their intersection illustrates that DNS may be close to flows encountered in real applications.

A major issue for DNS of reacting flows is the choice of a realistic scheme to

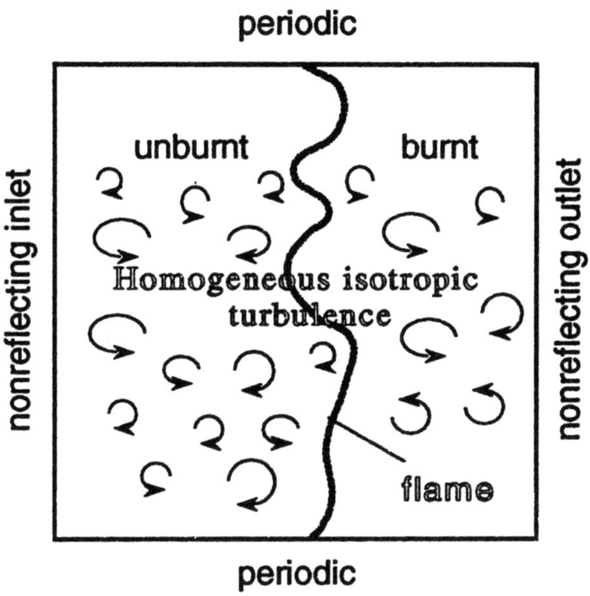

Figure 4: Schematic description of the computational domain with fresh gases entering from the left side and products exiting on the right side.

describe combustion kinetics. In most DNS studies, the description of chemistry is extremely simplified to be able to deal with the problem. Here we will concentrate on DNS where complete description of chemistry but also of transport coefficients and thermodynamic laws for multispecies flows have been used.

The main characteristics of the DNS-method used in our studies are summarized in [1,2]. This paper is devoted to recent work employing DNS with complex chemistry. A study of the flame structure of $H_2/O_2/N_2$ flames is presented in Section 2, turbulent methane-air flames are addressed in Section 3 and an investigation of quenching phenomena of methane-air by a cold wall in Section 4 concludes.

2 Structure of Turbulent $H_2/O_2/N_2$ Flames

While DNS with simple chemistry is commonly used to study the response of flames to turbulence, it is important to validate this approach and its results. DNS with detailed chemical reaction kinetics provides the necessary information on the structure of turbulent flames and the interaction of flames and turbulence to check the assumptions of simple chemistry DNS and to obtain further insights in turbulent combustion.

$H_2/O_2/N_2$ flames have been investigated in a first step for the simplicity of their chemical schemes compared to hydrocarbons. Furthermore the reaction mechanisms for hydrogen flames are quite well established to date and reasonable agreement between simulations and experiments (e.g. for ignition or planar, laminar flames) is reported. Moreover, a thorough understanding of the $H_2/O_2/N_2$ mecha-

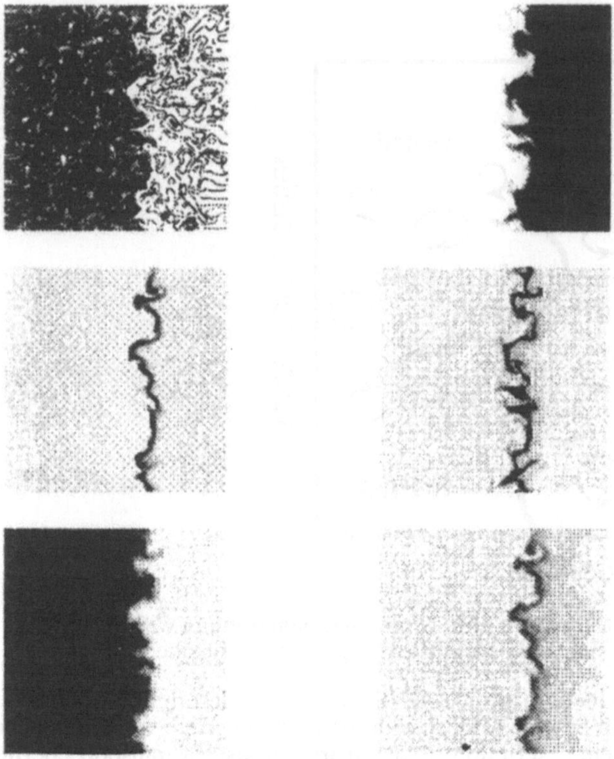

Figure 5: DNS of an H_2/O_2 flame with the Miller scheme: (a) vorticity contours, (b) temperature, (c) heat release, (d) H_2O_2 mass fraction, (e) H mass fraction, (f) OH mass fraction.

nism may be a guideline for the extension to hydrocarbons as their reaction schemes are based on the reaction mechanism for $H_2/O_2/N_2$ [3].

Main points of interest were the demonstration of the feasibility of DNS using detailed reaction kinetics and transport mechanisms and the comparison with results obtained from simple chemistry DNS. This comparison focused on the global and local statistics of the flame response to turbulence. The verification of the flamelet library approach was another important goal of this study.

A two dimensional configuration was considered (Fig. 4), where an initially planar laminar premixed flame evolves in a turbulent flow field. The chemical scheme of [4] (9 species, 19 reactions) has been retained to prescribe chemical reactions. The initial homogeneous isotropic turbulence field was given according to the energy spectrum proposed by Von Karmann-Pao [5]. Figure 5 presents an example of DNS – results are for a turbulent Reynolds number $u'L/\nu = 289$ in the fresh gas, a ratio of velocity fluctuations to laminar flame speed $u'/s_l = 30$ and an integral length scale to flame thickness $L/\delta_l = 1.4$. The size of the most energetic vortices L_i is much larger than the flame thickness: $L_i = 11 \times \delta_l$. The flame thickness based on the temperature gradient is 1 mm and the laminar flame speed 11 cm/s. The total box size is 2.5 cm in each direction.

Computations have been performed to cover a wide range of parameters (e.g. equivalence ratio, fresh gas temperature, turbulence level) and analyzed from a flamelet point of view. We refer to [2] for more details. The important results may be summarized as follows:

- Flames simulated with complex chemistry remain flamelet-like for most investigated cases. No quenching is observed along the flame front. This result confirms predictions obtained with simple chemistry models: In the absence of heat losses, localised quenching is difficult to achieve for these configurations.
- Flames simulated with complex chemistry align preferentially along extensive strain rates like simple chemistry flames. The curvature remains symmetric with near-zero mean value.

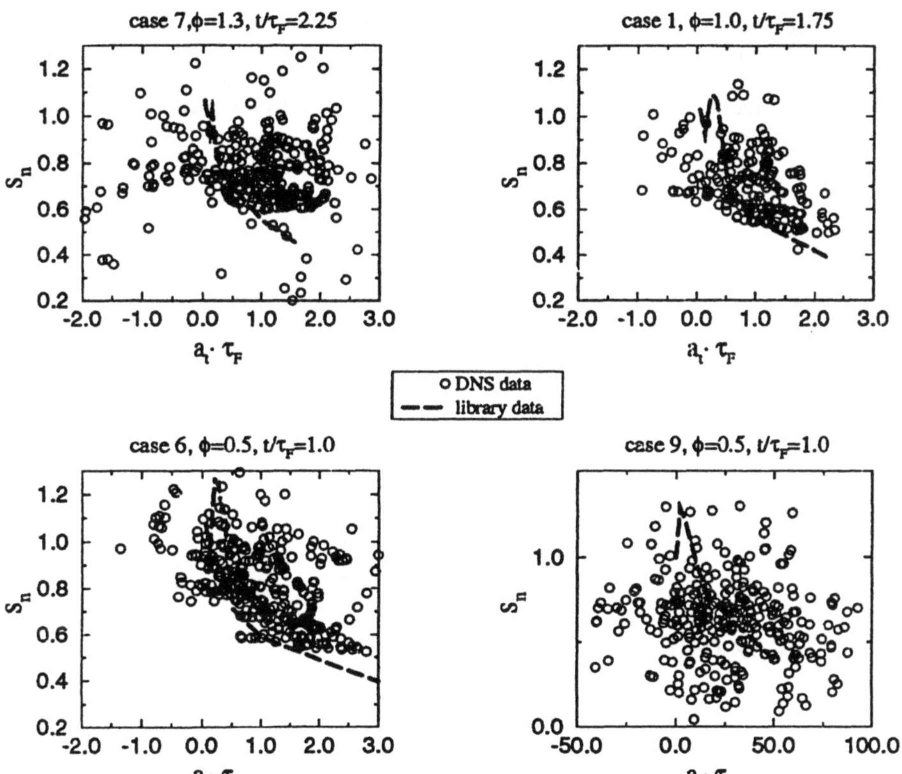

Figure 6: Correlation between strain rate and local consumption speed: comparison between DNS data and flamelet library. Cases 1, 6 and 7 are performed with $T_u = 800$ K, case 9 with $T_u = 300$ K.

- A transition in the behaviour of the flame is obtained for an equivalence ratio of 0.4. Below this value, the flame becomes unstable: correlations of flame speeds and maximum reaction rates along the flame front change sign when the equivalence ratio goes from rich to lean. A similar transition was observed for simple chemistry situations but for a Lewis number of unity. The definition of a Lewis number in a real flame is difficult but in this case, the trend observed in simple chemistry cases (and predicted by asymptotic theories) is confirmed by complex chemistry computations.
- A striking difference with simple chemistry results is that complex chemistry flames are more sensitive to strain than simple chemistry flames. Their thickness changes strongly with strain and the local consumption speed s_c correlates rather well with strain, a result which is opposite to the one obtained with simple chemistry computations. However, like simple chemistry flames, the space-averaged flame speed s^* remains of the order of the laminar flame speed s_L, except for unstable flames.
- Comparison of predictions of a flamelet library (based on planar steady strained flames with the same chemical and transport models as for the DNS) and DNS results was also given in Ref. 2. Despite the large scatter, it appears that the agreement is reasonable: the evolution of s_c with strain is correctly predicted by the library as shown in Fig. 6 even though the library overestimates the effects of strain on the flame (this effect is probably due to unsteady mechanisms). This result brings more credit to the idea of flamelet libraries than the simple chemistry computations did.

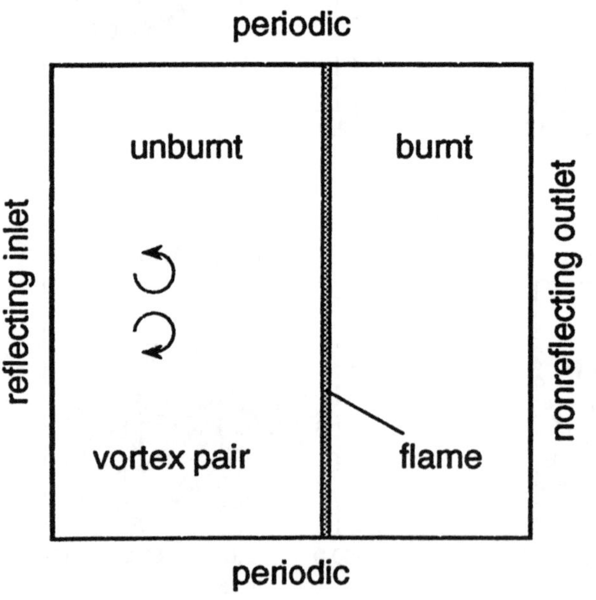

Figure 7: Schematic description of the computational domain with fresh gases entering from the left side and products exiting on the right side.

3 Turbulent Methane–Air Flames

While DNS of $H_2/O_2/N_2$ flames provided first insights in the structure of turbulent flames and their response to turbulence, they lack many characteristics of usual fuels for real engines (e.g. pollutant formation, unburnt hydrocarbons, etc.). Therefore the study presented in Section 2 has been extended to consider methane–air flames which incorporate many properties of usual fuels. Detailed reaction schemes (17 species, 52 reactions [6], and 17 species, 45 reactions [7]) were considered in two different configurations: (1) flame vortex interactions [8] and (2) flame turbulence interactions [9].

Following [10] the flame response to a turbulent flow can be considered as the sum of its responses to the vortices composing the turbulence spectrum in a first approximation. The principle regimes of interaction were identified by [10] as a function of two main parameters: the ratio of the maximum rotational speed of the vortices to laminar flame speed u_Θ/s_l and of the vortex core diameter to laminar flame thickness d/s_l. While most qualitative results were confirmed by experimental investigations [11,12], the simulations did not reproduce flame extinction observed in the experiments. Complex chemical kinetics phenomena are a possible explanation for flame extinction due to excessive strain as encountered during flame-vortex interaction.

Light emission during the flame-vortex interaction for a poor CH_4/Air flame was recently measured and a decrease of the luminous intensity, indicating possible local flame extinction, was noticed [13]. The current study permits a direct compar-

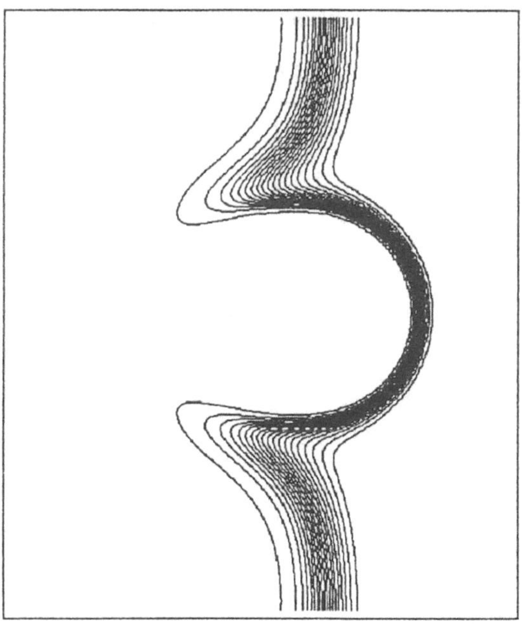

Figure 8: Isocontours of the instantaneous density field for a laminar premixed $CH_4/O_4/N_2$ interacting with a pair of vortices (Mixing ratio $\phi = 0.55$).

ison of instantaneous results from experiment and simulation for this configuration in order to investigate the modification of the local flame structure.

For the present simulation a two-dimensional configuration has been chosen. A rectangular computational domain (Fig. 7), discretized by a regular mesh was initialized with reactants on the left and products on the right, separated by a laminar premixed flame obtained by the PREMIX flame code [14]. We hereafter superimpose a pair of counter-rotating Oseen-vortices in the fresh gas region. Periodic boundary conditions are used in the vertical direction, while non-reflecting boundary conditions are applied along the inflow/outflow boundaries [15,16]. The detailed reaction scheme used in this calculation consisted of 17 species and 52 elementary reactions. An example of the density field at one instant of the interaction is given in Fig. 8. The initially planar flame front has been distorted by the vortex pair and a small quantity of H_2 has diffused into the vortex core. The initially relatively thick laminar flame is severely stretched on the centerline. As in the experiments of Ref. 13, the light emission from a premixed CH_4/Air flame computed according to [17]

$$I = A\, e^{-E_a/RT}\, [O][CO]\,, \tag{3}$$

decreases in the DNS as the vortex pair distorts the flame, Fig. 9. Nevertheless, as the flame surface increases, it is interesting to observe that the integrated heat release from the flame increases as well, but at a much slower rate. Figure 9 also indicates that light emission from CH radicals should increase while emissions from OH radicals should decrease during the interaction. Thus, emissions from CH radicals appear to be the only experimentally accessible indicator for the heat release rate of premixed methane-air flames. Luminous emission does not appear to follow

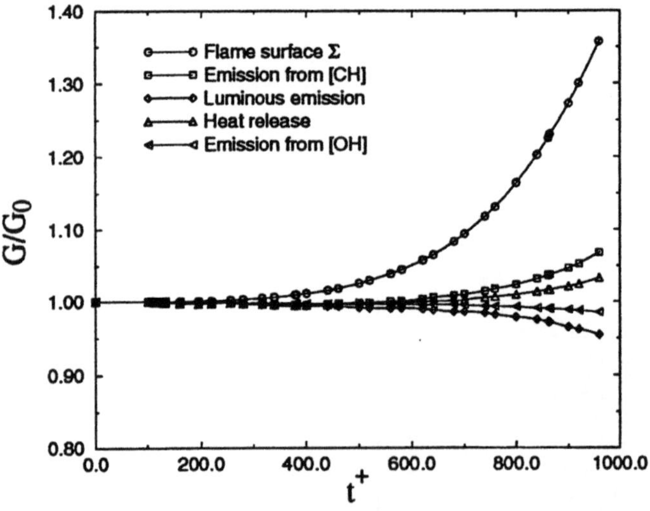

Figure 9: Evolution of flame surface, heat release and integral of [CH] and [OH] during flame-vortex interaction. All values are normalized by their initial value for an undisturbed planar flame and plotted against dimensionless time t^+.

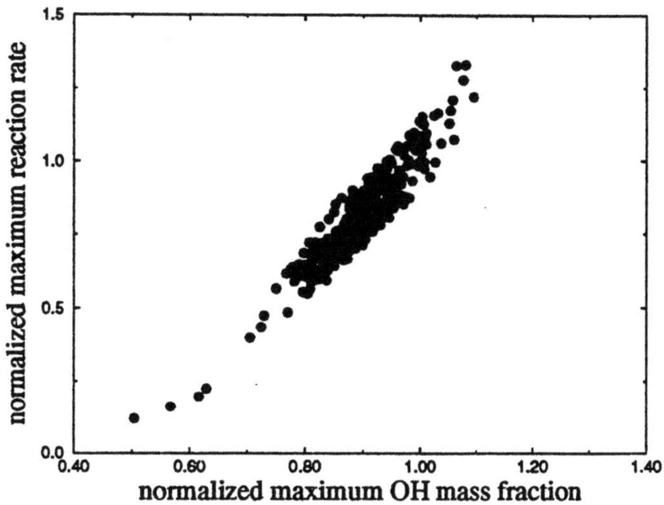

Figure 10: Correlation of the maximum heat release rate with maximum OH mass fraction along the flame front for a turbulent methane flame (equivalence ratio = 1.0; T_u = 298 K)

the heat release rate sufficiently close to allow the diagnostic of flame extinctions.

Statistical information on strain-rate, curvature, local burning ratio and displacement speed along the flame surface explored in these simulations support earlier findings of Ref. 10. We refer to Ref. 8 for more details.

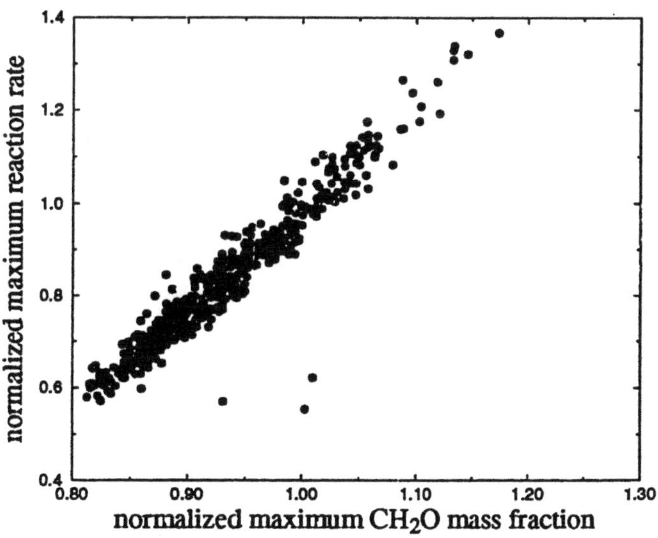

Figure 11: Correlation between the CH_2O concentration and maximum reaction rate along the flame front for a turbulent methane flame (equivalence ratio = 1.0; T_u = 298 K)

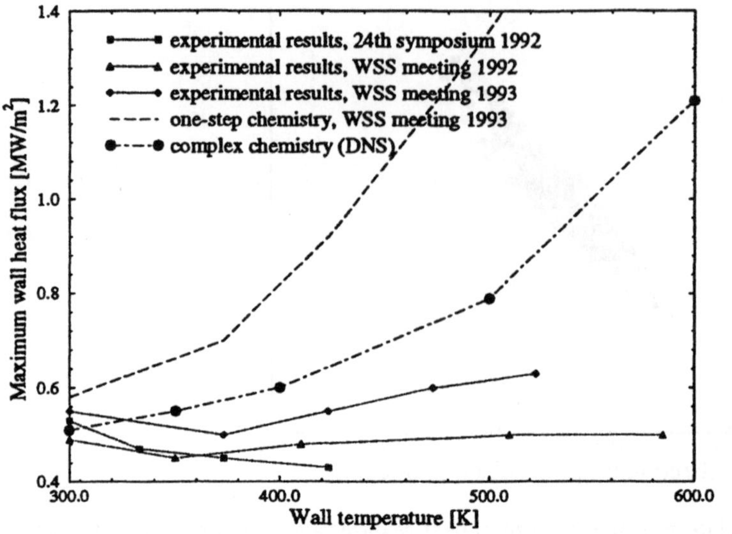

Figure 12: The maximum wall heat flux as a function of wall temperature. Results from experiments, one-step chemistry and detailed chemistry are shown.

Flame–turbulence interactions were studied in the same configuration as for the hydrogen flames in Section 2 (Fig. 4), but detailed chemical reaction kinetics for methane flames from [7] including NO_x formation were used. Main findings of the study presented in Section 2 were confirmed, especially [9]:

- The reactants (CH_4, O_2) as well as reaction intermediates (CH_3, CH_2O, HCO, H, HO_2, etc.) appear to be well represented by a flamelet model. Production and consumption of these species occurs in a reaction zone that is thin compared to the length scales associated with the turbulent flow. The formation of CO_2 and NO, however, proceeds through slower chemical reactions with their respective maximum concentration values relatively far behind the flame front.
- Correlations between the OH radical and the local maximum reaction rate (in planes perpendicular to the flame front) were found to be even more pronounced than for the hydrogen flames (Fig. 10) and other species like CH_2O turn out to correlate better with the maximum strain rate (Fig. 11).

4 Quenching of Methane–Air Flames Near a Wall

Experimental investigations in engines [18-20], as well as in much simpler environments such as constant volume combustion chambers [21,22], concerning the influence of combustion chamber wall insulation on engine efficiency have produced conflicting statements about the magnitude of the wall heat flux during quenching and its trend with respect to wall temperature. It seems thus desirable to have as an aid for future engine code design a complete understanding of the physical processes occurring in the flame-wall interaction zone as well as of the appropriate boundary conditions. We have studied the head-on quenching situation of laminar flames as it

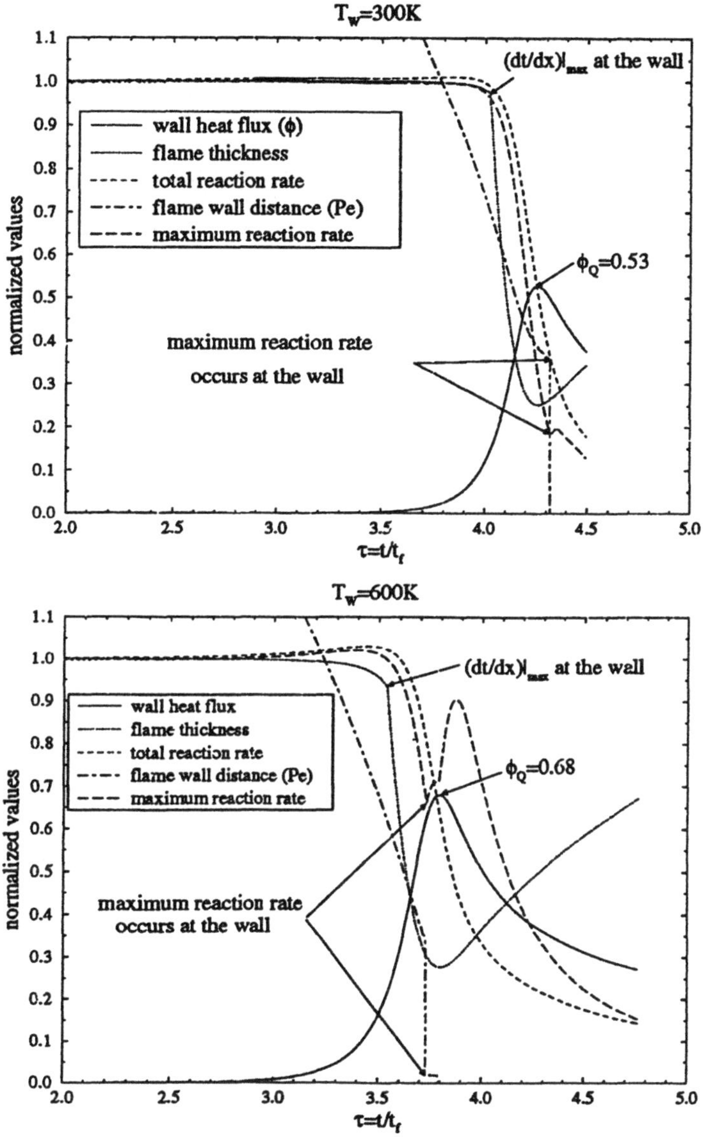

Figure 13: Global parameters of the flame wall interaction at 300 K (top) and 600 K (bottom).

can be found in constant-volume combustion chambers for a methane flame over a range of wall temperatures between 300 K and 600 K using DNS with detailed chemistry. This approach allows to compare various reaction schemes, to identify the most important species and reaction paths, and to investigate the influence of different modelling assumptions. The computational results (Fig.12) show that the dimensional wall heat flux increases with wall temperature over the whole range of wall temperatures studied; this agrees well with the most recent measurements in a precise experimental setup (Fig. 12). It is found that the wall can be modelled as chemically inert and that thermal diffusion processes are negligible for low wall temperatures between 300 K and 400 K. For higher temperatures, however, there are large discrepancies in the actual values of the maximum wall heat flux although they still show the same trend. Those differences are not any longer within the given uncertainty range of the experiment (about 11%). In addition, the bending in the profile at $T_W = 500$ K suggests that flame wall interaction for high temperatures might be dominated by other physical processes than at low temperatures.

In a first step to explain the increasing discrepancy between the numerical and experimental results with increasing wall temperature and to illustrate in more detail the physics of the quenching process we present at this point the most important global quenching parameters. Those parameters are frequently used in the literature [23-26] and demonstrate the global characteristics of this process. Figure 13 compares their time variation for wall temperatures of 300 K and 600 K, respectively. All parameters are normalized using their laminar undisturbed values except the wall heat flux and the Peclet number. As mentioned earlier the wall heat flux is normalized by the laminar flame power and the Peclet number yields the flame wall distance normalized by the laminar flame thickness based on the maximum temperature gradient. The flame wall interaction at different temperatures is following similar patterns. We may identify four phases:

1. The flame propagates steadily for some time before it starts sensing the wall (this corresponds in our case to about two and a half flame times). All flame characteristics are constant and identical to those of the laminar undisturbed flame.
2. The presence of the wall is felt once the flame enters the influence zone of the wall which has a width of about two flame thicknesses ($Pe_{infl,Tw=300K} \approx 1.8$, $Pe_{infl,Tw=600K} \approx 2.3$). The small increase in the total as well as the maximum reaction rate is the first indicator of the presence of the wall, followed by a small decrease in the flame thickness. The rise in the reaction rates lasts for a period larger than one flame time, till the maximum temperature gradient of the flame has reached the wall ($Pe_{Tw=300K} = 0.70$, $Pe_{Tw=600K} = 0.62$). At this point the wall heat flux has reached a value of about one fourth of its peak value, and the maximum reaction rate is still about one percent higher than in the undisturbed flame.
3. Once the maximum temperature gradient has reached the wall the interaction proceeds rapidly. The wall heat flux climbs up to its maximum value in only 0.2 flame times and the flame thickness reduces to 25% of its laminar undisturbed value indicating that the maximum temperature gradient increases by as much as a factor of four during the flame-wall interaction. The maximal wall heat flux Φ_Q occurs when the flame reaches its minimal thickness. Note that whereas in the 300 K case the total reaction rate has dropped at this moment to 50%, it is still 70% in the 600 K case.
4. For times larger than τ_Q (the instant where Φ_Q is reached) the wall heat flux and the reaction rates are decaying, yielding a maximum of the reaction rate that occurs at the wall. Note that in the high-temperature case the total reaction after τ_Q is much higher than in the 300 K case and is decaying much less rapidly.

Although the two graphs of Fig. 13 are very similar, an interesting difference exists between flame wall interaction at $T_W = 300$ K and $T_W = 600$ K: In the high temperature case the maximum reaction rate occurs at the wall before the wall suffers the maximum wall heat flux and the maximum in the reaction rate reaches values almost as high as in the undisturbed flame. At low wall temperatures the maximum of the reaction rate occurs after the maximum wall heat flux and the increase in the maximum reaction rate is almost negligible. This clearly indicates that important mechanisms are taking place directly at the wall and becoming very violent for high wall temperatures. Analysis of the reaction kinetics during flame quenching shows that dramatically increasing radical concentrations (H, O, OH) at the wall at higher wall temperatures lead to large heat release rates directly at the wall surface of the combustion chamber, and thus can not be neglected in the modelling of the quenching process. A complete discussion of the heat transfer and reaction mechanisms can be found in Ref. 27.

5 Conclusion

DNS of reacting flows with complex chemical schemes has become an efficient tool in the last three years to deal with problems where chemistry plays an important role and should not be simplified. Flame-wall interaction is a good example of such situations. Pollutant formation in turbulent flames will be another example of application for these techniques in the near future.

6 References

[1] M. Baum, *Etude de l'allumage et de la structure des flammes turbulentes*, Ph.D. thesis, Ecole Centrale Paris, (1994).

[2] M. Baum, T. Poinsot, D. Haworth, and N. Darabiha, J. Fluid Mech. **281**,1, (1994).

[3] J. Warnatz, Ber. Bunsenges. Phys. Chem. **87**,1008 (1983).

[4] H. P. Miller, R. Mitchell, M. Smooke, and R. Kee, 19th Symposium (Int.) on Combustion, pp. 181–196, The Combustion Institute, Pittsburgh (1982).

[5] J.O. Hinze, ***Turbulence*** (McGraw-Hill, 1975).

[6] N. Darabiha, S. Candel, and F.E. Marble, Combust. Flame **64**, 203 (1986).

[7] T. Coffee, Combust. Flame, **55**, 161 (1984).

[8] M. Hilka, D Veynante, M. Baum, and T. Poinsot, 10th Symposium on Turbulent Shear Flows, paper submitted, Pennsylvania State Univ. (1995).

[9] M. Hilka, M. Baum, T. Poinsot, and D. Veynante. in *Annual Research Briefs. Technip*, Centre de Recherche sur la Combustion Turbulente, Rueil Malmaison (1995).

[10] T. Poinsot, D. Veynante, and S. Candel, J. Fluid Mech. **228**, 561 (1991).

[11] W.L. Roberts and J.F. Driscoll, Combust. Flame **87**, 245 (1991).

[12] W.L. Roberts, J.F. Driscoll, M.C. Drake, and L.P. Goss, Combust. Flame **94**, 58 (1993).

[13] J.M. Samaniego, in *Proc. of the Summer Program. Center for Turbulence Research*, NASA Ames/Stanford Univ. (1993).

[14] R.J. Kee, J.F. Grcar, M. Smooke, and J.A. Miller, *PREMIX: A Fortran program for modeling steady laminar one-dimensional flames,* Technical Report SAND85-8240, Sandia National Laboratories (1985).

[15] T. Poinsot and S. Lele, J. Comput. Phys. **101** (1),104 (1992).

[16] M. Baum, T.J. Poinsot, and D. Thévenin, J. Comput. Phys. **116**, 247 (1994).
[17] C.T. Bowman, 15th Symposium (Int.) on Combustion, The Combustion Institute, Pittsburgh (1974).
[18] Y. Enomoto and S. Furuhama. JSME **29**(250) (1986).
[19] T. Morel, S. Wahiduzzaman, and E. Fort, *Heat transfer experiments in an insulated diesel engine,* Technical Report 880186, SAE (1988).
[20] G. Woschni, W. Spindler, and K. Kolesa, *Heat Insulation of Combustion Chamber Walls–A Measure to Decrease the Fuel Consumption of IC Engines?* , Technical Report 870339, SAE, (1987).
[21] L. Connelly, T. Ogasawara, D. Lee, R. Greif, and R.F. Sawyer, In Fall Meeting of the Western States Section, The Combustion Institute (Volume WSCI 93-077) Stanford, CA, USA (1993).
[22] W.M. Huang, S.R. Vosen, and R. Greif, 21st Symposium (Int.) on Combustion, pp. 1853-1860, The Combustion Institute, Pittsburgh (1986).
[23] A.P. Kurkov and W. Mirsky, 12th Symposium (Int.) on Combustion, pp. 615-624. The Combustion Institute, Pittsburgh (1968).
[24] S.R. Vosen, R. Greif, and C.K. Westbrook, 20th Symposium (Int.) on Combustion, pp. 76-83, The Combustion Institute, Pittsburgh (1984).
[25] J.H. Lu, O. Ezekoye, R. Greif, and F. Sawyer, 23rd Symposium (Int.) on Combustion, pp. 441-446, The Combustion Institute, Pittsburgh (1990).
[26] T. Poinsot, D. Haworth, and G. Bruneaux, Combust. Flame **95**(1/2),118, (1993).
[27] P. Popp, M. Baum, and T. Poinsot, Combust. Flame, submitted (1995).

Stability of Reaction Fronts

Vit.A. Volpert[1], Vl.A. Volpert[2], M. Garbey[1], and J.A. Pojman[3]
[1] Université Lyon 1, URA 740 CNRS, 69622 Villeurbanne, France
[2] Northwestern University, Evanston, IL 60208, U.S.A
[3] University of Southern Mississippi, Hattiesburg, MS 39406, U.S.A

Abstract
A review of recent experimental and theoretical results on stability of polymerization fronts is given. Different forms of instabilities are discussed as well as conditions of their appearance.

1 Frontal Polymerization

Propagation of reaction fronts is intensively studied in connection with combustion processes, self-propagating high-temperature synthesis, branching chain flames, frontal polymerization (FP), etc. One of the most interesting questions is the stability of propagating fronts. This paper provides a brief overview of recent studies, both theoretical and experimental, of the stability of reaction fronts, focusing mainly on polymerization fronts, where a localized polymerization zone, in which monomer is converted to polymer, propagates through the medium.

A systematic investigation of FP was initiated in [1-4], and followed by many others (see references in [5,6]). An important step in the experimental studies of FP was recently made by finding experimental conditions which allow to directly visualize the propagating fronts [7]. FP has much in common with combustion waves, which are much better studied than FP, since in both cases there are exothermic reaction waves propagating through a chemically active medium. FP differs from combustion processes by chemical kinetics, various phase transitions, hydrodynamical effects, and thermophysical properties of the substance. Typically polymerization waves are much slower and less exothermic than combustion waves.

In this section we discuss experimental observations of the instabilities of polymerization fronts. The next section provides a description of theoretical results.

1.1 Spinning Mode of Propagation

In this mode of propagation, which was first found in combustion [8], one or more high-temperature spots are observed to move in a helical fashion along the surface of the cylindrical sample. In FP the spinning mode was found in two cases, namely, in (i) anionic activated polymerization of ϵ-caprolactam [9] and (ii) methacrylic acid polymerization [10]. We describe briefly the experimental conditions and observations in both cases.

In the former case the reaction front propagates downwards through a vertical tube filled with a liquid monomer initially at 300 K. Adiabatic heat release due to polymerization is about 50 K. Since the adiabatic temperature is

Springer Series in Chemical Physics, Volume 61
Gas Phase Chemical Reaction Systems
Eds.: J. Wolfrum, H.-R. Volpp, R. Rannacher, and J. Warnatz

below the crystallization temperature of poly-ϵ-caprolactam, it crystallizes with an additional heat release of about 40 K. This phase transition appears to be important for the fronts stability (cf. Section 2).

The high-temperature spots change the conditions of crystallization of the polymer and form helical trajectories on the lateral surface of the cylindrical sample. Modes with two or three helical trajectories, i.e., with two or three high-temperature spots, were observed experimentally. Their number depends on the radius of cylinder. An increase in the radius leads to an increase in the number of helical trajectories (cf. Section 2). The direction of rotation of the high-temperature spots is determined by the initial conditions and can be either clockwise or counterclockwise.

In the latter case the reaction front also propagates downwards, but unlike ϵ-caprolactam polymerization, both the monomer and the polymer are solid. To maintain the monomer in the solid phase the tube with the monomer is held in an ice water bath. It causes an additional heat loss and keeps the monomer frozen. The distance between the front and the water bath was maintained constant by pulling the tube out of water. The monomer melts in the reaction zone, the reaction occurs in the liquid phase, and then solidification of the polymer takes place. Figures 1 and 2 show samples with one helical trajectory on the surface

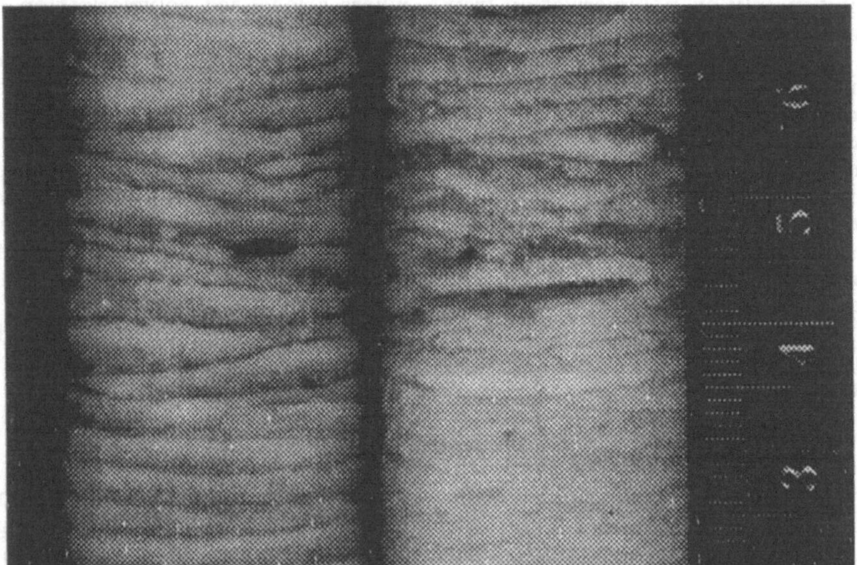

Figure 1: Samples obtained by the spinning mode of frontal polymerization.

and a curved front. We remark that the motion of hot spots is not necessarily regular – see Fig. 3 where a sample with multiple hot spots moving not in a helical fashion is displayed.

Figure 2: Curved front.

1.2 Convection

Natural convection plays an important role in FP. Since the reaction is exothermic, then in the case of an ascending front the cold monomer is heated from below. If the monomer is liquid, it can cause the convective instability. Convection increases the heat transfer from the reaction zone and can lead to extinction. This is why descending fronts were studied in [1-4,9-11].

Conditions of occurrence of natural convection caused by a reaction front are studied theoretically (see Section 2). Currently the experiments are being carried out to test the theory.

We mention also one more experimental observation concerning the interaction of reaction fronts and natural convection. If a planar descending front forming a solid product is considered, then the natural convection should not occur (see Section 2). However, if there is a heat loss through the side wall of the tube, the front is no longer planar. It becomes more advanced in the center. In this case the action of gravity can cause convective motion of the liquid under the front. This phenomenon was observed also in numerical simulations [12].

1.3 Hydrodynamic Instability

In Ref. 10 the fronts of methacrylic acid polymerization with AIBN initiator were studied. A solution of de-gassed monomer and initiator was placed into a 2.2×25 cm test tube. A front was initiated by the heat from a soldering iron. Propagating reaction waves were videotaped, and the position of the front as a function of time determined. Depending on experimental conditions two modes of frontal propagation were observed. In both the interface between the liquid monomer and the slowly growing (about 1 cm/min) region of solid polymer was essentially flat and travelled with essentially constant velocity. The main difference between the two modes of propagation was that convective motion of the monomer below the downward propagating front was observed in one mode, and not in the other.

Which of the two modes occurred was determined by experimental conditions, such as the initial concentration of the initiator, the ambient temperature and the external pressure. Uniformly propagating flat fronts without convection were observed when the pressure was sufficiently large, and/or the initial temperature and concentration of the initiator were sufficiently low. In the opposite case propagation of the front was accompanied by convection. An important

Figure 3: A sample with multiple hot spots.

distinction between the two modes of propagation is that the front accompanied by convection was also accompanied by the formation of gas bubbles in the monomer, while convection did not occur if gas bubbles were absent. The gas bubbles (carbon dioxide or nitrogen) are formed in the liquid monomer due to the decomposition of the initiator. High external pressures (100-150 psi) eliminate the bubbles and the front propagates without convection. Similarly, if the concentration of the initiator or the initial temperature are low, not much gas is released in the course of decomposition of the initiator, so that bubbles can be eliminated by lower pressures, and convection does not occur. Thus, the presence of gas bubbles appears to be a decisive factor in determining the mode of propagation of the front.

Figure 4 shows the propagating polymerization front with the liquid motion caused by the hydrodynamical instability under the front.

1.4 Fingering

Descent of polymer particles can be observed for the propagating downward fronts of methyl methacrylate polymerization [1-4]. This phenomenon is called "fingering". It occurs because the density of the solid polymer is greater than the density of the liquid monomer under the front. Fingering can initiate the reaction everywhere in the fresh monomer below the front, and destroy the FP. In Ref. 1, fingering was suppressed by high external pressure. Fingering has been studied in greater detail in Ref. 11.

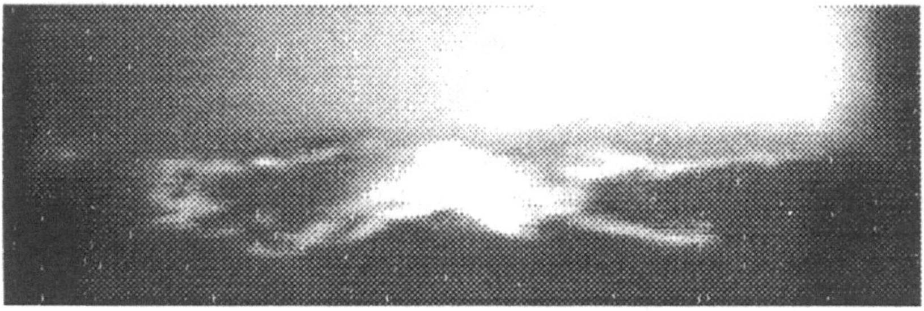

Figure 4: Hydrodynamical instability.

2 Modelling of Polymerization Fronts

2.1 Model

We consider a model which describes the propagation of reaction fronts through a condensed medium in the case of a one-step chemical reaction. It consists of the energy balance equation, mass balance equations, momentum equations and the continuity equation. The energy balance equation is a nonstationary nonlinear heat equation which includes a convective term, heat conduction term and a term describing liberation of heat in the chemical reaction. Mass balance equation is written for the concentration of the product of the reaction and includes convective and chemical reaction terms. Momentum equations are taken in the form of Navier-Stokes equations with a gravity term.

To complete the formulation of the problem we should specify how the density of the mixture depends on its temperature and pressure. Three different cases will be considered. The simplest one is the constant density approximation which assumes the density to be constant. In this case we eliminate all the phenomena connected with the hydrodynamics and can study the thermal instability in the pure form.

Next, we will employ the Boussinesq approximation in which case the density is assumed to be constant in all terms in the equations except for the buoyancy term. It allows to study the influence of the natural convection on the stability of reaction fronts.

Finally, we will discuss the formulation when the density depends on the temperature and on the concentration of the reaction product. Here we use a conventional isobaric approximation. This approximation assumes that density does not depend on pressure. As follows from the discussion below this approximation is not necessarily valid for FP.

2.2 Thermal Instability

If the density of the medium is constant, our model reduces to the heat equation and the equation for the concentration of the product of the reaction. This is a classical model of the condensed phase combustion studied in detail (see, e.g., [13-16]).

We describe briefly the main results. Stability of the front is determined by the Zeldovich number Z which is a nondimensional measure of the activation energy of the reaction. The front is stable if $Z < Z_{cr} = 8$, and unstable if $Z > Z_{cr}$. The instability occurs if heat production by the reaction is faster than removal of heat from the reaction zone. In this case the temperature in the reaction zone increases. This leads to further acceleration of the reaction rate and to further increase in temperature. This self-accelerating process causes the thermal instability which results in the oscillatory behavior of the reaction wave. It can be either one-dimensional or multidimensional modes of propagation. In the former case the front remains planar but the speed of propagation as well as all other characteristics of the reaction wave periodically oscillate in time [17]. Further increase in the activation energy leads to a sequence of period doubling bifurcations and eventually to transition to chaos [18].

Multidimensional modes of propagation depend on the geometry of the sample, e.g., spinning modes with a different number of high-temperature spots can appear in a circular cylinder. The number of spots depends on the radius of the cylinder [19] (cf. Section 1). We note that multiplicity of multidimensional modes can occur [20], i.e., different modes of propagation can exist for the same values of parameters. It can be for example spinning modes with one and two high-temperature spots. In this case initial conditions determine which mode is realized. Other modes for circular cylinders and for samples of a different geometry are described in [19,21]. We remark that in addition to the primary bifurcations described above, secondary and subsequent bifurcations occur as the Zeldovich number increases [20].

2.3 Convective Instability

If either the monomer or the polymer, or both, are liquid natural convection, caused by the heat liberated by the exothermic reaction, can occur. Consider first the case when the monomer is liquid and the polymer is solid (cf. Section 1). We will discuss separately upward and downward propagating fronts. If the front propagates upward, then the chemical reaction heats the monomer from below which reminds of the classical Rayleigh-Benard problem. If the Rayleigh number is sufficiently large, then the planar front loses its stability and stationary natural convection above the front occurs. For descending planar fronts there is no such convective instability. An approximate analytical approach allows one to find stability conditions for the propagating reaction front and to determine the modes which appear when the front loses stability [22].

The convective instability in pure form is observed when the Zeldovich number is sufficiently small. In this case there is no thermal instability. If Z is sufficiently large, then the thermal instability can occur. It is affected by hydrodynamics, which, in particular, changes the threshold of the thermal instability. For ascending fronts the critical value of the Zeldovich number becomes greater than that without hydrodynamics, i.e., the front is more stable. It happens due to an increased heat transfer from the reaction zone caused by the convective motion of the liquid. For descending fronts natural convection decreases the critical value of the Zeldovich number. The front becomes less stable. This analytical result can explain the occurrence of spinning modes of propagation of ϵ-caprolactam polymerization front which are experimentally observed for Z significantly less than Z_{cr}.

Consider now the case when both the monomer and the polymer are liquid. The principal difference between the cases is that now the medium can move not only ahead of but also behind the front. It significantly changes the results. Again, for the ascending fronts the convective instability occurs if the Rayleigh number, R exceeds the critical one. The liquid motion in this case becomes more complex than in the case of a solid product. Two vortices rather than one appear at the front. We note that convective motion of the liquid leads to the appearance of a high-temperature spot at the front. Thus, the origin of this hot spot is different from that in the case of the thermal instability.

If Z increases, then a bifurcation occurs, resulting in convective motion with an intensity which is periodic in time [23].

Unlike the case of FP forming a solid product, here the convective instability can also occur for descending fronts [23,24]. We note finally that natural

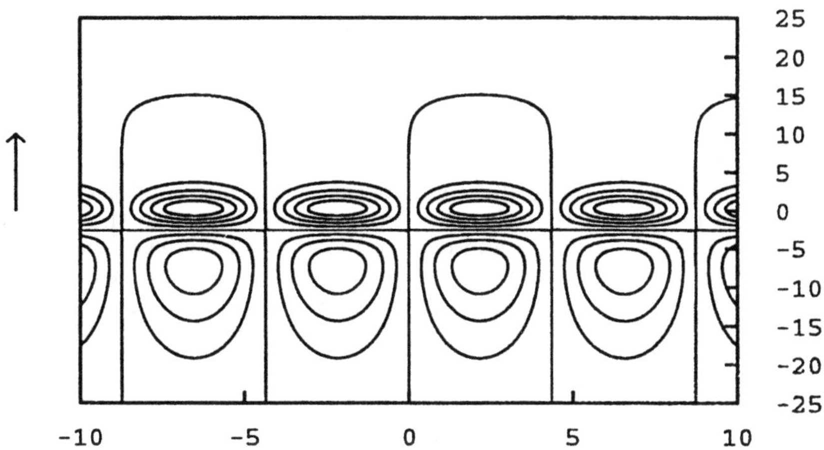

Figure 5: Level lines of the stream function for the convective instability.

convection also affects the threshold of the thermal instability in the case of a liquid polymer. However, this influence is the opposite to that in the case of a solid product. The ascending front becomes now less stable than the front without hydrodynamics, the descending front more stable. Figures 5 and 6 show the level lines of the stream function for different values of parameters.

2.4 Hydrodynamic Instability

Hydrodynamic instability (HI) of a reaction front was discovered by Landau in gaseous combustion [25] (see also Refs. 26 and 27). The physical reason for the instability is nonuniform thermal expansion of the medium near the reaction front. Unlike gaseous combustion, HI does not occur in FP processes forming solid polymer [28]. This result can be interpreted physically as follows. The heat release due to the reaction leads to expansion of the liquid near the front. If part of the liquid near the front expands at a greater rate than in the surrounding liquid, then this part of the liquid advances into the cold monomer. As a consequence, heat loss from this part of the liquid increases, which leads to a temperature drop,

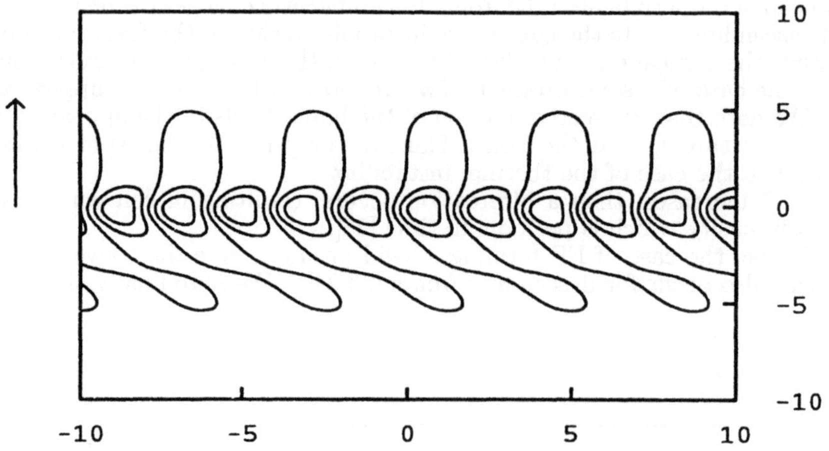

Figure 6: Level lines of the stream function for the thermal instability.

an increase of the density and consequently to a contraction of the liquid. Thus, the perturbation decays. The difference between gaseous combustion and FP forming a solid product is that thermal expansion of the gas in the combustion process occurs behind the front, and does not lead to an increased heat loss.

HI has been observed in FP experiments with solid polymer only in the case when gas bubbles are formed in the monomer (cf. Section 1). If gas bubbles are present, the processes in the wave become dependent on the external pressure through the influence of pressure on the bubbles. In this case the isobaric approximation cannot be used, and the analytical results [28] can no longer be applied. The bubbles can be effectively taken into account by allowing the density to depend on pressure as well as on temperature.

3 Acknowledgments

This work was supported by NSF grant CTS 9308708, NSF grant CTS 9319175 and NSF's Mississippi EPSCoR Program.

4 References

[1] N.M. Chechilo, R.Ya. Khvilivitsky, and N.S. Enikolopyan, Dokl. Phys. Chem. **204**, 512 (1972).

[2] N.M. Chechilo, and N.S. Enikolopyan, Dokl. Phys. Chem. **214**, 174 (1974).

[3] N.M. Chechilo, and N.S. Enikolopyan, Dokl. Phys. Chem. **221**, 391 (1975).

[4] N.M. Chechilo, and N.S. Enikolopyan, Dokl. Phys. Chem. **230**, 840 (1976).

[5] S.P. Davtyan, P.V. Zhirkov, and S.A. Vol'fson, Russ. Chem. Reviews **53**, 150 (1984).

[6] Vit.A. Volpert, and Vl.A. Volpert, Euro. J. Appl. Math. **5**, 201 (1994).

[7] J.A. Pojman, A.M. Khan, and W. West, Polym. Prepr. Am. Chem. Soc. Div. Polym. Chem. **33**, 1188 (1992).

[8] A.G. Merzhanov, A.K. Filonenko, and I.P. Borovinskaya, Dokl. Phys. Chem. **208**, 122 (1973).

[9] V.P. Begishev, V.A. Volpert, S.P. Davtyan,and A.Ya. Malkin, Dokl. Phys. Chem. **29**, 1057 (1984).

[10] J.A. Pojman, V.M. Ilyashenko, and A.M. Khan, Physica D **84**, 260 (1995).

[11] J.A. Pojman, R. Craven, A. Khan and W. West, J. Phys. Chem. **96**, 7466 (1992).

[12] A.I. Segal, private communication

[13] A.G. Merzhanov, and B.I. Khaikin, Prog. Ener. Combust. Sci. **14**, 1 (1988).

[14] G.M. Makhviladze, and B.V. Novozhilov, Zhurnal Prikl. Mech. Tech. Fiziki, No. **5**, 51 (1971) (in Russian).

[15] A.P. Aldushin, and S.G. Kasparyan, Soviet Physics - Doklady **24**, 29 (1979).

[16] B.J. Matkowsky, and G.I. Sivashinsky, SIAM J. Appl. Math.**35**, 465 (1978).

[17] K.G. Shkadinsky, B.I. Khaikin, and A.G. Merzhanov, Combustion, Explosion, and Shock Waves **7**, 15 (1971).

[18] A. Bayliss, B.J. Matkowsky, SIAM J. Appl. Math. **50**, 437 (1990).

[19] A.I. Volpert, V.A. Volpert, and V.A. Volpert, Translation of Math. Monographs **140** (AMS Providence 1994).

[20] Vit.A. Volpert, Vl.A. Volpert, S.P. Davtyan, I.N. Megrabova, and N.F. Surkov, SIAM J. Appl. Math. **52**, 368 (1992).

[21] S.B. Margolis, H.G. Kaper, G.K. Leaf, and B.J. Matkowsky, Combust. Sci. and Technol. **43**, 127 (1985).

[22] M. Garbey, A. Taik, and V. Volpert, Quart. Appl. Math. (to appear).

[23] M. Garbey, and V. Volpert, *Asymptotic and numerical computation of polymerization fronts*, Preprint No. 200, Unversité Lyon 1, Villeurbanne (1995).

[24] M. Garbey, A. Taik, and V. Volpert, Quart. Appl. Math. (to appear).

[25] L.D. Landau, and E.M. Lifshitz, *Fluid Mechanics* (Pergamon, Oxford UK/New York,1987).

[26] A.G. Istratov, and V.B. Librovich, J. Appl. Math. and Mech. **30**, 451 (1966). (in Russian)

[27] P. Clavin, Progr. Energy Comb. Sci. **11**, 1 (1985).

[28] Vit.A. Volpert, Vl.A. Volpert, J.A. Pojman, and S. Solovjov, Eur. J. Appl. Math. to appear

Modelling the Formation of N_2O and NO_2 in the Thermal De-NO_x Process

J.A. Miller
Combustion Research Facility, Sandia National Laboratories
Livermore, CA 94551-0969, USA

P. Glarborg
Department of Chemical Engineering,
Technical University of Denmark
2800 Lyngby, Denmark

Abstract

We have formulated a chemical kinetic model for the Thermal De-NO_x process that satisfactorily predicts the NO removed and the N_2O and NO_2 produced by the process over a range of temperatures and initial oxygen concentrations. The new feature of the mechanism is that NO_2 appears as an essential intermediate in the reaction scheme. It is formed as a consequence of NNH reacting with molecular oxygen,

$$NNH + O_2 \leftrightarrow N_2 + HO_2$$

$$HO_2 + NO \leftrightarrow NO_2 + OH,$$

and is converted back to NO by

$$NH_2 + NO_2 \leftrightarrow H_2NO + NO,$$

followed by $H_2NO \leftrightarrow HNO \leftrightarrow NO$. Nitrous oxide is produced by two different reactions,

$$NH_2 + NO_2 \leftrightarrow N_2O + H_2O$$

and

$$NH + NO \leftrightarrow N_2O + H.$$

The first is the primary source at high oxygen concentrations and the second is dominant for low O_2 levels. The branching fraction of the NH_2 + NO reaction (i.e. the fraction that produces NNH + OH) used in the model is $\alpha = 7.08 \times 10^{-4}\, T^{0.9}$, which above room temperature is somewhat higher than direct experimental determinations. The lifetime of NNH employed is $\tau_{NNH} = 10^{-7}$ sec, which is less than the upper limit set by experiment but still larger than the best theoretical prediction. All these points are discussed in detail.

1 Introduction

The Thermal De-NO_x process was developed by Richard Lyon at Exxon in the early 1970's and patented in 1975 [1]. It is one of three SNCR (selective non-catalytic reduction) schemes for nitrogen oxides (the others are RAPRENO$_x$, or cyanuric acid injection, and urea injection). Such after-treatment processes are commonly used on stationary combustion systems to control NO_x emissions. The Thermal De-NO_x process uses ammonia as the additive, and the complex reaction by which the ammonia reacts with nitric oxide has a number of fascinating properties that have prompted considerable research over the past 15 years or so.

Springer Series in Chemical Physics, Volume 61
Gas Phase Chemical Reaction Systems
Eds.: J. Wolfrum, H.-R. Volpp, R. Rannacher, and J. Warnatz

Some of the properties of interest are the following [1-7]:

1) The reaction requires oxygen. In the absence of oxygen the ammonia and nitric oxide do not react.
2) The reaction is self-sustaining, i.e. it does not require the addition of other fuel compounds to make it go.
3) Nitric oxide removal is possible only in a narrow temperature range centered at T $\approx$ 1250 K. At temperatures below 1100 K no reaction takes place, and at temperatures above 1400 K the ammonia is oxidized to form NO rather than destroy it.
4) If hydrogen (H_2) or hydrogen peroxide (H_2O_2) is added with the ammonia, the temperature window for NO removal moves to lower temperatures, but the width of the window remains unaltered.
5) The presence or absence of water has relatively little effect on the NO removal.
6) The reaction is not explosive [4]. It takes place relatively smoothly in the course of approximately 0.1 sec.

Miller, Branch, and Kee [5,6] were the first to attempt to explain these observations in terms of elementary reactions and to propose a quantitative mathematical model for the process. Miller and Bowman [7] subsequently updated, improved, and expanded upon the Miller-Branch-Kee model. Our point of departure is the Miller-Bowman mechanism. Their discussion of the important features of the mechanism is reasonably complete, so we shall give only a brief review here.

In combustion-exhaust products (i.e. with water present) ammonia is converted to NH_2 by reaction with hydroxyl,

$$NH_3 + OH \leftrightarrow NH_2 + H_2O.$$

Thus in order for the process to sustain itself, it must produce OH. Otherwise, what little OH might be present initially would be exhausted and the reaction would die out with little or no nitric oxide removed. It has been obvious from the outset that the NH_2 + NO reaction is key. This reaction must simultaneously remove NO and produce free radicals. It is clear both from experiment [21-27,31-33] and theory [10-14] that this reaction has two channels,

$$NH_2 + NO \leftrightarrow N_2 + H_2O$$

and

$$NH_2 + NO \leftrightarrow NNH + OH.$$

It is also clear from the argument given by Miller and Bowman that the radical channel must constitute at least 25% of the total reaction. This is because a hydrogen atom formed either directly or indirectly from NNH normally leads to three hydroxyls (note the need for O_2 to do so),

$$H + O_2 \leftrightarrow OH + O$$

$$O + H_2O \leftrightarrow OH + OH.$$

Thus the chain-terminating $N_2 + H_2O$ channel cannot be more than three times as fast as the chain-branching NNH + OH channel or the reaction would die out.

The process is limited at the high temperature end by the chain-branching cycle occurring too rapidly, thus producing hydroxyl concentrations that are so large that the reaction of NH_2 with OH is able to compete with the NH_2 + NO reaction. There ensues the sequence

$$NH_2 + OH \leftrightarrow NH + H_2O$$

$$NH + O_2 \leftrightarrow NO + OH$$

or

$$NH + O_2 \leftrightarrow HNO + O$$

followed by

$$HNO + M \leftrightarrow H + NO + M$$

$$HNO + OH \leftrightarrow NO + H_2O,$$

which produces NO. Therefore, at sufficiently high temperature (typically at $T \approx 1400$ K) there is a net increase in the nitric oxide.

The requirement that the reaction not be explosive is a very subtle one. Essentially it requires that the chain-carrier growth be self-limiting in some way. This constraint is related to the dependence of the NO removal on H_2 addition [4-7], water concentration [4-7], and O_2 level (this work) in that, when the reaction is explosive, these other effects also are not predicted accurately by the model. When NNH simply dissociates rapidly, i.e.

$$NNH \leftrightarrow N_2 + H,$$

chain-carrier growth is inevitably explosive. In the Miller-Bowman and Miller-Branch-Kee models the dominant NNH reaction is

$$NNH + NO \leftrightarrow N_2 + HNO.$$

As the reaction proceeds in time, the HNO initially dissociates and produces the branched-chain sequence. However, the hydroxyl concentration eventually reaches a point where the chain-terminating reaction,

$$OH + HNO \leftrightarrow H_2O + NO,$$

is able to compete with HNO dissociation. The branched-chain sequence is thus self-limiting, i.e. the faster the chain branching occurs, the sooner the chain termination step comes into play to slow it down.

The requirement from the modelling that NNH live long enough to react with NO places a lower limit on the NNH lifetime – according to Glarborg *et al.* [8], this limit is about 10^{-6} sec.

A great deal of our understanding of the Thermal De-NO_x process, and the NH_2 + NO reaction in particular, has come from *ab initio* electronic structure calcula-

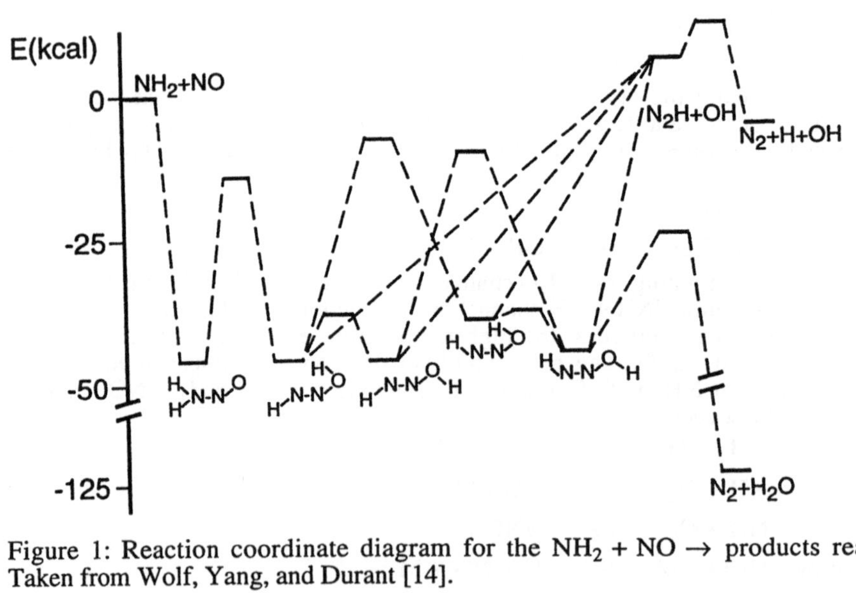

Figure 1: Reaction coordinate diagram for the NH_2 + NO → products reaction. Taken from Wolf, Yang, and Durant [14].

tions [9-17]. The pivotal contribution came from the BAC-MP4 calculations on NH_2 + NO by Melius and Binkley [12]. Subsequent calculations by different methods have produced some minor quantitative differences in the potential energy surface, but the basic conclusions to be drawn from the work are the same. Figure 1 shows a reaction coordinate diagram from the most recent calculations of Durant and co-workers [14]. The figure shows clearly the complex sequence of rearrangements required for the initial NH_2NO adduct to produce $N_2 + H_2O$ (the major channel) with no intrinsic energy barrier, and it shows the existence of a slightly endothermic radical channel, NNH + OH, which can occur through dissociation of any one of a number of intermediate adducts.

The properties of the potential energy surface can be used to explain, at least qualitatively, a number of features of the reaction observed experimentally. The rate coefficient is relatively large and decreases with increasing temperature, consistent with an addition-rearrangement reaction with no intrinsic energy barrier. The rate coefficient also is independent of pressure [18] from a few Torr to almost an atmosphere. Phillips [19] has calculated lifetimes of the intermediate complexes from a potential energy surface similar to that of Fig. 1 and has shown that they are $\sim 10^{-11}$ sec. This is considerably shorter than the time between collisions at pressures and temperatures of interest, so significant stabilization is impossible, consistent with the experimental observation. The branching fraction of the reaction α (i.e. the fraction that produces NNH + OH) is ~ 0.1 at room temperature and increases with increasing temperature [20-27]. This is consistent with the $N_2 + H_2O$ channel being favored energetically.

However, two important inconsistencies between the Miller-Bowman model and fundamental theory and experiment have arisen:

i) Direct experimental measurements of α at temperatures higher than room temperature generally indicate that $\alpha < 0.25$ at temperatures of interest for Thermal De-NO_x (see below). We shall not address this issue here except to point out that Diau *et al.* [28] have called into question much of the experimental work on the products of the NH_2 + NO reaction because of the potential influence of secondary reactions.

ii) Recent experimental [29] and theoretical [16,30] work indicates that τ_{NNH}, the lifetime of NNH, is less than 10^{-6} sec. This is the point we want to discuss.

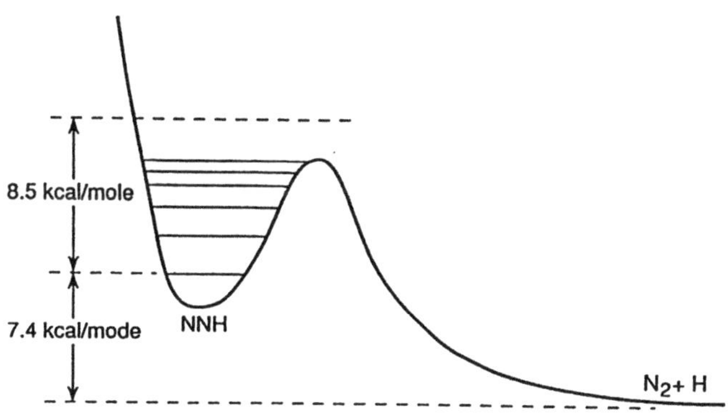

Figure 2: Reaction coordinate diagram for the dissociation of NNH. Drawn from calculations of Walch, Duchovic, and Rohlfing [16]. The uppermost dashed line indicates the energy of the ground vibrational state at the saddle point.

Walch and co-workers [16,17] have calculated a potential energy surface for the dissociation of NNH. A reaction-coordinate diagram based on their calculations is shown in Fig. 2. The radical dissociates exothermically and can do so even in the ground vibrational state by tunneling through the barrier shown in the diagram. This makes the lifetime very short. Curtiss *et al.* [15] were the first to call attention to this point and estimated the lifetime (in the ground vibrational state) to be $\tau_{NNH} \approx 10^{-11}$ sec. Walch *et al.* [16] made more accurate calculations of the potential for NNH and estimated the lifetime (by one-dimensional tunneling through an Eckart barrier defined by their potential) to be 5×10^{-9} sec. Koizumi *et al.* [30], using essentially the same potential as Walch *et al.*, have calculated $\tau_{NNH} = 3 \times 10^{-9}$ sec. from a complete quantum scattering calculation. Clearly, the best theoretical predictions indicate that τ_{NNH} is in the range 10^{-9}–10^{-8} sec.

Experimental results on τ_{NNH} are contradictory. Early experiments on the products of the NH_2 + NO reaction indicated that OH was a product but that H atom was not [31-33]. From the conditions of these experiments one can place lower limits on τ_{NNH}. In this way Miller and Bowman inferred $\tau_{NNH} \geq 10^{-4}$ sec. However, if NNH does not dissociate in these experiments (thus producing the hydrogen atom), it must react with NO (which is present in large excess) to produce HNO,

$$NNH + NO \leftrightarrow N_2 + HNO.$$

Unfried *et al.* [34] attempted to detect HNO in such experiments but were unsuccessful, thus creating a dilemma. However, the validity of all these experiments has recently been questioned by Diau *et al.* [28], and the fate of the NNH in such experiments appears still to be unresolved.

Selgren *et al.* [29] have taken a more direct approach, forming NNH by neutralizing a beam of NNH^+ ions. Their experiments indicate that NNH has a lifetime of less than 0.5 μsec, generally in agreement with the theoretical predictions described above. While the results of these experiments may be compromised somewhat by the formation of excited electronic states of NNH, it appears likely that τ_{NNH} is smaller than the lower limit given by Glarborg *et al.* [8].

In the Miller-Bowman model the only way that such a short lifetime for NNH might be accommodated (and an explosive reaction avoided) is if NNH reacts rapidly with a major stable species (not just with NO). The only legitimate candidate is molecular oxygen,

$$NNH + O_2 \leftrightarrow HO_2 + NO.$$

However, such a reaction leads to conversion of a large part of the NO to NO_2 via

$$HO_2 + NO \leftrightarrow NO_2 + OH,$$

in conflict with experimental observations that no NO_2 is produced in the Thermal De-NO_x process, at least under normal operating conditions [2-4].

Two recent developments have changed the situation just described. First, Glarborg *et al.* [35] have studied the oxidation of ammonia by NO_2 in the temperature range 850 K < T < 1400 K. They find that NO is the principal product of this reaction and propose that the oxidation takes place through a branched-chain sequence in which the key step is

$$NH_2 + NO_2 \leftrightarrow H_2NO + NO,$$

followed by conversion of the H_2NO to NO. This result suggests the possibility that NO_2 may be formed as a transient intermediate in Thermal De-NO_x experiments and converted back to NO by the mechanism proposed by Glarborg *et al.* The second development is that Kasuya *et al.* [36] have performed a series of Thermal De-NO_x experiments over a wide range (up to 50%) of initial oxygen concentrations. At the

Table 1: Selected rate coefficients from reaction mechanism, $k = A\, T^{\beta} \exp(-E/RT)$. Units are moles, cm^3, sec., K, and cal/mole.

	Reaction	A	β	E
61.	$NH_2 + NO \leftrightarrow NNH + OH$	8.92E12	-0.35	0.0
62.	$NH_2 + NO \leftrightarrow N_2 + H_2O$	1.26E16	-1.25	0.0
63.	$NH_2 + NO \leftrightarrow N_2 + H_2O$	-8.92E12	-0.35	0.0
68.	$NNH \leftrightarrow N_2 + H$	1.0E7	0.0	0.0
76.	$NNH + O_2 \leftrightarrow N_2 + HO_2$	2.0E14	0.0	0.0
77.	$NNH + O_2 \leftrightarrow N_2 + H + O_2$	5.0E13	0.0	0.0
78.	$H_2NO + O \leftrightarrow NH_2 + O_2$	2.5E14	0.0	0.0

highest oxygen levels they detect significant quantities of NO_2 in the products. They also measure the N_2O concentrations in the same set of experiments. Such a data set is ideal for testing our proposal that NO2 is an essential intermediate in the Thermal De-NO_x process.

The purpose of the present investigation now is clear. We wish to construct a chemical kinetic model for Thermal De-NO_x in which the lifetime of NNH is short, at least less than the upper limit set by the experiments of Selgren *et al.* We want to test this model against the experiments of Kasuya *et al.* for formation of NO_2 and N_2O, all as for NO removal.

2 Computational Details and Reaction Mechanism

In the present investigation we restrict our attention to modelling the experimental results of Kasuya *et al.* [36]. These investigators measured the concentrations of NO, NO_2, N_2O and NH_3 at the exit of a specially designed, quartz plug-flow reactor oper-

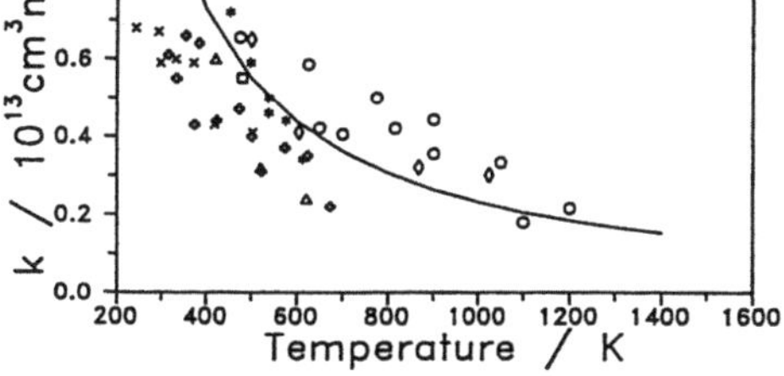

Figure 3: Rate coefficient for the NH_2 + NO reaction.

ating at atmospheric pressure. They employed a variety of initial NO, NH_3, and O_2 concentrations. Most importantly from our point of view, they varied the initial level of molecular oxygen from 0 to 50%. Because all their experiments were conducted at the same total mass flow rate, the residence time in the reactor was temperature-dependent, $\tau = 88K/T$ sec.

Our calculations simulating the experimental conditions of Kasuya *et al.* were performed with SENKIN [38], which runs in conjunction with CHEMKIN-II [39]. The thermodynamic data were taken from the Chemkin thermodynamic database [40]. The reaction time in the calculations was varied with temperature according to the expression given above.

The reaction mechanism used in the calculations is the same as that given by Glarborg *et al.* [35] with some significant exceptions. These exceptions are listed in Table I. The reaction numbers in the table correspond to the ordering of reactions in our Livermore computer file. The numbering in Table I of Ref. 35 is different, but the rate coefficients of all the significant chemical reactions are the same.

As discussed in the Introduction, the key reaction in the Thermal De-NO_x process is the reaction between NH_2 and NO. The total rate coefficient, k_{tot}, now has been measured many times, although relatively few studies have involved any significant range of temperatures. In Fig. 3 we plot the experimental data for k_{tot} from the most recent studies along with the rate expression used by Miller and Bowman [7] in their review. Clearly, the Miller-Bowman expression is still a very good representation of the experimental data available. Consequently we adopt it here. The rate coefficient may be expressed as

$$k_{tot} = 1.26 \times 10^{16}\ T^{-1.25}\ cm^3/mole\text{-}sec.$$

Of more importance to the Thermal De-NO_x mechanism than k_{tot} is the branching fraction α, i.e. the fraction of the total rate coefficient in which the NH_2+NO reaction

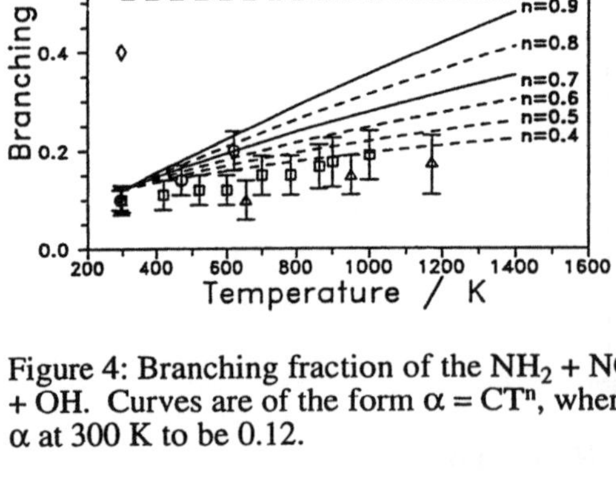

Figure 4: Branching fraction of the NH_2 + NO reaction, i.e. fraction that forms NNH + OH. Curves are of the form $\alpha = CT^n$, where C is determined by fixing the value of α at 300 K to be 0.12.

produces NNH + OH. Figure 4 is a plot of a number of experimental results for α as a function of temperature. Also plotted is a family of curves of the form:

$$\alpha = CT^n, \quad \text{(II.1)}$$

where, for any value of n, C has been chosen so that $\alpha = 0.12$ at 300 K. This value of α, or something very close to it, is reasonably well established now at room temperature. The experimental results for α appear to be most consistent with a value of $n \approx 0.4$. However, as discussed by Glarborg, *et al.* [8], any value of $n < 0.7$ results in model predictions that show no NO removal at Thermal De-NO_x temperatures (1100 K < T < 1400 K), in disagreement with a large body of flow reactor data. For the present investigation we have adopted n = 0.9. How we arrived at this value is explained in the Results and Discussion section below.

For any value of n the rate coefficient k_{61} for reaction (R61),

$$NH_2 + NO \leftrightarrow NNH + OH, \quad \text{(R61)}$$

is expressed as

$$k_{61} = \alpha k_{tot}, \quad \text{(II.2)}$$

and the rate coefficient for the $N_2 + H_2O$ channel is $k_{tot} - \alpha\, k_{tot}$. This latter expression is introduced into Chemkin as two separate reactions, (R62) and (R63), one with a positive rate coefficient and one with a negative rate coefficient (the same thing is done for the $NH + NO \leftrightarrow N_2O + H$ reaction, (R41) and (R42)). Any sensitivity coefficient for the reaction $NH_2 + NO \leftrightarrow N_2 + H_2O$, or the instantaneous rate of the reaction, is obtained by adding algebraically the corresponding sensitivity coefficients or rates for the individual CHEMKIN reactions (R62) and (R63). This statement is obviously true for rates. Although we shall not prove it here, it is also true for sensitivity coefficients, but perhaps not obviously so.

Two other reactions that play an important role in the present model are

$$NNH + O_2 \leftrightarrow N_2 + HO_2 \quad \text{(R76)}$$

and

$$NNH + O_2 \leftrightarrow N_2 + H + O_2. \quad \text{(R77)}$$

Reaction (R77) is viewed as being a special case of (R76) in which an HO_2^* is formed with enough vibrational energy to dissociate spontaneously, $HO_2^* \rightarrow H + O_2$. Reaction (R77) is approximately 8 kcal/mole exothermic. The values of k_{76} and k_{77} used in the model are closely tied to the value of k_{68},

$$NNH \leftrightarrow N_2 + H, \quad \text{(R68)}$$

where $\tau_{NNH} = 1/k_{68}$ is the lifetime of NNH. This point is discussed in detail below.

The last reaction listed in Table I is reaction (R85),

$$O + H_2NO \leftrightarrow NH_2 + O_2. \quad \text{(R85)}$$

In our calculations this reaction goes in the reverse direction. It plays a role only as a source of chain carriers for the cases with very high initial O_2 concentration.

Recently Hennig *et al.* [41] have studied the $NH_2 + O_2$ reaction in a shock tube in the temperature range 1450–2300 K. They deduced rate coefficients for the three different channels,

$$NH_2 + O_2 \leftrightarrow H_2NO + O$$

$$NH_2 + O_2 \leftrightarrow HNO + OH$$

$$NH_2 + O_2 \leftrightarrow NO + H_2O,$$

concluding that the molecular channel ($NO + H_2O$) is dominant. We have attempted to use their results in our model. However, when we replace our reaction (R85) with their three-channel process, including their rate coefficients, we find that NO is produced rather than destroyed at all temperatures in our flow-reactor calculations with 50% initial O_2. This is in complete disagreement with the experiments. Nitric oxide is produced in the calculations because the $NH_2 + O_2$ reactions are so fast that they dominate the $NH_2 + NO$ reactions in the competition for NH_2. Consequently we have not adopted the rate coefficients of Hennig *et al.* in the present model. However, the value of k_{85} used here is somewhat larger than we have used in previous modelling.

3 Results and Discussion

Figures 5, 6, and 7 summarize the results of the present investigation. Figure 5 compares our model predictions for NO with the experimental results of Kasuya *et al.* Fig. 6 is a similar comparison for NO_2, and Fig. 7 is for N_2O. The comparisons shown are for the single set of initial concentrations for which Kasuya *et al.* obtained a complete set of data. All three figures contain (a) and (b) parts. In each case the (a) figure is the experimental data and the (b) figure is the model prediction. Plotting the results in this way facilitates the comparison of changes in the temperature dependence of product (NO, NO_2, N_2O) concentrations with changes in initial oxygen level.

From Fig. 5 one can see that the "initiation temperature", i.e. the temperature at which NO first begins to disappear (to be precise, we shall define it as the lowest temperature at which 10% of the NO is removed) decreases in a systematic fashion as the initial oxygen concentration increases. At the same time, the maximum amount of NO removed and the temperature at which this occurs both decrease with increasing O_2 concentrations. In other words, the minima in the NO_f/NO_i curves move upward and to the left with increasing levels of molecular oxygen. All these trends are predicted accurately by the model. The initiation temperatures are so sensitive to the critical parameters in the kinetic model that we have used these temperatures in a systematic manner to determine the kinetic parameters given in Table I.

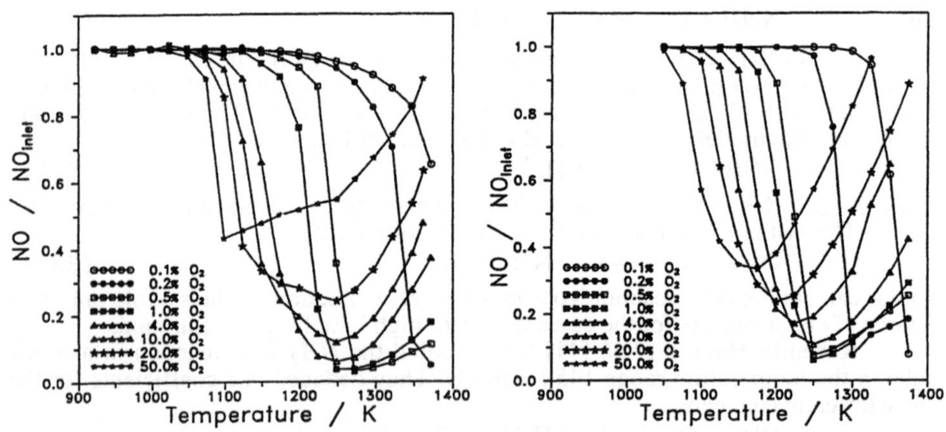

Figure 5: Experimental and theoretical NO concentrations as a function of temperature for various initial O_2 concentrations (flow reactor conditions of Kasuya *et al.* [36]). Initial conditions: NO = 500 ppm, NH_3 = 1000 ppm, H_2O = 5%, O_2 concentrations as indicated, balance N_2. Residence time = 88 K/T sec. (a) experimental results, (b) theoretical predictions.

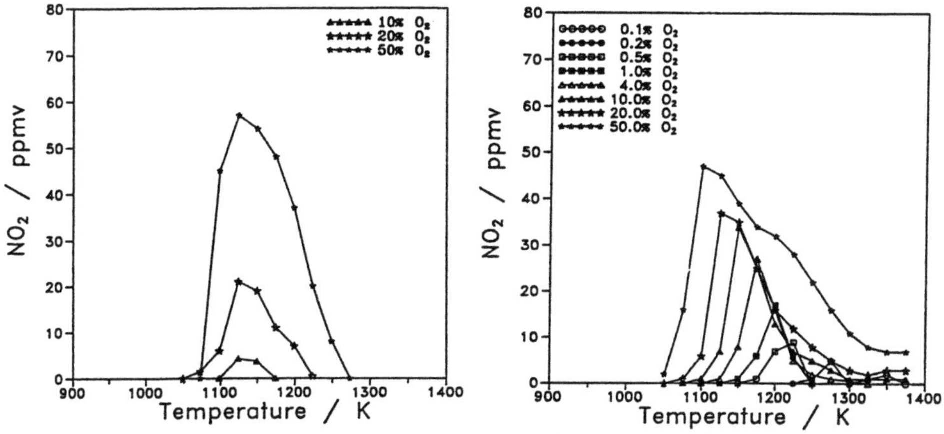

Figure 6: Experimental and theoretical NO_2 concentrations as a function of temperature for various initial O_2 concentrations (flow reactor conditions of Kasuya *et al.* [36]). Initial conditions: NO = 500 ppm, NH_3 = 1000 ppm, H_2O = 5%, O_2 concentrations as indicated, balance N_2. Residence time = 88 K/T sec. (a) experimental results, (b) theoretical predictions.

As discussed in the Introduction, a primary objective of the present investigation is to develop a kinetic model in which the lifetime of NNH is as close as possible to recent experimental and theoretical determinations. In particular, we want to restrict the lifetime of NNH to values smaller than 0.5 μsec, the upper limit placed on it by the experiments of Selgren *et al.* [29] and to get as close as possible to the value of

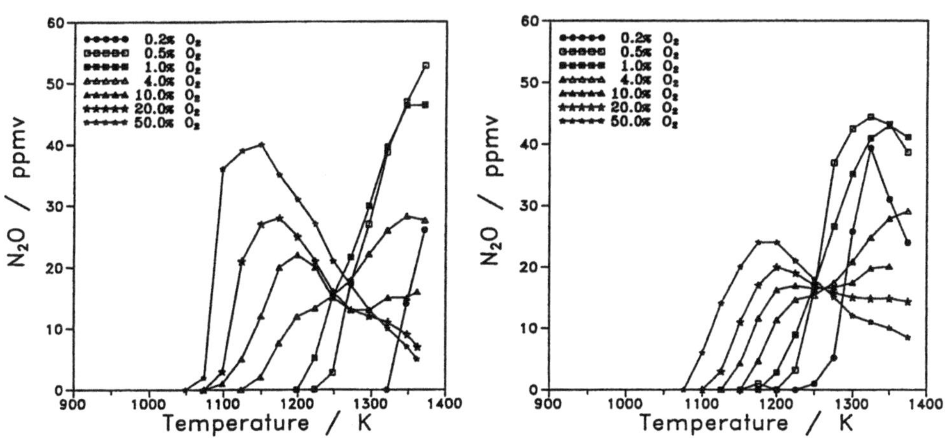

Figure 7: Experimental and theoretical N_2O concentrations as a function of temperature for various initial O_2 concentrations (flow reactor conditions of Kasuya *et al.* [36]). Initial conditions: NO = 500 ppm, NH_3 = 1000 ppm, H_2O = 5%, O_2 concentrations as indicated, balance N_2. Residence time = 88 K/T sec.; (a) experimental results, (b) theoretical predictions.

$\tau_{NNH} = 3 \times 10^{-9}$ sec obtained theoretically by Koizumi *et al.* [30]. To accomplish this objective requires a very fast reaction between NNH and O_2,

$$NNH + O_2 \leftrightarrow N_2 + HO_2. \qquad (R76)$$

Molecular oxygen is the only species present in sufficient quantities, under normal Thermal De-NO_x conditions, to react with NNH and compete with spontaneous dissociation,

$$NNH \leftrightarrow N_2 + H. \qquad (R68)$$

Spontaneous dissociation leads to unconstrained chain-carrier growth and a branched-chain "explosion". Such a property is not observed experimentally [4] and, when it is present in the model, many features of the overall reaction cannot be predicted, e.g. dependence on initial H_2 and H_2O concentrations [4-7] and, in the present investigation, dependence on initial oxygen levels. Therefore we have assumed a very large value for $k_{76} = 2 \times 10^{14}$ cm^3/mole-sec. (very few reactions between neutral species are faster). In fact, such a large value of the rate coefficient is probably to be expected for such an exothermic ($\Delta H° \approx -58$ kcal/mole) hydrogen abstraction.

Even with this large value of k_{76}, in the 0.1% – O_2 case of Fig. 5 virtually all the NNH dissociates. Consequently, the initiation temperature for this case is determined completely by the branching fraction of the NH_2 + NO reaction, and we can use this case to determine α, or n in Eq. (II.1) to be more precise. In this way we determined n = 0.9. Even a value of n as high as 0.80 results in an initiation temperature for the 0.1% – O_2 case in excess of 1400 K, in disagreement with the experiments.

Once we have established n and k_{76} the competition between (R76) and (R68) determines the rate at which the initiation temperature drops with increasing oxygen concentration, at least for the smaller values of initial O_2. Using this observation we determined $k_{68} = 10^7$/sec., or $\tau_{NNH} = 10^{-7}$ sec. With this value of k_{68} virtually all the NNH reacts with O_2 by the time the initial oxygen concentration is as high as 4%. To illustrate the sensitivity of the initiation temperatures to τ_{NNH} we have computed product distributions for the 1% - O_2 case with $\tau_{NNH} = 10^{-7}$ sec. and with $\tau_{NNH} = 10^{-8}$ sec. With the longer lifetime the predicted (and experimental) initiation temperature is approximately 1180 K. With the shorter lifetime the initiation temperature drops dramatically to 1080 K.

The rate coefficients for reactions (R85) and (R77) are important only for the high-oxygen cases, and their values were chosen to obtain good agreement for the initiation temperatures in the cases of initial O_2 concentration between 4% and 50%. These are "softer" numbers than the others given in Table I, and to a limited extent it is possible to increase one and decrease the other without modifying the results too drastically.

With all the rate parameters chosen, we can depict the reaction mechanism diagrammatically as in Fig. 8. The most important change in the mechanism from that described by Miller and Bowman [7] is that NO_2 is now an essential intermediate in the process. Nitrogen dioxide is formed by the sequence,

$$NNH + O_2 \leftrightarrow N_2 + HO_2 \qquad (R76)$$

$$HO_2 + NO \leftrightarrow NO_2 + OH \qquad (R25),$$

and converted back to NO through the fast reaction,

$$NH_2 + NO_2 \leftrightarrow H_2NO + NO, \qquad (R30)$$

followed by

$$H_2NO \rightarrow HNO \rightarrow NO,$$

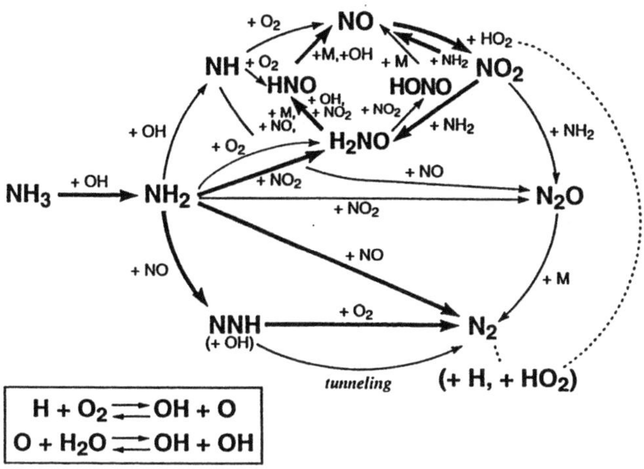

Figure 8: Reaction path diagram for the Thermal De-NO_x process. The bold arrows represent the dominant paths.

as shown in Fig. 8. The chain branching (i.e. H-atom production) required to sustain the process comes from several sources: the spontaneous dissociation of NNH, the reaction between NNH and O_2, (R77) NNH + $O_2 \leftrightarrow N_2 + H + O_2$, the dissociation of HNO, and the dissociation of H_2NO. An "explosive" branched-chain reaction is avoided as before by reaction (R79),

$$HNO + OH \quad \leftrightarrow \quad NO + H_2O, \qquad (R79)$$

and to a lesser extent by

$$H_2NO + OH \leftrightarrow \quad HNO + H_2O. \qquad (R90)$$

As the reaction progresses in time the radical concentrations, particularly that of OH, grow to the point where reactions (R79) and (R90) are competitive with HNO and H_2NO dissociation, thus inhibiting the chain carrier growth. As before, the "turn-up" in the NO_f/NO_i curves of Fig. 5 comes as a consequence of the OH concentrations reaching a level where reaction (R57),

$$NH_2 + OH \quad \leftrightarrow \quad NH + H_2O, \qquad (R57)$$

can compete with the NH_2 + NO reaction for the NH_2. This results in NO formation, rather than destruction, as shown in Fig. 8.

As noted above, increasing levels of molecular oxygen cause the minima of the NO_f/NO_i curves in Fig. 5 to move up and to the left. This trend is predicted extremely well by the model. The reduction of the peak NO-removal efficiency with increased oxygen concentration is a direct consequence of the participation of NO_2 in the process. As the initial O_2 level is increased, there is more NO_2 formed and a greater likelihood that NH_2 is oxidized to NO by NO_2, i.e.

$$NH_2 + NO_2 \leftrightarrow H_2NO + NO, \qquad (R29)$$

followed by the sequence $H_2NO \rightarrow HNO \rightarrow NO$, rather than reacting with NO to produce N_2.

The agreement between the NO_2 concentrations predicted by the model and those observed experimentally is quite good for the 50% – O_2 and 20% – O_2 cases, as shown in Fig. 6. For 10% O_2 and less, the predicted NO_2 substantially exceeds that shown in the experimental results of Fig. 6. However, this is a little misleading. For the cases where there is significant disagreement between the model and experiment, the experimental results (and in most instances the predicted results as well) are significantly smaller than the uncertainty in the measurements themselves, i.e. ± 20 ppm. Consequently, we conclude that the overall agreement is quite good.

An important point to note about the NO_2 experimental results is that, as a function of temperature, the NO_2 concentrations rise as the NO concentrations fall, and *vice versa*. Actually, the peak NO_2 concentration occurs generally at a temperature slightly lower than the minimum in the corresponding NO_f/NO_i curve. These results coupled with the relatively large peak NO_2 concentration observed in the 50% – O_2 case (60 ppm NO_2 with 500 ppm NO present initially) suggest strongly that the NO_2 formation mechanism is intimately tied to the NO removal process. The good agreement between model and experiment in this regard is strong support for the proposed mechanism.

The dependence of the N_2O results on temperature and initial O_2 level is even more enlightening than that for NO_2. There are two possible sources of N_2O in the mechanism,

$$NH + NO \leftrightarrow N_2O + H \qquad \text{(R41, R42)}$$

and

$$NH_2 + NO_2 \leftrightarrow N_2O + H_2O. \qquad \text{(R29)}$$

For large initial O_2 concentrations reaction (R41, R42) is dominated by the NH + O_2 reaction,

$$NH + O_2 \leftrightarrow HNO + O \qquad \text{(R39)}$$

$$NH + O_2 \leftrightarrow NO + OH, \qquad \text{(R40)}$$

in the competition for NH, and consequently N_2O formation from the NH + NO reaction is not very important. Under these conditions N_2O is formed predominantly by the reaction of NH_2 with NO_2, and the N_2O concentrations in Fig. 7 track those of NO_2 in Fig. 6 fairly closely. In these cases the N_2O concentration peaks in the vicinity of the minimum in the corresponding NO_f/NO_i curve.

For small initial O_2 levels, less NO_2 is formed than in the high-O_2 cases. This restricts the amount of N_2O that can be formed by the $NH_2 + NO_2$ reaction. At the same time reaction (R41, R42), $NH + NO \leftrightarrow N_2O + H$, can compete more favorably with the NH + O_2 reaction than it can in the high-O_2 cases and becomes the dominant source of N_2O. However, N_2O formation in these cases becomes NH-limited, i.e. N_2O is formed only when NH is formed. From Fig. 8 one can see that NH is produced only through the oxidation sequence that causes the "turn-up" in the NO_f/NO_i curves. Consequently, significant quantities of N_2O are produced only at temperatures where this oxidation sequence is effective, i.e. at temperatures higher than that at which the minimum in the NO_f/NO_i curve occurs. One can see this effect by comparing Figs. 5 and 7.

All the effects discussed above are predicted by the model. The various correlations among the NO_f/NO_i, NO_2, and N_2O results are intimately tied to the details of the model. The quantitative prediction of these correlations is strong evidence that the basic structure of the mechanism is correct and that even the individual rate coefficients are generally accurate.

4 Concluding Remarks

We have developed a chemical kinetic model for the Thermal De-NO_x process (based on the previous mechanism of Miller and Bowman [7] and the NH_3/NO_2 mechanism of Glarborg *et al.* [35]) that satisfactorily predicts the NO_2 and N_2O produced, as well as the NO removed, in the process over a range of temperatures and initial oxygen concentrations. The new feature of the mechanism is that NO_2 appears as an essential intermediate in the reaction under most conditions. Nitrogen dioxide is formed through the sequence,

$$NNH + O_2 \leftrightarrow N_2 + HO_2 \qquad (R76)$$

$$HO_2 + NO \leftrightarrow NO_2 + OH, \qquad (R25)$$

and is rapidly converted back to NO by reaction with NH_2,

$$NH_2 + NO_2 \leftrightarrow H_2NO + NO, \qquad (R30)$$

followed by $H_2NO \rightarrow HNO \rightarrow NO$. Thus relatively little NO_2 accumulates except under high oxygen concentrations.

There are two sources of nitrous oxide in the mechanism. Under high oxygen concentrations N_2O is formed from NO_2,

$$NH_2 + NO_2 \leftrightarrow N_2O + H_2O, \qquad (R29)$$

and under low oxygen concentrations N_2O is formed from the NH + NO reaction,

$$NH + NO \leftrightarrow N_2O + H. \qquad (R41, R42)$$

At intermediate O_2 levels both reactions contribute to the N_2O produced.

The lifetime of NNH used in the model is $\tau_{NNH} = 10^{-7}$ sec., which is less than the upper limit of 0.5 μsec. set by the experiments of Selgren *et al.* [29] but larger than the best theoretical estimate of 3×10^{-9} sec. [30]. In the kinetic model the only way that a smaller value of τ_{NNH} can be used successfully is with a larger value of the rate coefficient for the NNH + O_2 reaction. The value of this rate coefficient is already extremely large, $k_{76} = 2 \times 10^{14}$ cm^3/mole-sec. Although significant progress has been made in resolving the difference between theoretical predictions of τ_{NNH} and values used in Thermal De-NO_x modelling (at one point the best theoretical value was 10^{-11} sec. and the value used in modelling was 10^{-4} sec.) there is a remaining discrepancy of between one and two orders of magnitude. However, the lifetime for such a "hanging potential well" is normally quite sensitive to the height of the potential barrier to dissociation (and perhaps to other features of the potential as well). It is quite possible that an error in the barrier height as small as 1 or 2 kcal/mole could be the main source of the present discrepancy.

5 Acknowledgement

This work was supported by the United States Department of Energy, Office of Basic Energy Sciences, Division of Chemical Sciences and by the CHEC (Combustion and Harmful Emission Control) program. The CHEC program is co-funded by the Danish Technical Research Council, Elsam (the Jutland Funen Electricity Consortium), Elkraft (The Zealand Electricity Consortium) and the Danish Ministry of Energy.

6 References

[1] R. K. Lyon, U. S. Patent 3,900,554 (1975).
[2] R. K. Lyon, Int. J. Chem. Kinet. **8**, 315 (1976).
[3] R. K. Lyon and D. Benn, *Seventeenth Symposium (International) on Combustion,* pp 601, The Combustion Institute, Pittsburgh, PA (1979).
[4] R. K. Lyon and J. E. Hardy, Ind. Eng. Chem. Fundam. **25**, 19 (1986).

[5] J. A. Miller, M. C. Branch, and R. J. Kee, "A Chemical Kinetic Model for the Selective Reduction of Nitric Oxide by Ammonia", Sandia National Laboratories Report SAND 80-8635 (1980).
[6] J. A. Miller, M. C. Branch, and R. J. Kee, Comb. Flame **43**, 81 (1981).
[7] J. A. Miller and C. T. Bowman, Prog. Energy Comb. Sci. **15**, 287 (1989).
[8] P. Glarborg, K. Dam-Johansen, J. A. Miller, R. J. Kee, and M. E. Coltrin, Int. J. Chem. Kin. **26**, 421 (1994).
[9] C. J. Casewit and W. A. Goddard, J. Am. Chem. Soc. **104**, 3280 (1982).
[10] J. A. Harrison, R. G. A. R. Maclagan, and A. R. Whyte, J. Phys. Chem. **91**, 6683 (1987).
[11] H. Abou-Rachid, C. Pouchan, and M. Chaillet, Chem. Phys. **90**, 243 (1984).
[12] C. F. Melius and J. S. Binkley, *Twentieth Symposium (International) on Combustion,* p. 575, The Combustion Institute, Pittsburgh, PA (1985)
[13] S. P. Walch, J. Chem. Phys. **99**, 5295 (1993).
[14] M. Wolf, D. L. Yang, and J. L. Durant, J. Photochem. Photobiol. A. Chem. **80**, 85 (1994).
[15] L. A. Curtiss, D. L. Drapcho, and J. A. Pople, Chem. Phys. Lett. **103**, 437 (1984).
[16] S. P. Walch, R. J. Duchovic, and C. M. Rohlfing, J. Chem. Phys. **90**, 3230 (1989).
[17] S. P. Walch, J. Chem. Phys. **93**, 2384 (1990).
[18] R. Lesclaux, P. V. Khé, P. DeZauzier, and J. C. Soulignac, Chem. Phys. Lett. **35**, 493 (1975).
[19] L. F. Phillips, Chem. Phys. Lett. **135**, 269 (1987).
[20] L. J. Stief, W. D. Brobst, D. F. Nava, R. P. Borkowski, and J. V. Michael, J. Chem. Soc. Faraday Trans. II **78**, 1391 (1982).
[21] B. Atakan, A. Jacobs, M. Wahl, R. Weller, and J. Wolfrum, Chem. Phys. Lett. **155**, 609 (1989).
[22] B. Atakan, J. Wolfrum, and R. Weller, Ber. Bunsenges. Phys. Chem. **94**, 1372 (1990).
[23] P. Pagsberg, B. Sztuba, E. Ratajczak, and A. Sillesen, Acta Chemica Scandinavica **45**, 329 (1991).
[24] D. A. Dolson, J. Phys. Chem. **90**, 6714 (1986).
[25] V. P. Bulatov, A. A. Ioffee, V. A. Lozovsky, and O. M. Sarkisov, Chem. Phys. Lett. **161**, 141 (1989).
[26] J. L. Hall, D. Zeitz, J. W. Stephens, J. V. V. Kasper, G. P. Glass, R. F. Curl, and F. K. Tittel, J. Phys. Chem. **90**, 2501 (1986).
[27] J. W. Stephens, C. L. Morter, S. K. Farhat, G. P. Glass, and R. F. Curl, J. Phys. Chem. **97**, 8944 (1993).
[28] E. W. Diau, T. Yu, M. A. G. Wagner, and M. C. Lin, J. Phys. Chem. **98**, 4034 (1994).
[29] S. F. Selgren, P. W. McLoughlin, and G. I. Gellene, J. Chem. Phys. **90**, 1624 (1989).
[30] H. Koizumi, G. C. Schatz, and S. P. Walch, J. Chem. Phys. **95**, 4130 (1991).
[31] P. Andresen, A. Jacobs, C. Kleinermanns, and J. Wolfrum, *Nineteenth Symposium (International) on Combustion*, p. 11, The Combustion Institute, Pittsburgh, PA (1983).
[32] J. A. Silver and C. E. Kolb, J. Phys. Chem. **86**, 3249 (1982).
[33] J. A. Silver and C. E. Kolb, J. Phys. Chem. **91**, 3713 (1987).
[34] K. G. Unfried, G. P. Glass, and R. F. Curl, Chem. Phys. Lett. **173**, 337 (1990).

[35] P. Glarborg, K. Dam-Johansen, and J. A. Miller, "The Reaction of Ammonia with Nitrogen Dioxide in a Flow Reactor: Implications for the $NH_2 + NO_2$ Reaction" Int. J. Chem. Kinet. (in press).
[36] F. Kasuya, P. Glarborg, J. E. Johnsson, and K. Dam-Johansen, Chem. Eng. Sci **30**, 1455-1466 (1995).
[37] W. Hack, H. Schacke, H. Schroeter, and H. Gg. Wagner, *Seventeenth Symposium (International) on Combustion*, p. 505, The Combustion Institute, Pittsburgh, PA (1979).
[38] A. Lutz, R. J. Kee, and J. A. Miller, "SENKIN: A Fortran Program for Predicting Homogeneous Gas Phase Chemical Kinetics with Sensitivity Analysis", Sandia National Laboratories Report SAND87-8248 (1987).
[39] R. J. Kee, F. M. Rupley, and J. A. Miller, "CHEMKIN-II: A Fortran Chemical Kinetics Package for the Analysis of Gas-Phase Chemical Kinetics", Sandia National Laboratories Report SAND89-8009 (1989).
[40] R. J. Kee, F. M. Rupley, and J. A. Miller, "The Chemkin Thermodynamic Database", Sandia National Laboratories Report SAND 87-8215B (1990).
[41] G. Hennig, M. Klatt, B. Spindler, and H. Gg. Wagner, Ber. Bunsenges Phys. Chem. **99**, 651 (1995).

Simplifying Chemical Kinetics Using Intrinsic Low-Dimensional Manifolds

U. Maas
Konrad-Zuse-Zentrum für Informationstechnik
Heilbronner Str. 10, 10711 Berlin, Germany

Abstract

During the last years the interest in the numerical simulation of reacting flows has grown considerably. Numerical methods are available which allow to couple chemical kinetics with flow and molecular transport. However, the use of detailed physical and chemical models, involving more than 100 chemical species, is restricted to very simple flow configurations with very simple geometries, and models are required which simplify chemistry without sacrificing accuracy. As early as one hundred years ago Bodenstein observed that some chemical reactions are so fast that some chemical species in the reaction system are in a quasi-steady state. This observation has been the basis for practically all attempts to simplify the description of chemical reaction systems. We discuss a mathematical method, which can be used for the simplification of chemical kinetics. The method is simply based on local time scale analyses of chemical reaction systems. In this way the fast (and thus not rate limiting) chemical processes are identified and decoupled, and the chemistry can be described in terms of a small number of governing reaction progress variables. Examples for reacting flow calculations are shown and verify the approach.

1 Introduction

During the last years the interest in the numerical simulation of reacting flows has grown considerably in a variety of different applications, such as for example combustion processes [1,2], hypersonic flows [3], and chemistry in the atmosphere [4]. Some of the approaches use very detailed descriptions of the underlying physical and chemical processes, thus forcing a restriction to very simple flow configurations like one- or two dimensional systems with simple geometries. Other approaches, which are typical for industrial applications, handle very complex, i.e. realistic three-dimensional flow geometries, but, on the other hand, usually over-simplify the governing physical and chemical processes (e.g., by using one-step chemistry). Thus, methods are needed, which simplify the description of the physical and chemical processes without sacrificing accuracy, but nevertheless allow the mathematical models to be applied in realistic computations of three-dimensional reacting flows of practical interest.

As early as one hundred years ago Bodenstein [5] observed that some chemical reactions are so fast that some chemical species in the reaction system are in a quasi-steady state. This means that after a very fast relaxation process to the quasi-steady state the concentrations of the quasi-steady state species are explicitly known, if the concentrations of the remaining species are known. The consequence for the mathematical simulation is that instead of solving differential

Springer Series in Chemical Physics, Volume 61
Gas Phase Chemical Reaction Systems
Eds.: J. Wolfrum, H.-R. Volpp, R. Rannacher, and J. Warnatz

equations for all the chemical species, rate equations have only to be solved for the species which are not in steady state (usually a much smaller number).

Based on the ideas of Bodenstein, many attempts have been made to develop simplified descriptions of chemical reaction systems, e.g., for the simulation of complex combustion processes, and a variety of different approaches can be found in the literature (see, e.g., [6–8] for references considering combustion processes). In principle two ways of simplifying the chemical kinetics can be distinguished. One is to use the knowledge about the reaction system, i.e. the information on which species are in quasi-steady state or which reactions are in partial equilibrium (see [6, 7] for references). The other is to extract exactly this information from the detailed reaction mechanism based on mathematical methods (see e.g., [8–11]).

Recently we developed a method, the method of intrinsic low-dimensional manifolds (ILDM) [10], which provides a simplification of the chemical kinetics starting from a detailed reaction mechanism. The method is based on the dynamic systems theory. All the method does is to identify the fast chemical time scales (corresponding to species in quasi-steady state or reactions in partial equilibrium) and to decouple them. Besides the detailed reaction mechanism, the only input to the scheme is the desired accuracy, corresponding to the maximum time scale which is assumed to be fast enough to be decoupled. Sample calculations for a perfectly stirred reactor [12], laminar flames [13, 14], turbulent flames [15, 16], and hypersonic flows [17] have shown that the method yields good results, and is able to describe the coupling of chemical reaction with physical processes like mixing or molecular transport. In the following we outline the method and its implementation in reacting flow calculations.

2 Simplification of Chemical Kinetics

2.1 Intrinsic Low-Dimensional Manifolds

The dynamics of a chemical reaction system with n_s species is governed by the sytem of conservation equations for mass, momentum, energy, and species masses. For the moment let us restrict to a simple closed adiabatic homogeneous system at constant pressure, where the chemical kinetics is governed by the n_s rate equations

$$\frac{\partial \underline{w}}{\partial t} = F(\underline{w}) \qquad \underline{w}(t=0) = \underline{w}^0, \tag{1}$$

where t denotes the time, $\underline{w} = (w_1, w_2, \ldots, w_{n_s})^T$ the vector of species mass fractions, $\underline{F} = (M_1\omega_1/\rho, M_2\omega_2/\rho, \ldots, M_{n_s}\omega_{n_s}/\rho)^T$ the vector describing the chemical rates of change, ρ the density, and ω_i is the molar rate of formation and M_i the molar mass of the species i.

If we assume that we have a detailed reaction mechanism which is valid all over the range of different compositions, we can calculate the time evolution starting from any given initial composition $\underline{w}^0$. Let us now look at the chemical kinetics from a geometrical point of view. Every composition $\underline{w}$ is a point in the n_s-dimensional composition space which is the coordinate system spanned by the mass fractions w_i as coordinates. Starting from a point $\underline{w}^0$ the chemistry determines completely how the system eveolves, i.e. how $\underline{w}(t)$ changes, and thus how we move along in the composition space. This means that chemical reaction corresponds to trajectories in the composition space starting at $\underline{w}^0$ and finally

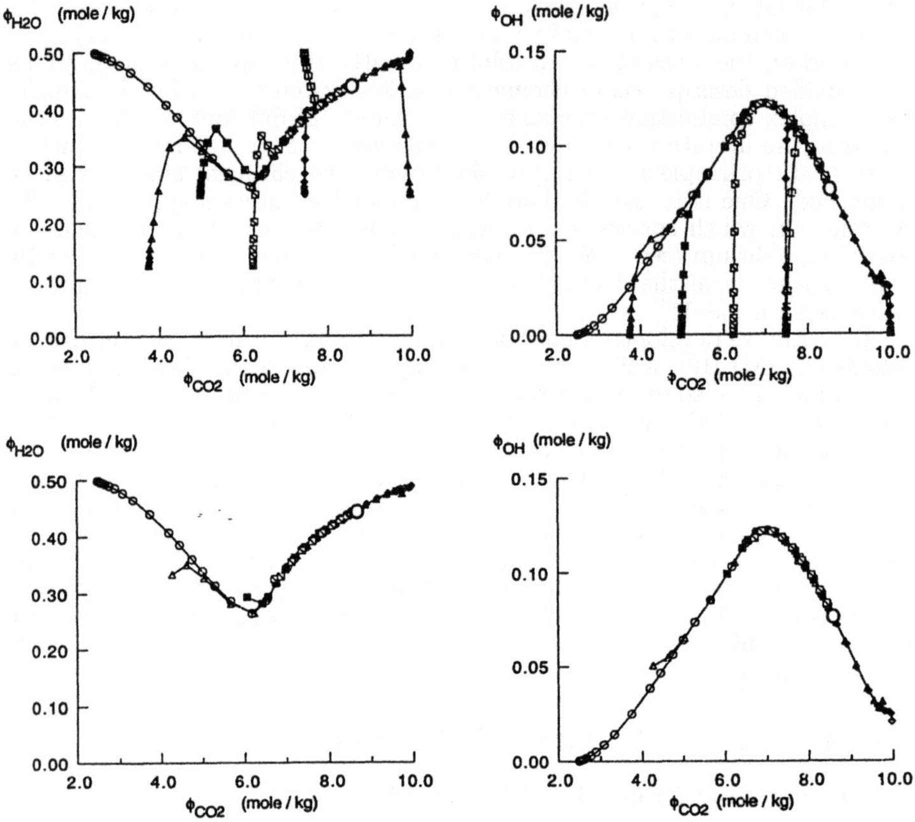

Figure 1: Trajectories of the chemical reaction for a $CO - H_2$-air system ($\phi_i = w_i/M_i$); $\circ$ denotes the equilibrium value; projection into the CO_2-H_2O- and the CO_2-OH-planes. Upper part: complete reaction process, lower part: trajectories after a relaxation time of 100 μs.

(for $t \to \infty$) reaching the equilibrium value $\underline{w}^e$ which is a function of the element composition, enthalpy, and pressure only.

Now let us consider a specific example, namely the CO-H_2-O_2-N_2-system. If we neglect NO-formation, the system consists of 13 different species (N_2, CO, H_2, O_2, CO_2, H_2O, H, O, OH, H_2O_2, HO_2, CH_2O, CHO) involved in 67 elementary reactions [10]. Of course, there is no way to visualize a 13-dimensional space. Thus let us simply look at projections into the CO_2-H_2O and CO_2-OH-planes, respectively. In Fig. 1 sample trajectories for different initial values $\underline{w}_0$ (however, for the same element composition, enthalpy, and pressure) are plotted [18]. The trajectories start at the initial values and end at the equilibrium value, which is denoted by the large circle. These plots can reveal insight into the dynamics of the chemical reaction. In the case of the CO_2-H_2O- plane the first phase of the reaction corresponds to an increase of H_2O whereas the amount of CO_2 remains almost constant. This, e.g., means that the hydrogen is consumed prior to the carbon monoxide. It can be seen from Fig. 1 that the chemical kinetics leads to a quite complicated dynamics in the composition space. The reason is that the dynamics considered so far includes all chemical processes, i.e., also all the fast equilibration processes causing species to obtain quasi-steady states. On the other hand the picture becomes quite simple if we are not interested in the dynamics of the chemical system at all. Then we can approximate the chemical system by its equilibrium value, which is only one single point in Fig. 1. The question is now, how to get rid of the fast relaxation processes while still retaining the essential dynamics of the reaction system.

Let us assume that we are not interested in the dynamics which takes place with time scales smaller than, say, 100 μs. Then all we have to do is to cut off the parts of the trajectories, which correspond to the time interval $[0, 100\mu s]$. The result is shown in Fig. 1, too. We obtain a very simple behavior: All the chemical states in the composition space are described by a simple line. The reason is that the fast relaxation processes lead to species being in quasi-steady state or reactions in partial equilibrium, thus introducing correlations among the various species. Roughly spoken this behavior states the following: After 100 μs we do no longer need n_s species mass fractions to characterize the system, but (for a given element composition, enthalpy, and pressure) only one species. Then all the other species mass fractions are explicitely known. Instead of solving all the n_s rate equations, we only have to solve the evolution equation for one species (say, CO_2).

Thus, the fast chemical relaxation processes cause the existence of low-dimensional attractors (intrinsic low-dimensional manifolds) in the composition space. Depending on the time which is allowed for the relaxation to the low-dimensional manifold, we obtain different dimensions of the manifolds. If we allow a very long time (5ms in the example above), we obtain a point (0-dimensional manifold), namely the equilibrium value, if we allow 100 μs, we obtain a line (1-dimensional manifold), if we only allow even smaller relaxation times, we obtain manifolds of higher dimension, and finally, if we do not allow any relaxation processes at all to take place instantanously, we end up with an n_s-dimensional manifold, namely the composition space itself, which corresponds to detailed chemical kinetics.

The idea of the method presented in the following is to identify the low-dimensional manifolds describing the slow processes, then to extract the dynamics within those manifolds, and finally to couple this simplified chemical model with other processes like flow or molecular transport.

2.2 Mathematical Method

The mathematical model is described in detail elsewhere [10–12,18]. We shall only outline the basic ideas. For a reaction system governed by the n_s-dimensional equation system (1) local eigenvector analyses of the Jacobians F_w ($(F_w)_{ij} = \partial F_i / \partial w_j$) reveal that there are n_s eigenvalues (i.e., time scales) associated with n_s eigenvectors (characteristic directions). Typically there are n_e zero eigenvalues which correspond to conserved quantities, namely the n_e chemical elements. Among the remaining $n_r = n_s - n_e$ eigenvalues there are some positive or moderately negative eigenvalues and (at least in many reacting flow systems) many negative eigenvalues large in magnitude. Those large negative eigenvalues correspond to fast relaxation processes, namely reactions in partial equilibrium or species in steady state. Of course these eigenvalues (i.e. fast relaxation processes) depend on the regime, i.e. they are defined locally in the state space. If we want to identify the slow manifolds, we simply have to determine the points in state

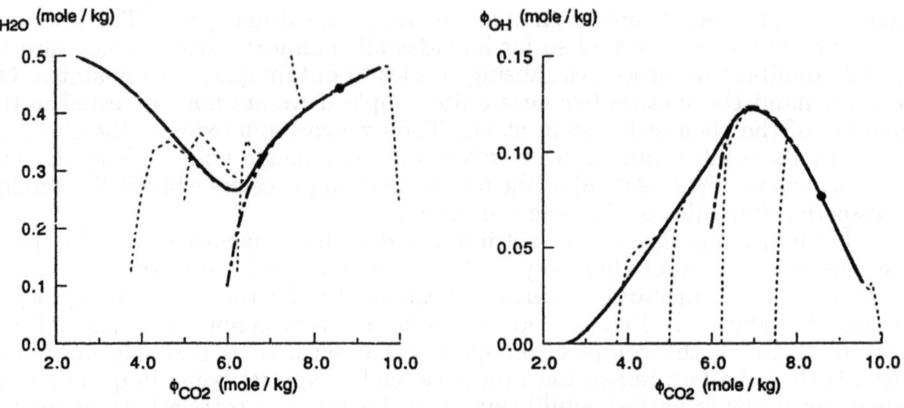

Figure 2: Low-dimensional manifolds and trajectories of the chemical reaction for a $CO - H_2$-air system ($\phi_i = w_i/M_i$); projection into the CO_2-H_2O- and the CO_2-OH-planes. Thin lines: trajectories, ——: intrinsic one-dimensional manifold, — - —: one-dimensional manifold obtained from steady state and partial equilibrium assumptions, • equilibrium value (0-dimensional manifold)

space, where the fast processes are in local equilibrium. This means that the low-dimensional manifolds are composed of those points where the rate vanish in direction of the n_f eigenvectors which correspond to large negative eigenvalues. In this case these n_f conditions define an $m = (n_r - n_f)$-dimensional manifold in the n_r dimensional reaction space according to:

$$\tilde{V}_f \left(F_w(\underline{w})\right) F(\underline{w}) = 0, \tag{2}$$

where $\tilde{V}_f$ denotes the $n_f \times n$ matrix of left eigenvectors with corresponding smallest real part of the eigenvalue. The manifolds, which are defined by (2) can be computed numerically [10,18].

Figure 2 shows plots of trajectories together with computed one-dimensional manifolds. The intrinsic low-dimensional manifold (ILDM) is obtained using the analysis outlined above. The other low-dimensional manifold has been obtained by explicitly specifying partial equilibrium or steady state conditions. Whereas the ILDM describes the system well in large parts of the domain, the other manifold fails if the system is far from equilibrium. The reason for this behavior is that the specified assumptions are only valid near equilibrium. At lower temperatures (smaller amount of CO_2) the reaction mechanism changes and the partial equilibrium and steady state assumptions valid near equilibrium can no longer be used in this part of the state space.

After the ILDM has been identified, the chemical reaction system governed by the n_s-dimensional system of conservation equations (1) can be described by a lower dimensional equation system describing the dynamics within the N-dimensional manifold.

$$\frac{\partial}{\partial t}\underline{\theta} = \underline{S}(\underline{\theta}), \tag{3}$$

where $\underline{\theta}$ is the N-dimensional vector of reaction progress variables and $\underline{S}(\underline{\theta})$ is the rate of change of the reaction progress variables. The reaction progress variables

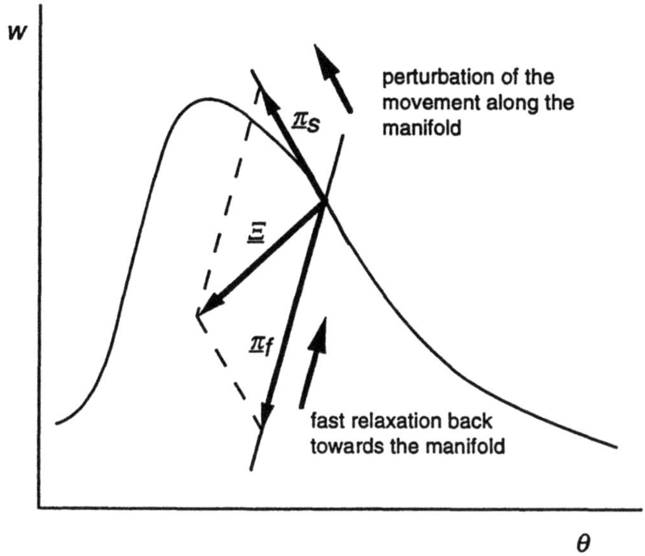

Figure 3: Schematical illustration of a perturbation of the chemistry by some physical process.

can be chosen quite arbitrarily [10, 12] and the rate of change can be computed explicitly from the chemical reaction rates $\underline{F}(\underline{w})$.

2.3 Coupling with Physical Processes

It was shown in the previous section that intrinsic low-dimensional manifolds of chemical systems can be used to simplify the chemical kinetics. However, the interesting case in practical applications is the coupling with flow and molecular transport. In principle the concept of the ILDM is still valid in systems with flow and molecular transport, but the physical processes act as perturbations of the chemical system, i.e. they tend to move the system off the manifold. If the perturbations occur with time-scales larger than the time scales of the relaxation towards the attracting manifold, then the fast chemical processes move the system back to the manifold instantaneously, and we can still use the manifold to simplify the kinetics.

A spectrum of time scales, say, in a combustion process, reveals that there are many chemical time scales which are smaller than the typical physical time scales. From the consideration explained above, it is obvious that all the chemical time scales which are sufficiently smaller than the physical time scales can be decoupled. The remaining question of how to account for the physical perturbations can be answered by a simple idea (Fig. 3).

Let us assume that we are at a point on the manifold and that there is some perturbation $\underline{\Xi}$. The perturbation can be decomposed into its components in the local eigenvector basis, i.e. into two parts – one part, describing the rate of change in the slow subspace and one describing the rate of change in the fast subspace.

Now let us assume that the time scale of the perturbation is of the order of the time scales of the slow movement within the manifold, i.e. much smaller than those of the fast relaxing time scales. The components of the perturbation in the direction of the fast subspace will have a minor effect on the chemical reaction system, because chemistry (fast equilibration processes) relaxes the perturbation back to the manifold. The components of the perturbation in the slow subspace instead, will directly couple with the time scales of the chemical reaction and thus move the state within the manifold. That means if we project the perturbation locally onto the slow subspace, we can account for the interaction of the physical processes with the slow time scales, whereas we neglect all processes which perturb the chemistry, but are "equilibrated" by the chemistry within a very short time (of the order of the fast time scales). Using these ideas, we can formulate a projection operator P, which depends on the local characteristics of the manifold and which projects a physical perturbation $\underline{\Xi}$ onto a perturbation $\underline{\Gamma}(\underline{\theta})$ within the manifold (see [12]). This means that all we have to know in order to solve the coupled problem are the rates of change of the parameters $\underline{S}(\underline{\theta})$, the composition space $\underline{w}(\underline{\theta})$, and the local projection matrices $P(\underline{\theta})$ as functions of the parameters $\underline{\theta}$. This method will work and yield good approximations as long as the time scales of the perturbations are slow compared with the fast relaxing time scales which had been decoupled by the construction of the intrinsic low-dimensional manifold. A detailed mathematical description of the ideas outlined above can be found in [10–13, 18].

3 Example

Various calculations of reacting flows, such as perfectly stirred reactors [12], laminar flames [13,14], turbulent flames [15,16], and hypersonic flows [17] have verified the approach presented above. Due to space limitation we shall only present one example, namely a premixed laminar flat flame calculation [13]. It provides a nice, simple test case for the verification of the model. The specific example is a syngas (40 Vol. % CO, 30 Vol. % H_2, 30 Vol. % N_2)-air system at $p = 1$ bar, and with a temperature of 290 K in the unburnt gas. The fuel/air ratio is 6/10. The influence of simplified transport models is described elsewhere [13]. Here, for the sake of simplicity, only systems with equal diffusivity shall be considered. In this case a three-dimensional manifold with enthalpy and two reaction progress variables as parameters has been calculated, i.e. the chemistry has been simplified to a two-step reaction scheme. Figure 4 shows profiles of CO_2, H_2O, OH and H in the flame both for reduced and detailed kinetics [13]. Good agreement is obtained in the flame front as well as behind the flame front, where the slow equilibration process takes place. Differences are observed at the beginning of the flame front. This is caused by the fact that at the low temperatures ($T < 1000$ K) the decoupled time scales are of the order of the time scales of diffusion and heat conduction. However, if an additional reaction progress variable is used, the results will become much more accurate. The small deviations behind the flame front can be attributed to discretization errors in the numerical simulation (the points reflect directly the mesh used in the computations).

Comparisons of cpu-times needed for laminar flame calculations show that the computational demand is reduced considerably if simplified chemical kinetics is used. Even for the CO-H_2-N_2-N_2 system consisting of 13 chemical species, we observe a speed-up factor of 10 if simplified chemistry with 2 reaction progress variables is used. The reason is that not only the number of equations to be

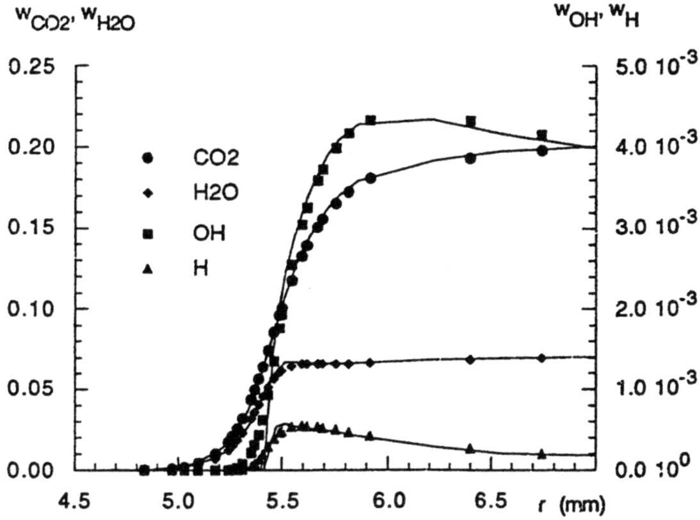

Figure 4: Calculated structure of a syngas–air flame; points: reduced mechanism, line: detailed mechanism.

solved is reduced considerably, but also much of the stiffness of the differential equation system. It can be estimated that a reduction of cpu-time by a factor of more than 1000 is possible for more complicated systems like combustion of higher hydrocarbons.

4 Conclusions

The method of simplifying chemical kinetics by constructing intrinsic low-dimensional manifolds can be used successfully in the simulation of reacting flows which are governed by a coupling of chemical reaction, flow, and molecular transport. Comparisons with detailed chemistry calculations show a very good agreement and verify the approach. The computational effort in the reacting flow calculations is reduced considerably, and it can be estimated that the method can speed up calculations of flows with complex chemistry (like, e.g., combustion of higher hydrocarbons) by a factor of more than 1000.

5 Acknowledgements

The author would like to thank D. Schmidt and S. B. Pope for interesting discussions.

6 References

[1] S. B. Pope, *23th Symp. (Intl.) Combustion.* The Combustion Institute, Pittsburgh, p. 591 (1990).

[2] J. Warnatz, *24th Symp. (Intl.) Combustion.* The Combustion Institute, Pittsburgh, p. 553 (1992).

[3] H. Oertel, H. Körner (Ed.), *Orbital Transport, Technical, Meteorological and Chemical Aspects.* Springer, Berlin (1993).

[4] Proc. Eurotrac 1994, *Transport and Transformation of Pollutants in the Troposphere*, Garmisch-Partenkirchen, Germany (1994).

[5] M. Bodenstein, S. C. Lind, *Z. phys. Chem.* **57**, 168 (1906).

[6] M. D. Smooke (Ed.), *Reduced Kinetic Mechanisms and Asymptotic Approximations for Methane-Air Flames.* Lecture Notes in Physics 384, Springer, Berlin Heidelberg New York (1991).

[7] N. Peters, B. Rogg, *Reduced Kinetics Mechanisms for Applications in Combustion Systems.* Springer, Berlin (1993).

[8] A. S. Tomlin, M. J. Pilling, T. Turanyi, J. H. Merkin, J. Brindley, *Combust. Flame* **91**, 107 (1992).

[9] S. H. Lam, D. A. Goussis, *22nd Symp. (Intl.) Combustion.* The Combustion Institute, Pittsburgh, p. 931 (1990).

[10] U. Maas, S. B. Pope, *Combust. Flame* **88**, 239 (1992).

[11] U. Maas, *Appl. Math.* **40**, 249 (1995).

[12] U. Maas, S. B. Pope, *24th Symp. (Intl.) Combustion.* The Combustion Institute, Pittsburgh, p. 103 (1992).

[13] U. Maas, S. B. Pope, *25th Symp. (Intl.) Combustion.* The Combustion Institute, Pittsburgh, p. 1349 (1994).

[14] U. Riedel, D. Schmidt, U. Maas, J. Warnatz , Laminar Flame Calculations Based on Automatically Simplified Chemical Kinetics *Proc. Eurotherm Seminar # 35, Compact Fired Heating Systems*, Leuven, Belgium (1994).

[15] A. Wölfert, M. Nau, U. Maas, J. Warnatz, University of Heidelberg, Technical Report 94-69 (1994).

[16] A. Norris, *The Application of PDF Methods to Turbulent Diffusion Flames.* Dissertation, Cornell University, Ithaca, NY, USA, (1993).

[17] D. Schmidt, U. Maas, J. Warnatz, Simplifying Chemical Kinetics for the Simulation of Hypersonic Flows Using Intrinsic Low-Dimensional Manifolds *Proc. 5th International Symposium on Computational Fluid Dynamics*, Sendai, Japan (1993).

[18] U. Maas, *Automatische Reduktion von Reaktionsmechanismen zur Simulation reaktiver Strömungen*, Habilitationsschrift, Universität Stuttgart (1993).

Index of Contributors

Springer Series in Chemical Physics

Editors: Vitalii I. Goldanskii Fritz P. Schäfer J. Peter Toennies

Springer Series in Chemical Physics

Editors: Vitalii I. Goldanskii   Fritz P. Schäfer   J. Peter Toennies

1 Atomic Spectra and Radiative Transitions
By I. I. Sobelman

2 Surface Crystallography by LEED
Theory, Computation and Structural Results. By M. A. Van Hove, S. Y. Tong

3 Advances in Laser Chemistry
Editor: A. H. Zewail

4 Picosecond Phenomena
Editors: C. V. Shank, E. P. Ippen, S. L. Shapiro

5 Laser Spectroscopy
Basic Concepts and Instrumentation
By W. Demtröder   3rd Printing

6 Laser-Induced Processes in Molecules   Physics and Chemistry
Editors: K. L. Kompa, S. D. Smith

7 Excitation of Atoms and Broadening of Spectral Lines   By I. I. Sobelman, L. A. Vainshtein, E. A. Yukov

8 Spin Exchange
Principles and Applications in Chemistry and Biology
By Yu. N. Molin, K. M. Salikhov, K. I. Zamaraev

9 Secondary Ion Mass Spectrometry SIMS II   Editors: A. Benninghoven, C. A. Evans, Jr., R. A. Powell, R. Shimizu, H. A. Storms

10 Lasers and Chemical Change
By A. Ben-Shaul, Y. Haas, K. L. Kompa, R. D. Levine

11 Liquid Crystals of One- and Two-Dimensional Order
Editors: W. Helfrich, G. Heppke

12 Gasdynamic Lasers   By S. A. Losev

13 Atomic Many-Body Theory
By I. Lindgren, J. Morrison

14 Picosecond Phenomena II
Editors: R. M. Hochstrasser, W. Kaiser, C. V. Shank

15 Vibrational Spectroscopy of Adsorbates   Editor: R. F. Willis

16 Spectroscopy of Molecular Excitons
By V. L. Broude, E. I. Rashba, E. F. Sheka

17 Inelastic Particle-Surface Collisions
Editors: E. Taglauer, W. Heiland

18 Modelling of Chemical Reaction Systems   Editors: K. H. Ebert, P. Deuflhard, W. Jäger

19 Secondary Ion Mass Spectrometry SIMS III
Editors: A. Benninghoven, J. Giber, J. László, M. Riedel, H. W. Werner

20 Chemistry and Physics of Solid Surfaces IV   Editors: R. Vanselow, R. Howe

21 Dynamics of Gas-Surface Interaction
Editors: G. Benedek, U. Valbusa

22 Nonlinear Laser Chemistry
Multiple-Photon Excitation
By V. S. Letokhov

23 Picosecond Phenomena III
Editors: K. B. Eisenthal, R. M. Hochstrasser, W. Kaiser, A. Laubereau

24 Desorption Induced by Electronic Transitions DIET I   Editors: N. H. Tolk, M. M. Traum, J. C. Tully, T. E. Madey

25 Ion Formation from Organic Solids
Editor: A. Benninghoven

26 Semiclassical Theories of Molecular Scattering   By B. C. Eu

27 EXAFS and Near Edge Structures
Editors: A. Bianconi, L. Incoccia, S. Stipcich

28 Atoms in Strong Light Fields
By N. B. Delone, V. P. Krainov

29 Gas Flow in Nozzles
By U. G. Pirumov, G. S. Roslyakov

30 Theory of Slow Atomic Collisions
By E. E. Nikitin, S. Ya. Umanskii

31 Reference Data on Atoms, Molecules, and Ions   By A. A. Radzig, B. M. Smirnov

32 Adsorption Processes on Semiconductor and Dielectric Surfaces I
By V. F. Kiselev, O. V. Krylov

33 Surface Studies with Lasers
Editors: F. R. Aussenegg, A. Leitner, M. E. Lippitsch

34 Inert Gases
Potentials, Dynamics, and Energy Transfer in Doped Crystals   Editor: M. L. Klein

35 Chemistry and Physics of Solid Surfaces V   Editors: R. Vanselow, R. Howe

36 Secondary Ion Mass Spectrometry SIMS IV   Editors: A. Benninghoven, J. Okano, R. Shimizu, H. W. Werner

37 X-Ray Spectra and Chemical Binding
By A. Meisel, G. Leonhardt, R. Szargan

38 Ultrafast Phenomena IV
By D. H. Auston, K. B. Eisenthal

39 Laser Processing and Diagnostics
Editor: D. Bäuerle

MIX
Papier aus verantwortungsvollen Quellen
Paper from responsible sources
FSC® C105338
FSC
www.fsc.org

If you have any concerns about our products,
you can contact us on
ProductSafety@springernature.com

In case Publisher is established outside the EU,
the EU authorized representative is:
Springer Nature Customer Service Center GmbH
Europaplatz 3, 69115 Heidelberg, Germany

Printed by Libri Plureos GmbH
in Hamburg, Germany